Thermodynamics

Paulo Cesar Philippi

Thermodynamics

From Fundamentals to Multiphase and Multicomponent Systems

PUCPR
GRUPO MARISTA

Springer

Paulo Cesar Philippi
Mechanical Engineering Graduate Program
Pontifical Catholic University of Paraná
Curitiba, Paraná, Brazil

ISBN 978-3-031-49356-0 ISBN 978-3-031-49357-7 (eBook)
https://doi.org/10.1007/978-3-031-49357-7

Jointly published with Associação Paranaense de Cultura
The print edition is not for sale in China (Mainland). Customers from China (Mainland) please order the print book from: Associação Paranaense de Cultura.

This Springer imprint is published by the registered company Springer Nature Switzerland AG
The registered company address is: Gewerbestrasse 11, 6330 Cham, Switzerland

Paper in this product is recyclable.

To Raquel

Foreword

This new book on thermodynamics, *Thermodynamics: From Fundamentals to Multiphase and Multicomponent Systems* came, originally, from a graduate/undergraduate education which Paulo spoke to me during a stay at the Pontifical Catholic University of Paraná in Curitiba in 2018.

But this book goes far beyond classic academic lessons!

What challenges first of all is a very high-quality iconography, with historical images and many educational illustrations to implement the two fundamental principles. This iconography also addresses thermodynamic systems in non-equilibrium conditions, with the result of numerical simulations carried out by the author and his colleagues. The book presents a wide variety of approaches with, on the one hand, the formal aspects such as the introduction of the Legendre transform to specify the different thermodynamic potentials and on the other hand the more physical aspects with for example the Marangoni effect, going even toward chemistry with the notion of affinity between species. A great variety of exercises stimulates the student to consolidate and mature his learning.

The book is divided into three main parts: *The Principles of Thermodynamics*, *Thermodynamic Equilibrium*, and an *Introduction to Non-equilibrium Thermodynamics*. Hence a succession of chapters, from the First Principle of Thermodynamics to Multicomponent Systems in non-equilibrium conditions, the book ends with a chapter dedicated to the kinetic standpoint of non-equilibrium thermodynamics.

Of course, this mass of knowledge requires exposure time and Paulo Philippi's work includes nearly 350 ages and offers around 200 bibliographical references.

In summary, the book *Thermodynamics: From Fundamentals to Multiphase and Multicomponent Systems* is a reference work for students of course, but also for academic scholars and professionals, in the long tradition of books from Maxwell, Callen, Huang, or Landau and Lifchitz.

Good reading!

August 2023

Francois Dubois
Conservatoire National des Arts et Métiers
Paris, France

Preface

This book on the fundamentals of thermodynamics was written intended to be a textbook for advanced undergraduate and graduate students, who are interested in a more in-depth understanding of the influence of irreversibility on thermodynamic processes between equilibrium states and to acquire a better knowledge of the physical mechanisms underlying the macroscopic behavior of multiphase and multicomponent systems in equilibrium and non-equilibrium states.

Part I deals with the laws of thermodynamics as developed by outstanding nineteenth-century researchers and philosophers such as Benoît Clapeyron, Sadi Carnot, James Prescott Joule, Rudolf Clausius, Willian Thomson, Hermann von Helmholtz, and Max Planck. The first principle of thermodynamics was formulated based on the idea that "heat is motion," i.e., that "heat is nothing else than the mechanical energy of the particles that form a body," an atomistic concept that, at that time, was not universally accepted by the scientific community.

The second principle is based on the idea that "nature has an arrow" or that in all natural phenomena some "vis viva" is being converted into heat, and not the inverse. This idea was first pictured by Carnot in his 1824 book *Réflexions sur la puissance motrice du feu* and, as admitted by Willian Thomson, Lord Kelvin, in 1848, was the main difficulty to be solved for the acceptance of Joule's proposal of the equivalence between heat and work. Carnot's book is a *chef d'oeuvre* of scientific thought and had a large influence on the scientific community in the first half of the nineteenth century. In this book, Carnot bases his approach on the Lavoisier theory on the conservation of the caloric and the caloric theory is more akin to the dissipative nature of the physical processes than Joule´s equivalence principle.

It is important, for young students—supposed to be the actors who will be asked to overcome new challenges in the future—to understand how all these old challenges were overcome in the nineteenth century, giving rise to the recognition of the principle of energy conservation and the birth of thermodynamics as a new science. It is also important for their education that proper reference be made for the authors of the knowledge they are acquiring and to be conscious of the amount of effort that was employed to construct this knowledge.

Most of the development of thermodynamics in the nineteenth century was motivated by the eighteenth-century steam atmospheric engines. The challenge was to find the key to increasing the conversion efficiency of these power machines disseminated throughout Britain and being responsible for a true revolution in manufacturing and transportation.

Nowadays—to mention only a few engineering applications—thermodynamics is an invaluable tool for the design of power plants, refrigerating systems, heat pumps, solar and photovoltaic plants, fuel cells and capillary pumps, and the heat pipes that are, presently, wildly employed in computers.

Nevertheless, besides being born to solve engineering problems, thermodynamics is, also, a unified science that tries to understand nature and the ubiquitous processes around us, being fundamental to the scientific thought in several branches of human knowledge such as geology, atmospheric science, chemistry, biomedical, and health sciences.

Part I of this book has two chapters.

Chapter 1 deals with the first law and begins with a brief history of the science of heat. Clapeyron's equation is not only a suitable equation of state for elastic (or ideal) gases—as frequently understood by students—but has a 200 years history (if we start from the first Galileu's experiments with the thermoscope) and means the recognition of the absolute temperature and the first indication of the existence of an unbridgeable absolute zero, firstly pictured by Gay-Lussac.

Chapter 2 is dedicated to the second law and begins with the meaning of irreversibility. The work delivered in irreversible expansions of gases cannot be written as $\int PdV$. In fact, it does not make any sense to talk about a single pressure for a non-equilibrium thermodynamic system. Nevertheless, in certain simple situations, the effective work can be computed and compared with the work that would be delivered in a reversible expansion. In this chapter, I also try to clarify the fundamental role of James Joule in demonstrating the principle of energy conservation. Some academic exercises are proposed at the end of each part.

The second law received different formulations by Clausius, Kelvin, and Planck. Demonstrating the equivalence between different statements is important because this equivalence shows that the underlying principle behind these statements is the same. Indeed, all the formulations of the second law state that energy flows in one direction, "from vis viva to heat." In other words, dissipation is a common attribute of all natural phenomena.

Classical mechanics, the groundwork of the mechanical theory that thermodynamicists advocate for heat is, however, reversible. In effect, the previous states of a given mechanical system of particles can always be retrieved when the current state of this system is known. So, how do conciliate a reversible science—classical mechanics—with a science that has irreversibility as its one of its underlying principles? The answer was given by Ludwig Boltzmann in the second half of the nineteenth century and a brief summary of this outstanding development for human knowledge is given at the end of Chap. 3.

Part II of this book is dedicated to the concept of thermodynamic equilibrium.

Chapter 3 deals with homogeneity, Legendre transforms (leading to the Helmholtz and Gibbs energies and the concept of enthalpy) followed by Maxwell relations and minimum principles and establishing the framework that is needed for the theoretical analysis of thermodynamic systems in equilibrium. A thermodynamic system in equilibrium is homogeneous, meaning that its intensive properties do not vary from point to point inside the system. This property leads to an Euler's relationship between its extensive variables and the Gibbs–Duhem equation. This equation found, simultaneously, by Josiah Willard Gibbs and Pierre Maurice Marie Duhem enables a Lagrangian description of thermodynamic systems, of crucial importance in the analysis of open systems in non-equilibrium states. Minimum principles state that when a restriction that prevents a system from reaching equilibrium is removed, this system will attain an equilibrium state given by the minimization of one of its energies (internal, Helmholtz, Gibbs, or enthalpy), in accordance with the thermodynamic variables that are maintained constant during the equilibration process.

In Chap. 4, we begin by following the efforts of Maxwell to show that the attractive forces between molecules were responsible for vapor condensation, at temperatures below the critical temperature found by Andrews for CO_2. In his 1875 paper "*On the dynamical evidence of the molecular constitution of bodies*," Maxwell recognizes the impact of the 1873 Doctoral Thesis of Johannes Diderik van der Waals for clearing the role of the interaction of real molecules on phase transition and does not omit to mention the important work of Willard Gibbs who, in Maxwell's words, "has given us a remarkably simple and thoroughly satisfactory method of representing the relations of the different states of matter by means of a model."

Chapter 5 is dedicated to multicomponent systems. In this kind of thermodynamic system, there is a competition between the attraction forces between identical molecules and the forces between molecules belonging to different species. When the thermodynamic system is such that the former forces are dominant the mixture will segregate leading to a system with distinct phases. When the latter forces are dominant, the different components form a non-ideal homogeneous system, which is the subject of this chapter. Equilibrium calculations are based on the Unifac method, a semiempirical method for the prediction of non-electrolyte activity in non-ideal mixtures that use the functional groups, present on the molecules that make up the mixture to calculate the activity coefficients. The Unifac method was first published in 1975 by Fredenslund, Jones, and Prausnitz, from the University of California.

In Chapter 6, we deal with fluid interfaces trying to give a deeper insight into the molecular mechanisms that are responsible for the interface tension and for the adhesion of liquids on solid surfaces.

In Part III, we give an introduction to non-equilibrium thermodynamics.

We start in Chapter 7 with the continuum mechanics hypothesis of local equilibrium, when each point of the space is considered to have a characteristic length much greater than the molecular dimensions and, at the same time, much smaller than the characteristic length of the whole system. So, each point can be thought of as an elementary system in local thermodynamic equilibrium, although the local thermodynamic properties at this point differ from the ones at the neighboring points.

This local equilibrium hypothesis is the key to finding the mass conservation law and the balance equations for the linear momentum and internal energy in their differential forms, enabling to write a closed set of differential equations for the density, velocity, and temperature fields in non-equilibrium systems. This chapter ends with the balance equation for entropy. When we consider a material volume, the rate with which the entropy varies with time is dependent on the heat flow through its boundary and on the internal production of entropy $\mathscr{P}(s)$, due to internal non-equilibrium and related to the Clausius non-compensated heat. By the second law, $\mathscr{P}(s)$ cannot be negative, $\mathscr{P}(s) \geq 0$, and is always positive in non-equilibrium systems.

In Chap. 8, we study single-component two-phase systems where two approaches are presented: the singular and the diffuse interface approaches. In the singular interface approach, the transport equations are solved in each phase and the interface is considered as a boundary, where the field variables from the two phases are linked. In the second approach, the interface is considered to be diffuse, with a continuous variation of the thermodynamic properties from one phase to another. In the diffuse interface approach, the law of increasing entropy, $\mathscr{P}(s) \geq 0$, is used to find the influence of surface forces on the pressure tensor and heat flux.

The source of entropy is written as a product of fluxes and forces, and this form is used throughout Chap. 9 as a method for obtaining a closed system of differential equations for non-equilibrium thermodynamic systems, especially for systems involving non-ideal fluids and their mixtures. For multicomponent systems, this is the key to relating the macroscopic properties of the system with the molecular sources of segregation. Indeed, miscibility is not only an intrinsic property of a pair of liquids but is dependent on the thermodynamic state of their mixture. Two liquids that can mix in all proportions forming homogeneous solutions at a given temperature can partially segregate when the temperature is decreased. The amount of solute that remains in the solution at this lower temperature is a measure of the solubility, and this property is dependent on the temperature and pressure. In this chapter, the physical underlying processes that are at the origin of mixing and segregation in non-ideal mixtures of partially miscible fluids are investigated.

The book ends with a chapter dedicated to exposing the kinetic standpoint of non-equilibrium thermodynamics. The kinetic theory is the branch of science to which the *mechanical theory of heat* of Clausius, Maxwell, and Boltzmann has evolved after a system of particles ruled by deterministic mechanics was considered from a probabilistic point of view. Furthermore, the complex behavior of non-ideal fluids and solutions requires downscaling to the molecular scale. The Boltzmann equation is nevertheless restricted to rarefied gases, and this restriction requires the introduction of kinetic models that, in the reverse sense, make use of the available information in the macroscopic scale. Kinetic models for the Boltzmann collision term and for non-ideal fluids and their mixtures are presented in this chapter.

The numerical simulation of kinetic models for thermodynamic systems in non-equilibrium conditions has had a rapid growth after 1990, and most of the simulation examples presented throughout the book were performed with discrete forms of kinetic models in the lattice Boltzmann framework.

I wish to give my thanks to Pablo Javier Blanco from the National Laboratory of Scientific Computation (LNCC) for his careful revision of the text and to François Dubois from the Conservatoire National des Arts et Métiers de Paris for his stimulating words in the foreword. Most of the material presented in this book resulted from my almost 30 years of research work and my thermodynamics graduate/undergraduate courses and I thank my colleagues and my former graduate and doctoral students with whom I had, during these years, many fruitful discussions and a very friendly relationship.

Curitiba, Brazil
September 2023

Paulo Cesar Philippi

Contents

List of Figures

List of Tables

Part I
The Principles of Thermodynamics

Chapter 1
The First Principle

Abstract The first principle of thermodynamics is considered to be one of the greatest achievements of scientific thought when heat, or the *internal energy* of a thermodynamic system, was recognized as a kind of mechanical energy. In this chapter, we will try to understand the main reasons that led to this outstanding discovery. We start with the main seventeenth and eighteenth century findings, from the Mariotte discovery of atmospheric pressure to James Watt steam engines. While the seventeenth century was the century of the discovery of vacuum by Pascal, the thermoscope of Galileu, the gas thermometer of the Grand Duke of Tuscany, Ferdinand II, and the discovery of the properties of elastic gases by Robert Boyle, the eighteenth century saw the born of the steam engines and the industrial revolution they produced, atmospheric gases were isolated and the Stahl phlogiston theory was overthrown by Lavoisier's mass conservation principle. At the end of the eighteenth century, Lavoisier, controversially, also proposed the conservation of caloric principle, in the reverse sense with the principle of energy conservation sustained by important philosophers and further postulated by Helmholtz, based on the pioneer experimental works of Benjamin Thompson, Joule, and Mayer.

1.1 The Laws of Gases

1.1.1 Pressure

The term '*pressure*' as the normal force per unit area acting on a given surface in contact with a fluid was introduced by Pascal. But, pressure is not only a force per unit area acting on a given surface, but the result of the momentum variation of the molecules that are reflected by the wall. Indeed, in accordance with Newton's second law, the force **F** that a single molecule of mass m exerts on a surface is the rate of change of its momentum before and after the collision of this molecule with the surface and during the time interval Δt this molecule is in contact with the surface

$$\frac{m\mathbf{v}_{in} - m\mathbf{v}_{out}}{\Delta t} = \mathbf{F}.$$

P. C. Philippi, *Thermodynamics*, https://doi.org/10.1007/978-3-031-49357-7_1

Fig. 1.1 Measurement of the atmospheric pressure by Evangelista Torricelli

Considering the molecules to be material points, this collision will be elastic, and the tangential components of $\mathbf{v}_{in}$ and $\mathbf{v}_{out}$ will be identical in modulus and opposed. So, the resulting force will be normal to the surface.

Pressure is not only a normal force but it is, also, isotropic, in the sense that, in a given point $\mathbf{x}$, it has the same value independently of the surface normal direction with which the fluid (a gas or a liquid) is in contact, at this point x. Therefore the pressure inside a spherical steel vessel filled with vapor is the result of the collisions of the vapor molecules with the inner surface of the vessel shell and the liquid pressure inside a pressure cooker is the pressure of the vapor above it.

In 1594 Galileo obtained the patent for a machine to pump water from a river for the irrigation of land. Galileo found that 10 m was the limit to which the water would rise in the suction pump, but had no explanation for this phenomenon. Knowing this curious fact, Galileo encouraged his former pupil Evangelista Torricelli to investigate this supposed limitation. Torricelli reasoned, that this limitation was the result of the pressure exerted on the liquid by the surrounding air.

To prove this theory, he filled a long glass tube, sealed at one end, with mercury and upended it into a dish also containing mercury. The column of mercury invariably fell to about 760 mm, leaving an empty space above its level. Torricelli attributed the cause of the phenomenon to a force on the surface of the earth, without knowing, where it came from. He also concluded that the space on top of the tube is empty, that nothing is in there, and called it '*vacuum*' (Fig. 1.1).

From about 1645 Blaise Pascal (1623–1662) heard about the experiments of Torricelli and was searching for the reasons for Galileo's and Torricelli's findings. He came to the conviction that the force, that keeps the column at 760 mm, is the weight of the air above. Thus, on a mountain, the force must be reduced by the weight of

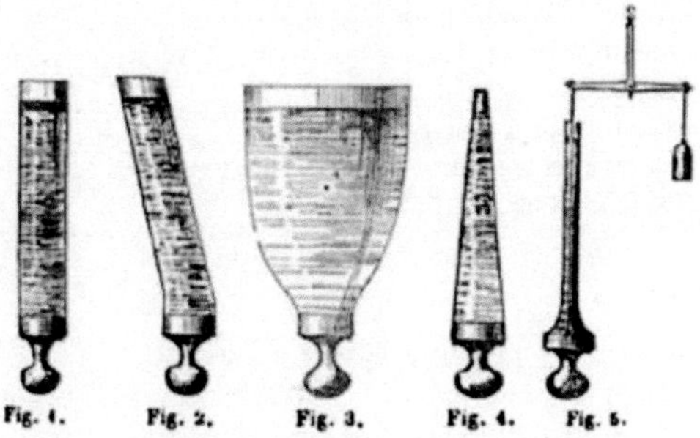

Fig. 1.2 Pascal's drawing showing the law of pressure for liquids

the air between the valley and the mountain. He predicted that the height of the column would decrease which he proved with his experiments at the mountain Puy de Dome in central France. From the decrease, he could calculate the weight of the air. Pascal also formulated that this force, he called it '*pressure*', is acting uniformly in all directions.[1] Descartes visited Pascal on 23 September. His visit only lasted two days and the two argued about the vacuum which Descartes did not believe in. In August of 1648 Pascal observed that the pressure of the atmosphere decreases with height and deduced that a vacuum existed above the atmosphere.[2]

From May 1653 Pascal worked on mathematics and physics and wrote the book *Traité de l' équilibre des liqueurs*[3] in which he explains Pascal's law of pressure (Fig. 1.2). This treatise is a complete outline of a system of hydrostatics, the first in the history of science, it embodies his most distinctive and important contribution to physical theory.

Robert Boyle (1627–1691) was an alchemist, but he is most known by its important work on *elastic* gases, showing the constancy of the product of the pressure by the volume,

$$PV = const.$$

published in his 1660 book[4] (Fig. 1.3) and meaning that the volume of a fixed amount of an elastic gas is inversely proportional to the pressure. This finding was also discovered by Edme Mariotte (1620–1684) and published in his 1679 book.[5]

[1] B. Pascal (1647). *Expériences nouvelles touchant le vide*. O'Euvres de Blaise Pascal, II. La bibliothèque libre.

[2] B. Pascal (1648) *Récit de la grande expérience de l'équilibre des liqueurs*. Bibliothèque Nationale de France. Available in http://gallica.bnf.fr/ark:/12148/bpt6k105083f/f1.image.

[3] B. Pascal (1653) *Traité de l'équilibre des liqueurs*. A new edition with comments was produced by Arvensa Éditions. Paris.

[4] Boyle R (1660) *New experiments Physico-Mechanical touching the spring of the air and its effects*. H. Hall, Oxford, [19].

[5] E. Mariotte, *Essais de Physique, ou mémoires pour servir á la science des choses naturelles*. Paris: E. Michallet, 1679 [111].

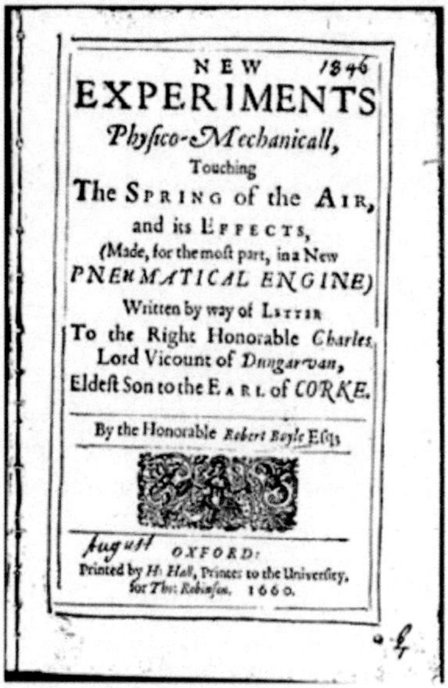

NEW
EXPERIMENTS
Phyſico-Mechanicall,
Touching
The SPRING of the AIR,
and its EFFECTS,
(Made, for the moſt part, in a New
PNEUMATICAL ENGINE)
Written by way of LETTER
To the Right Honorable Charles
Lord Vicount of Dungarvan,
Eldeſt Son to the EARL of CORKE.
By the Honorable Robert Boyle Eſq;
OXFORD:
Printed by H: Hall, Printer to the Univerſity,
for Tho: Robinſon. 1660.

Fig. 1.3 Front page of the book *New experiments Physico-Mechanical touching the spring of the air and its effects*, published by Boyle in 1660

1.1.2 Temperature

Galileo Galilei in the end of the sixteenth century was able to gauge the relative "*coldness*" or "*hotness*" of air, using a rudimentary thermoscope (Fig. 1.4), but the first thermometer was invented by Santorio Santorio in 1612. Santorio's instrument was an air thermometer. Its accuracy was poor as the effects of varying air pressure on the thermometer were not understood at that time. The sealed liquid-in-glass thermometer was first produced in 1654 by the Grand Duke of Tuscany, Ferdinand II (1610–1670). His thermometer had an alcohol filling. Although this was a significant development, his thermometer was inaccurate and there was no standardized scale in use. Gabriel Fahrenheit (1686–1736) was the first person to make a thermometer using mercury in 1714. The more predictable expansion of mercury combined with improved glass working techniques led to a much more accurate thermometer. In 1742 a Swedish scientist named Anders Celsius (1701–1744) devised a thermometer scale dividing the freezing and boiling points of water into 100 °C. Celsius chose 0 °C for the boiling point of water, and 100 °C for the freezing point [119]. One year later, Jean Pierre Cristin (1683–1755) inverted the Celsius scale to produce the Centigrade scale used today (freezing point 0 °C, boiling point 100 °C).

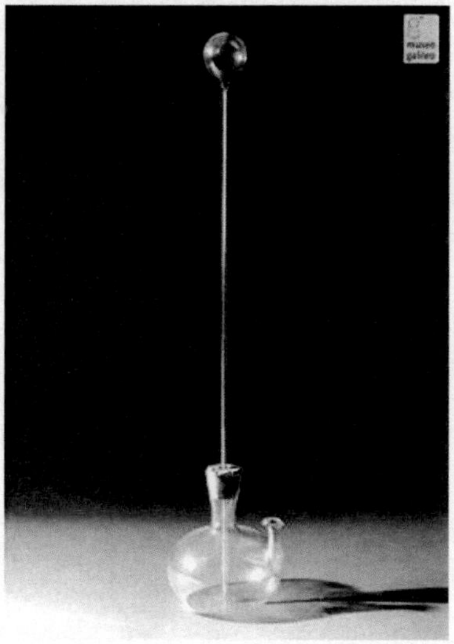

Fig. 1.4 A replica of Galileo's thermoscope exposed in Galileo's museum (Florence)

1.2 From Stahl to Lavoisier

> There have been few, if any, revolutions in science so great, so sudden, and so general, as the prevalence of what is now usually termed the new system of chemistry or that of the Antiphlogistians, over the doctrine of Stahl, which was at one time thought to have been the greatest discovery that had ever been made in science. I remember hearing Mr. Peter Woulfe, whose knowledge of chemistry will not be questioned, say, that there had hardly been anything that deserved to be called a discovery subsequent to it. Though there had been some who occasionally expressed doubts of the existence of such a principle as that of phlogiston, nothing had been advanced that could have laid the foundation of another system before the labors of Mr. Lavoisier and his friends, from whom this new system is often called that of the French. (Joseph Priestley, 1796[6])

It is almost impossible to fully understand the history of thermodynamics without the help of some knowledge of the main achievements in chemistry during the eighteenth century, when the chemical elements were discovered, establishing the framework for the atomistic theory.

In 1667, Johann Joachim Becher published his book *Physica subterranea*, in which was the first mention of what would become the phlogiston theory. Traditionally, alchemists considered that there were four classical elements: fire, water, air, and earth. In his book, Becher eliminated fire, water, and air from the classical element model and replaced them with three forms of earth: *terra lapidea*, *terra fluida*, and *terra pinguis*. Becher believed that *terra pinguis* was a key feature of combustibles and was released when combustible substances were burned.

---

[6] Priestley J (1796) *Considerations on the doctrine of phlogiston and the decomposition of water*. Thomas Dobson, Philadelphia, [139].

In 1703 Georg Ernst Stahl, professor of medicine and chemistry at Halle, proposed a variant of the theory which he renamed Becher's *terra pinguis* to *phlogiston*, and it was in this form that the theory probably had its greatest influence.

Carbon dioxide was the first gas to be described as a discrete substance. In about 1640, the Flemish chemist Jan Baptist van Helmont observed that when he burned charcoal in a closed vessel, the mass of the resulting ash was much less than that of the original charcoal. His interpretation was that the rest of the charcoal had been transmuted into an invisible substance he termed a *gas* or *wild spirit* (*spiritus sylvestre*).

The properties of carbon dioxide were studied more thoroughly in the 1750s by the Scottish physician Joseph Black. He found that limestone (calcium carbonate) could be heated or treated with acids to yield a gas he called *fixed air*. He observed that the fixed air was denser than air and supported neither flame nor animal life. Black also discovered the magnesium and the latent and specific heat.

In his article *On Fractious airs*, published in 1766,[7] Henry Cavendish (1731–1810) reports the discovery of hydrogen (*flammable air*). He speculated that the *flammable air* was in fact identical to the hypothetical substance called *phlogiston* and in a further finding in 1781, he discovered that the gas produces water when burned. In 1783, Antoine Lavoisier gave the element the name hydrogen (from the Greek *hydro* meaning water and *genes* meaning generator) when he and Laplace reproduced Cavendish's finding that water is produced when hydrogen is burned.

The discovery of oxygen is attributed to Joseph Priestley (1733–1804) in 1774. Priestley was an eighteenth century English theologian, dissenting clergyman, natural philosopher, chemist, educator, and a liberal political theorist who published over 150 works.

During his lifetime, Priestley's considerable scientific reputation rested on his invention of soda water, his writings on electricity, and his discovery of several *airs* (gases), the most famous being what Priestley dubbed *dephlogisticated air* (oxygen). Many of the friends of Priestley were members of the *Lunar Society*, a group of manufacturers, inventors, and natural philosophers who assembled monthly to discuss their work. The core of the group included men such as the manufacturer Matthew Boulton, the chemist and geologist James Keir, the inventor and engineer James Watt, and the botanist, chemist, and geologist William Withering. Priestley was asked to join this unique society and contributed much to the work of its members. However, Priestley's determination to defend the phlogiston theory and to reject what would become the chemical revolution eventually left him isolated within the scientific community.

Priestley's science was integral to his theology, and he consistently tried to fuse Enlightenment rationalism with Christian theism. In his metaphysical texts, Priestley attempted to combine theism, materialism, and determinism. Priestley, who strongly believed in the free and open exchange of ideas, advocated toleration and equal rights for religious Dissenters, which also led him to help found Unitarianism in England.

---

[7] Cavendish H (1766) *Three Papers, Containing Experiments on Factitious Air, by the Hon. Henry Cavendish, F. R. S.* Philos Trans R Soc London 56:141–184 [28].

The controversial nature of Priestley's publications combined with his outspoken support of the French Revolution aroused public and governmental suspicion; he was eventually forced to flee, in 1791, first to London, and then to the United States after his home and church were burned. He spent the last ten years of his life living in Northumberland County, Pennsylvania.

Antoine Laurent de Lavoisier was an outstanding chemist, considered to be the *father of modern chemistry*. Lavoisier's chemical research between 1772 and 1778 was largely concerned with developing his own new theory of combustion.

In 1777, Antoine Lavoisier wrote the *Mémoire sur la combustion en général*, the first of what proved to be a series of attacks on phlogiston theory[8],[9]; it was against these attacks that Priestley responded in 1796. While Priestley accepted parts of Lavoisier's theory, he was unprepared to assent to the major revolutions Lavoisier proposed: the overthrow of phlogiston, a chemistry-based conceptually on elements and compounds, and a new chemical nomenclature. Priestley's original experiments on *dephlogisticated air*, combustion, and water provided Lavoisier with the data he needed to construct much of his system; yet Priestley never accepted Lavoisier's new theories and continued to defend the phlogiston theory for the rest of his life. Lavoisier's system was based largely on the quantitative concept that mass is neither created nor destroyed in chemical reactions (i.e., the conservation of mass).

> Mais si tout s'explique en chimie d'une manière satisfaisante sans le secours du phlogistique, il est par cela seul infiniment probable que ce principe n'existe pas; que c'est un être hypothétique, une supposition gratuite; et, en effet, il est dans les principes d'une bonne logique de ne point multiplier les êtres sans nécessité. Peut-être aurais-je pu m'en tenir à ces preuves négatives, et me contenter d'avoir prouvé qu'on rend mieux compte des phénomènes sans phlogistique qu'avec le phlogistique; mais il est temps que je m'explique d'une manière plus précise et plus formelle sur une opinion que je regarde comme une erreur funeste à la chimie, et qui me paraît en avoir retardé. (Antoine Lavoisier, 1783, [101])

1.2.1 Gay-Lussac and Clapeyron

Louis Joseph Gay-Lussac (1778–1850) measured the dilation of ideal gases as affected by the temperature in 1802.[10] His apparatus is sketched in Fig. 1.5. It includes a balloon full of a dried gas, isolated from the atmosphere through a mercury stopper placed in a graduated capillary tube. It is submitted at melting ice and the temperature is increased till boiling water temperature. The volume of the gas was measured by reading the mercury drop position.

One of Gay-Lussac's conclusions was: "*For the permanent gases, the increase of volume received by each of them between the temperature of melting ice and that*

[8] Lavoisier A (1777) *Mémoire sur la combustion en général*. Mémoires l'Académie des Sci 592–600.

[9] Lavoisier A (1783) *Réflexions sur le phlogistique, pour servir de suite á la théorie de la combustion et de la calcination* publiée en 1777. Mémoires l'Académie des Sci 505–538, [101].

[10] Gay-Lussac LJ (1802) *L'expansion des gaz par la chaleur*. Ann Chim 43:137, [58].

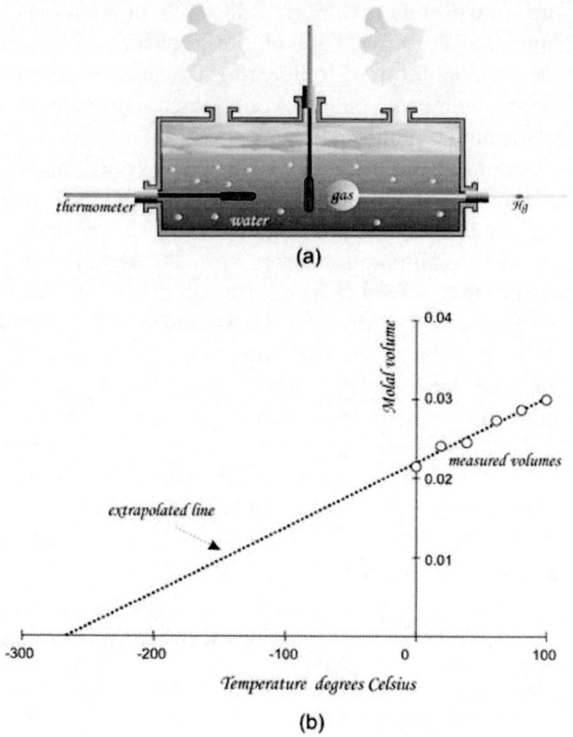

Fig. 1.5 **a** Gay-Lussac experiment for measuring the dilation of gases with the temperature. **b** Results found for the molar volume dependence on the temperature at atmospheric pressure

of boiling water is equal to (...) 100/266.66 of the same volume for the centigrade thermometer".

So, he found that for all '*permanent*', or ideal gases, the relative variation of the volume $\frac{\Delta v}{v_0}$ between the melting ice and boiling water was constant,

$$\frac{\Delta v}{v_0} = \frac{100}{267}.$$

For an arbitrary temperature θ in °C, this relationship can be written as

$$\frac{v(\theta)}{v_0} = \frac{\theta + 267}{267} \quad \text{or} \quad v(\theta) = const. \times (\theta + 267)$$

and, although Gay-Lussac did not report this finding, the value $\theta = -267\,°C$ was the first estimation of the zero absolute temperature.[11]

[11] Presently, the absolute zero is considered to be $\theta = -273.16\,°C$.

The law of gas expansion is presently known as the Charles and Gay-Lussac law. Indeed, in his 1802 paper, Gay-Lussac wrote: "*Before going further, I must jump ahead. Although I had recognized on many occasions that the gases oxygen, nitrogen, hydrogen, carbonic acid, and atmospheric air all expand identically, citizen Charles had noticed the same property in these gases 15 years ago; however, since he never published his results, it is only by great luck that I knew it*".

An important conclusion in his article was that the law of gas expansion could not be applied to vapors—or condensable gases—except when these vapors were in the 'elastic state': "*Since all gases are equally expansible by heat and equally compressible, and since these two properties depend on each other, as I shall show in another place, the vapors which are equally expansible with the gases should also be equally compressible, but I may mention that this last conclusion cannot be true except so long as the compressed vapors remain entirely in the elastic state, and this requires that their temperature shall be sufficiently elevated to enable them to resist the pressure which tends to make them assume the liquid state*".

The equation of state for ideal gases as it is presently known was only enunciated in 1834, 32 years after Gay-Lussac, in a historical article by Benoît émile Clapeyron (1799–1864), published in the Journal de l'Ecole Royale Polytechnique[12] (Fig. 1.6). In this article, Clapeyron combined the Gay-Lussac's law with Boyle-Mariotte's law, reasoning that when the temperature increases from θ_0 to θ at a constant pressure P_0, then the molar volume $v\,(\theta, P_0)$ will follow

$$\frac{v\,(\theta, P_0)}{v\,(\theta_0, P_0)} = \frac{\theta + 267}{\theta_0 + 267}.$$

Therefore, by inserting this relationship into Boyle-Mariotte's law

$$pv\,(\theta, P) = p_0 v\,(\theta, P_0) = \frac{p_0 v\,(\theta_0, P_0)}{\theta_0 + 267}\,(\theta + 267)\,,$$

or

$$\frac{pv\,(\theta, P)}{\theta + 267} = \frac{p_0 v\,(\theta_0, P_0)}{\theta_0 + 267} = cte = \mathcal{R},$$

where $\mathcal{R}$ is the well-known universal gas constant nowadays.

[12] Clapeyron BE (1834) *Mémoire sur la Puissance Motrice de la Chaleur.* J l'Ecole R Polytech 23:153–190, [33].

Soit v_0 le volume du gaz sous la pression p_0 et à la température t_0; soient p et v le volume et la pression du même poids du gaz à la température t du corps A. La loi de Mariotte, combinée avec celle de Gay-Lussac, établit entre ces quantités diverses la relation

$$pv = \frac{p_0 v_0}{267 + t_0}(267 + t),$$

ou posant, pour simplifier, $\frac{p_0 v_0}{267 + t_0} = \mathrm{R}$:

$$pv = \mathrm{R}(267 + t).$$

Fig. 1.6 Clapeyron's equation published at page 164 of the *Mémoire sur la Puissance Motrice de la Chaleur*

1.3 The Steam Engines and Carnot

> L'étude de ces machines est du plus haut intérêt, leur importance est immense, leur emploi s'accroît tous les jours. Elles paraissent destinées à produire une grande révolution dans le monde civilisé. Déjá la machine à feu exploite nos mines, fait mouvoir nos navires, creuse nos ports et nos rivières, forge le fer, façonne les bois, écrase les grains, file et ourdit nos étoffes, transporte les plus pesans fardeaux, etc. Elle semble devoir un jour servir de moteur universel et obtenir la préférence sur la force des animaux, les chutes d'eau et les courants d'air. Elle a, sur le premier de ces moteurs, l'avantage de l'économie; sur les deux autres, l'avantage inappréciable de pouvoir s'employer en tous temps et en tous lieux, et de ne jamais souffrir d'interruption dans son travail. (Sadi Carnot, 1824, [27])

Readers are referred to Robert Henry Thurston[13] book '*A History of the Growth of the Steam-Engine*', [175], for a complete and detailed history of steam engines beginning with Hero in Greece. This book was re-edited several times and a pdf is available for downloading.

The steam engine was first used to pump water out of coal mines. In 1712 Thomas Newcomen's atmospheric engine became the first commercially successful engine using the principle of the piston and cylinder, which was the fundamental type of steam engine used until the early twentieth century (Fig. 1.7).

Steam produced by boiling water pushes the piston upward. When the piston is at the top of the cylinder, a valve opens allowing a spray of cold water to be admitted into the cylinder, condensing the steam and, so, creating a partial vacuum below the piston. The piston is, then, forced to move downward, completing the cycle, by the action of the atmospheric pressure above it. In the Newcomen engine, the steam generator and the condenser were, then, in the same place.

While working as an instrument maker at the University of Glasgow, James Watt became interested in the technology of steam engines. He realized that contemporary engine designs wasted a great deal of energy by repeatedly cooling and reheating the cylinder. Watt introduced a design enhancement, the separate condenser, which avoided this waste of energy and radically improved the power, efficiency,

[13] Robert Henry Thurston (1839–1903) was an American engineer, and the first Professor of Mechanical Engineering at Stevens Institute of Technology.

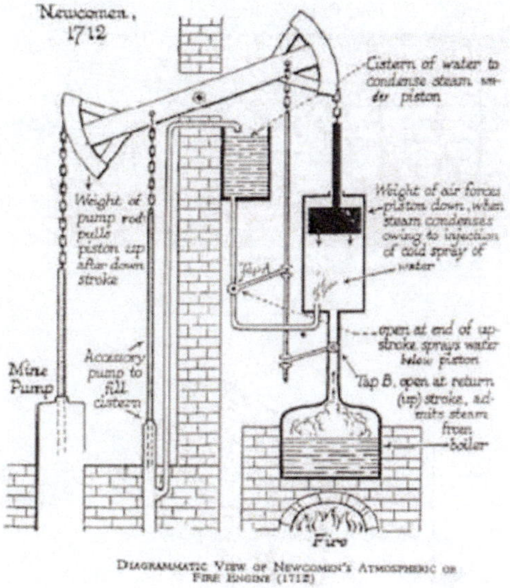

Fig. 1.7 Newcomen steam engine

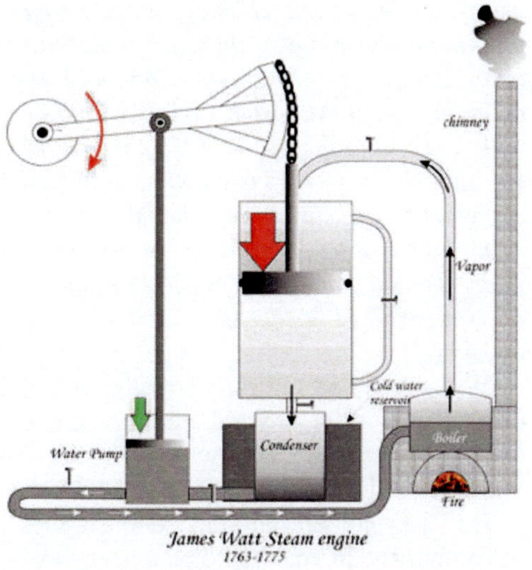

Fig. 1.8 First stage of the James Watt steam engine

and cost-effectiveness of steam engines. He, also, invented the double-acting cylinder (Figs. 1.8 and 1.9). These improvements more than doubled the efficiency of the engine with respect to Newcomen's engine.

In Watt's words,[14] "*I had gone to take a walk on a fine Sabbath afternoon. I had entered the Green by the gate at the foot of Charlotte Street and had passed the old*

[14] Reminiscences of James Watt, Robert Hart, *Transactions of the Glasgow Arch Theological Society*, 1709.

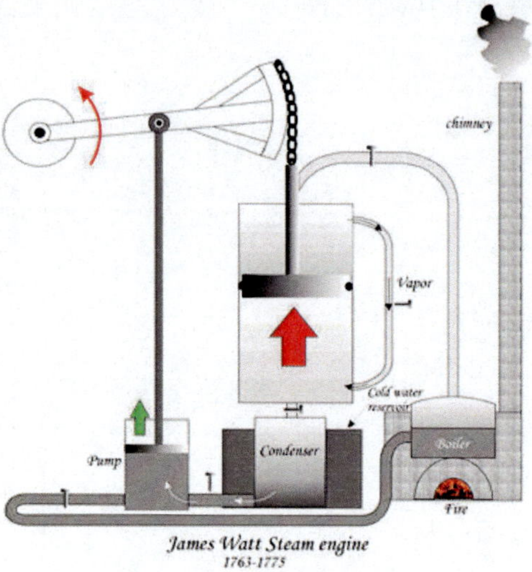

Fig. 1.9 Second stage of the James Watt steam engine

washing house. I was thinking upon the engine at the time and had gone as far as the herd's house, when the idea came into my mind that, as steam was an elastic body, it would rush into a vacuum, and, if communication were made between the cylinder and an exhausted vessel, it would rush into it and might be there condensed without cooling the cylinder. I then saw that I must get rid of the condensed steam and injection water if I used a jet, as in Newcomen's engine. Two ways of doing this occurred to me: First, the water might be run off by a descending pipe, if an offlet could be got at the depth of 35 or 36feet, and any air might be extracted by a small pump. The second was, to make the pump large enough to extract both water and air. I had not walked farther than the Golf house when the whole thing was arranged in my mind"

Watt attempted to commercialize his invention but experienced great financial difficulties until he entered a partnership with Matthew Boulton in 1775. The new firm of Boulton and Watt was highly successful and Watt became a wealthy man. Watt's invention was a key point in the Industrial Revolution.

Both Newcomen and Watt steam engines depend on a source of energy—the heat of combustion of coal—and on a heat sink—the cold water.

When Sadi Carnot [27], began working on his '*Réflexions sur la puissance motrice du feu*', published in 1824, steam engines had achieved widely recognized economic and industrial importance, but there had been no real scientific study of them. Engineers in Carnot time had tried to use highly pressurized steam to improve the efficiency of engines. In these early stages of engine development, the efficiency of a typical engine—the useful work it was able to do when a given quantity of fuel was burned—was only about 3%.

In the first part of his *Réflexions*Carnot advocates that *everywhere where we have a difference in temperature it is possible to generate motive power*, a central idea in thermodynamics. A corollary of this idea is that *it is impossible to produce motive power unless we dispose, simultaneously, of a hot and a cold body*, which is a first enunciate of the second principle of thermodynamics. Carnot based his *réflexions* on Lavoisier's caloric theory, comparing the flow of caloric from the high to the low temperature to the flow of water through a hydraulic machine, producing work in its passage: the higher the height of the waterfall the higher the delivered work. He concluded: "*The production of motive power is, therefore, due in steam engines not to actual consumption of caloric but to its transportation from a warm body to a cold body*".

1.4 The Principle of Energy Conservation

> Believing that the power to destroy belongs to the Creator alone I affirm...that any theory which, when carried out, demands the annihilation of force is necessarily erroneous. (James Prescott Joule, 1843)

1.4.1 Joseph Black

The concept of '*heat*' or the internal energy of a thermodynamic system as the result of the mechanical energy of its individual molecules is very old and was supported by philosophers and scientists. "*Heat is motion*" in the words of Francis Bacon (around 1600). "*Heat is nothing else but the motion of the particles that form a body*" (Thomas Hooke, around 1650). Nevertheless, the caloric theory and mechanics evolved as independent sciences since the invention of the thermoscope by Galileu in 1600, the built-up of the mercury thermometer and the establishment of measurement methods for the specific and latent heat in terms of heat unity, the calorie (*the amount of heat that raises the temperature of 1 g of water by 1°C*).

Joseph Black (1728—1799) was a Scottish physicist and chemist, known for his discoveries of magnesium and carbon dioxide, latent and specific heat. He was Professor of Anatomy and Chemistry at the University of Glasgow for 10 years from 1756, and then Professor of Medicine and Chemistry at the University of Edinburgh from 1766, teaching and lecturing there for more than 30 years.

Black was the first to illustrate that heat and temperature are different things. Although thermometers existed, there were several different ideas as to what they were measuring. One way Black demonstrated the distinction was to take a bucket of ice and monitor its temperature constantly. The ice gradually melted, yet the temperature stayed the same. This finding credited Black with the discovery of latent heat. From this Black developed the idea of'specific heat' by measuring the amount of heat required to raise the temperature of a substance by a specified number of degrees.

In his *Lectures on the Elements of Chemistry delivered at the University of Edinburgh* [12] Black gives a specific example involving the measurement of the specific heat of mercury (or quicksilver, as mercury was known at that time): "*To make this plainer by an example in numbers, let us suppose the water to be at the 100th degree of heat, and that an equal measure of warm quicksilver at the 150th degree, is suddenly mixed and agitated with it. We know that the middle temperature between 100 and 150 is 125, and we know that this middle temperature would be produced by mixing the cold water at 100 with an equal measure of warm water at 150; the heat of the warm water being lowered by 25 degrees, while that of the cold is raised just as much. But when warm quicksilver is used in place of warm water, the temperature of the mixture turns out 120 degrees only instead of 125. The quicksilver, therefore, is become less warm by 30 degrees, while the water has become warmer by twenty degrees only; and yet the quantity of heat that the water has gained is the very same quantity that the quicksilver has lost*".

In [12] Black proposed in his lecture '*On the General Effects of Heat*': (i) to ascertain what he means by the word *heat*; (ii) to explain the meaning of the term *cold*, and ascertain the real difference between cold and heat; (iii) to mention some of the attempts that have been made to discover the nature of heat, or to form an idea of what may be the immediate cause of it; (iv) to describe the sensible effects produced by heat on the bodies to which it is communicated.

He writes in page 25 of [12] "*When we consider the communication of heat from hot bodies to colder ones, the first question that may naturally occur to our mind is, in what manner have these two bodies acted, the one on the other, on this occasion? Has one of them lost something, which the other has gained? The vulgar opinion is, that the hot body has lost something which has been added to the other. And those who have attempted to reason more profoundly on the nature of heat, have agreed with the multitude in this point and have supposed that heat is a positive quality and depends either upon an exceedingly subtle and active matter, introduced into the pores of bodies, or upon a tremor or vibration excited among their particles*".

Therefore, although Black does not deny the idea of heat as a *subtle fluid*, the concept of heat or internal energy as the mechanical energy of the *particles* that form a body is present in Black's writings.

1.4.1.1 Lavoisier and the Caloric Theory

In Sect. 1.2 we saw that Lavoisier showed the inconsistency of Stahl's phlogiston theory of combustion with his own experimental results leading him to postulate the *principle of mass conservation*. Lavoisier did early research in physical chemistry and thermodynamics in joint experiments with Laplace and, controversially, also postulated the *principle of caloric conservation*. Caloric was considered by him as a kind of *weightless fluid* and, according to this theory, the quantity of this substance is constant throughout the universe and flows from warmer to colder bodies.

He writes in his *Réflexions sur le phlogistique: on the nature of heat [101]:*

> When an ordinary body—solid or fluid—is heated, that body increases in dimensions in all directions it occupies a larger and larger volume. If the cause of heating ceases, the body retreats through the same degrees of extension that it reached, at the same rate as it cools. Finally, if it is returned to the same temperature that it had at the first instant, it will clearly return to the same volume as it had before. Hence the corpuscles of matter do not touch each other, there exists between them a distance that heat increases and that cold decreases. One can scarcely conceive of these phenomena except by admitting the existence of a subtle fluid, the accumulation of which is the cause of heat and the absence of which is the cause of coldness. No doubt it is this fluid that lodges between the particles of matter, which spreads them apart and which occupies the space left between them. Along with the majority of physicists, I name this fluid (whatever it is) igneous fluid, the matter of heat and fire.
>
> I do not deny that the existence of this fluid is, up to a certain point, hypothetical. But supposing it is a hypothesis (that it is not rigorously proven), it is the only one that I am obliged to posit. The partisans of the phlogiston doctrine are no more sophisticated than me on this point and, if the existence of the igneous fluid is a hypothesis, it is common to both their system and my own.
>
> We see that in such a state of affairs, the corpuscles would not have any connection with each other, that there would be no solid bodies if they were not held together by another force—by attraction—which is a general principle of nature to which all matter seems to submit, whatever the cause.
>
> According to this first glimpse, all natural bodies obey two forces: igneous fluid (the matter of fire), which tends to continually separate corpuscles, and the attraction that counter-balances this force. So long as the latter force—attraction—is victorious, bodies remain in a solid state. When these two forces are in a state of equilibrium, the substance becomes liquid. Finally, when the expansive force of heat carries it off, the substance passes into the aeriform state. But if only these two forces existed, at the moment the bodies ceased to be in the solid state the slightest increase in heat that they received would be sufficient to vaporize them. Not only would they pass quickly to the aeriform state, but their corpuscles would also be carried away endlessly further and further. But there is a third force, which stops this effect from taking place—this is the weight of the atmosphere. Without this pressure, at the moment when water ceases to be ice (at zero degrees) it would become an aeriform fluid, whereas in fact this effect only occurs at a temperature of 100 C, at a pressure of one atmosphere.

In present days, the conservation of caloric principle would be written as

$$\delta Q = dU, \tag{1.1}$$

meaning that the amount of caloric U of a body—or the mechanical energy of their molecules, as we consider today—can only be changed by the addition of a certain amount of heat δQ from an external source of caloric. For an isolated thermodynamic system

$$U = \text{constant}.$$

Therefore, following Lavoisier, the total amount of caloric U in the universe was constant and was not to be confused with *force*, which was the designation of *energy* at the time, but should be considered as a *subtle fluid* that was transferred from warm to cold bodies.

1.4.2 Joule's Experiments

The text in the following is based on the notes about Joule by Ingo Muller [119] and on the Wikipedia article[15] (https://en.wikipedia.org/wiki/James_Prescott_Joule).

The son of a wealthy brewer, Joule was tutored as a young man by the famous scientist John Dalton[16] and was strongly influenced by chemist William Henry[17] and Manchester engineers Peter Ewart and Eaton Hodgkinson. He was fascinated by electricity. As an adult, Joule managed the brewery. Science was merely a serious hobby. Sometime around 1840, he started to investigate the feasibility of replacing the brewery's steam engines with the newly invented electric motor. He discovered Joule's first law in 1841, that the heat which is evolved by the proper action of any voltaic current is proportional to the square of the intensity of that current, multiplied by the resistance to conduction which it experiences.[18]

Joule's interest diverted from the narrow financial question to that of how much work could be extracted from a given source, leading him to speculate about the convertibility of energy. In 1843 he published results of experiments showing that the heating effect he had quantified in 1841 was due to the generation of heat in the conductor. This was a direct challenge to the caloric theory which held that heat could neither be created nor destroyed. The caloric theory has dominated thinking in the science of heat since introduced by Antoine Lavoisier in 1783. Lavoisier's prestige and the practical success of Sadi Carnot's caloric theory of the heat engine

---

[15] James P. Joule received several biographies. See for instance:

Bottomley, J. T. (1882).*James Prescott Joule*. Nature 26 (678): 617–620. Bibcode:1882Natur..26..617B. https://doi.org/10.1038/026617a0.

Cardwell, D. S. L. (1991). *James Joule: A Biography*. Manchester University Press. ISBN 0-7190-3479-5.

Forrester, J. (1975). *Chemistry and the Conservation of Energy: The Work of James Prescott Joule*. Studies in the History and Philosophy of Science 6 (4): 273–313. https://doi.org/10.1016/0039-3681(75)90025-4.

Fox, R, *James Prescott Joule, 1818–1889*, in North, J. (1969). Mid-nineteenth-century scientists. Elsevier. pp. 72–103. ISBN 0-7190-3479-5.

Reynolds, Osbourne (1892). *Memoir of James Prescott Joule*. Manchester, England: Manchester Literary and Philosophical Society.

[16] The most important of all Dalton's investigations are those concerned with the beginning of atomic theory in chemistry. The main points of Dalton's atomic theory were: (i) elements are made of extremely small particles called atoms; (ii) atoms of a given element are identical in size, mass, and other properties; atoms of different elements differ in size, mass, and other properties; (iii) atoms cannot be subdivided, created, or destroyed; (iv) atoms of different elements combine in simple whole-number ratios to form chemical compounds; (v) in chemical reactions, atoms are combined, separated, or rearranged.

[17] Henry W (1803) *Experiments on the Quantity of Gases Absorbed by Water, at Different Temperatures, and under Different Pressures*. Philos Trans R Soc London 93:29–274. https://doi.org/10.1098/rstl.1803.0004 [69].

[18] Joule JP (1841) *On the heat evolved by metallic conductors of electricity, and in the cells of a battery during electrolysis*. Philos Mag Ser 3 19:260–277. https://doi.org/10.1080/14786444108650416 [79].

since 1824 ensured that the young Joule, working outside either academia or the engineering profession, had a difficult road ahead.

Further experiments and measurements led Joule to estimate the mechanical equivalent of heat as 4.1868 J per calorie of work to raise the temperature of a pound of water by one degree Fahrenheit. He announced his results at a meeting of the chemical section of the British Association for the Advancement of Science in Cork in August 1843 and was met by silence.

Joule was undaunted and started to seek a purely mechanical demonstration of the conversion of work into heat. By forcing water through a perforated cylinder, he could measure the slight viscous heating of the fluid. He obtained a mechanical equivalent of 770 ft.lbf/Btu (4.14 J/cal). The fact that the values obtained both by electrical and purely mechanical means were in agreement to at least one order of magnitude was, to Joule, compelling evidence of the reality of the convertibility of work into heat.

Joule now tried a third route. He measured the heat generated against the work done in compressing a gas. He obtained a mechanical equivalent of 798 ft.lbf/Btu (4.29 J/cal). In many ways, this experiment offered the easiest target for Joule's critics but Joule disposed of the anticipated objections by clever experimentation. Joule read his paper to the Royal Society on 20 June 1844, however, his paper was rejected for publishing by the Royal Society and he had to be content with publishing in the Philosophical Magazine in 1845.[19]

In June 1845, Joule read his paper *On the Mechanical Equivalent of Heat* to the British Association meeting in Cambridge. In this work, he reported his best-known experiment, involving the use of a falling weight, in which gravity does the mechanical work, to spin a paddle wheel in an insulated barrel of water which increased the temperature (Fig. 1.10). He now estimated a mechanical equivalent of 819 ft.lbf/Btu (4.41 J/cal). He wrote a letter to the Philosophical Magazine, published in September 1845 describing his experiment.[20]

In 1850, Joule published a refined measurement of 772.692 ft.lbf/Btu (4.159 J/cal).[21]

Joule was proposing a mechanical theory of heat, and this required a conceptual leap: *if heat was a form of molecular motion, why the motion of the molecules did not gradually die out?* Joule's ideas required one to believe that the collisions of molecules were perfectly elastic. We should also remember that the very existence of atoms and molecules was not widely accepted for another 50 years.

Although it may be hard today to understand the allure of the caloric theory, at the time it seemed to have some clear advantages. Carnot's successful theory of heat engines had also been based on the caloric assumption, and only later was it proved

---

[19] Joule JP (1845) *On the changes of temperature produced by the rarefaction and condensation of air*. Philos Mag Ser 3 26:369–383. https://doi.org/10.1080/14786444508645153 [78].

[20] Joule, J.P. (1845) *On the Mechanical Equivalent of Heat*, Brit. Assoc. Rep., trans. Chemical Sect, p.31, read before the British Association at Cambridge, June 1845.

[21] Joule, J.P. (1850). *On the Mechanical Equivalent of Heat*. Philosophical Transactions of the Royal Society of London 140: 61–82. https://doi.org/10.1098/rstl.1850.0004 [80].

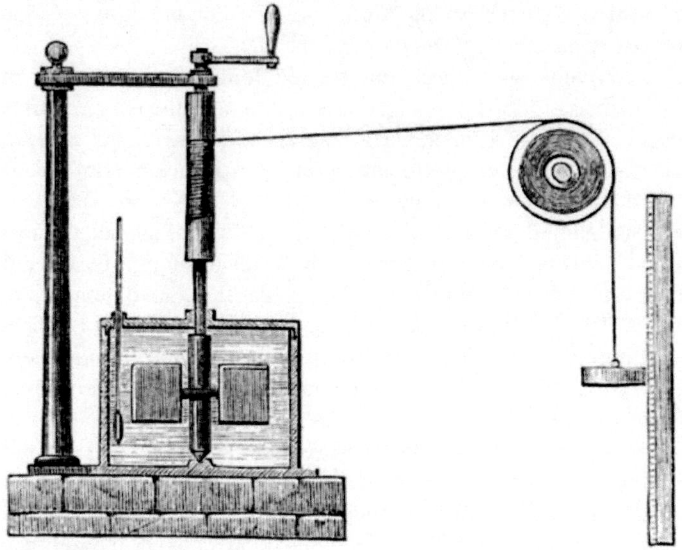

Fig. 1.10 Joule apparatus for measuring the mechanical equivalent of heat

by Lord Kelvin that Carnot's mathematics were equally valid without assuming a caloric fluid.

However, in Germany, Hermann Helmholtz became aware both of Joule's work and the similar 1842 work of Julius Robert von Mayer. Though both men had been neglected since their respective publications, Helmholtz's definitive 1847 declaration of the conservation of energy credited them both.[22]

Also in 1847, another of Joule's presentations at the British Association in Oxford was attended by George Gabriel Stokes, Michael Faraday, and the precocious and maverick William Thomson, later to become Lord Kelvin, who had just been appointed professor of natural philosophy at the University of Glasgow. Stokes was *inclined to be a Joulite* and Faraday was *much struck with it* though he harbored doubts. Thomson was intrigued but skeptical.

Though Thomson felt that Joule's results demanded a theoretical explanation, he retreated into a spirited defense of the Carnot-Clapeyron school. In 1848, Thomson wrote that *the conversion of heat (or caloric) into mechanical effect is probably impossible, certainly undiscovered*—but a footnote signaled his first doubts about the caloric theory, referring to Joule's *very remarkable discoveries*. Surprisingly, Thomson did not send Joule a copy of his paper but when Joule eventually read it he wrote to Thomson on 6 October, claiming that his studies had demonstrated conversion of heat into work but that he was planning further experiments. Thomson replied on the 27th, revealing that he was planning his own experiments and hoping for a reconciliation of their two views. Though Thomson conducted no new experiments,

[22] von Helmholtz, HLF. *Über die Erhaltung der Kraft*. Reimer, Berlin 1847 [68].

over the next two years he became increasingly dissatisfied with Carnot's theory and convinced of Joule's. In his 1851 paper, Thomson was willing to go no further than a compromise and declared *the whole theory of the motive power of heat is founded on two propositions,*[23] *due respectively to Joule, and to Carnot and Clausius.*

As soon as Joule read the paper, he wrote to Thomson with his comments and questions. Thus began a fruitful, though largely epistolary, collaboration between the two men, Joule conducting experiments, Thomson analyzing the results and suggesting further experiments. The collaboration lasted from 1852 to 1856, and its discoveries including the Joule-Thomson effect and the published results did much to bring about general acceptance of Joule's work and the kinetic theory.

1.4.3 Exercise

In his 1850 paper [80], Joule wrote: "*For a long time it had been a favorite hypothesis that heat consists of a force or power belonging to bodies, but it was reserved for Count Rumford to make the first experiments decidedly in favor of that view. That justly celebrated natural philosopher demonstrated by his ingenious experiments that the very great quantity of heat excited by the boring of cannon could not be ascribed to a change taking place in the calorific capacity of the metal; and he therefore concluded that the motion of the borer was communicated to the particles of metal, thus producing the phenomena of heat. It appears to me, he remarks, extremely difficult, if not quite impossible, to form any distinct idea of anything, capable of being excited and communicated, in the manner the heat was excited and communicated in these experiments, except it be a renovation of the most important parts of Count Rumford's paper, though one to which little attention has hitherto been paid, is that in which he makes an estimate of the quantity of mechanical force required to produce a certain amount of heat. Referring to his third experiment, he remarks that the total quantity of ice-cold water which, with the heat actually generated by friction, accumulated in 2h30, might have been heated* 180 °*F, or made to boil 26.58 lbs. In the next page, he states that the machinery used in the experiment could easily be carried round by the force of one horse (though to render the work lighter, two horses were actually employed in doing it). Now the power of a horse is estimated by Watt as 33,000 foot-pounds per minute.*"

Using the data from Count Rumford and the Watt estimation of the power of a horse, find the mechanical equivalent of heat, and compare it with the 772 lbf.foot/btu finding of Joule.

[23] The first and second principle of thermodynamics.

1.4.4 The Principle of Energy Conservation

The principle of mechanical energy conservation was first evidenced by Galileo with his experiments on falling bodies and on the inclined plane. Gottfried Leibniz (1646–1716) extended the work of his mentor Huygens on the importance of the product of mass times velocity squared which had been originally investigated by Huygens and which Leibniz called *vis viva*, the living force. He believed the *vis viva* to be the real measure of force, as opposed to Descartes's force of motion (equivalent to mass × velocity, or momentum). That *vis viva* was the most fundamental conserved quantity comes extremely close to an early statement of the principle of conservation of energy in mechanics. Since, however, the conservation of momentum had become one of the pillars of Cartesian natural philosophy, his *vis viva* was seen as rivaling the conservation of momentum championed by Newton in England and by Descartes in France; hence academics in those countries tended to neglect Leibniz's idea. In reality, both energy, and momentum are conserved, so the two approaches are equally valid.

In present days, the principle of mechanical energy conservation would be written as

$$\delta W = dE_c + dE_p, \tag{1.2}$$

meaning that the total mechanical energy $E_c + E_p$ of a system of particles is conserved unless it is affected by the work performed by an external force. So, if we neglect the air friction and consider a single body at rest, under the action of gravity and suspended at an height h from the ground level, the velocity v this body will acquire at the ground level when falling from a height h will be found by solving

$$\underbrace{\frac{1}{2}m\mathsf{v}^2}_{E_c \text{ at ground level}} + \underbrace{0}_{E_p \text{ at ground level}} = \underbrace{0}_{E_c \text{ at heigth } h} + \underbrace{mgh}_{E_p \text{ at heigth } h} .$$

The solution of this equation gives $\mathsf{v} = \sqrt{2gh}$, which was the expression for the velocity found by Galileu for falling bodies: *the velocity of a falling body is independent of its weight*. This was an important finding, because contrary to Aristotle's early idea that heavy objects fall faster than lighter ones, *in direct proportion to weight*.

Drawing on the earlier work of Sadi Carnot, Benoît Paul émile Clapeyron, James Prescott Joule, and Faraday, Helmholtz [68] postulated a relationship between mechanics, heat, light, electricity and magnetism by treating them all as manifestations of a single force: energy in modern terms.

When heat is recognized as energy, the principle of energy conservation postulated by Helmholtz would be written, in a purely mechanical context, as the sum of Eqs. (1.1) and (1.2)

$$\delta Q + \delta W = dU + dE_c + dE_p. \tag{1.3}$$

Therefore, when a thermodynamic system performs a cycle, starting from and returning to the same thermodynamic state

$$\oint (\delta Q + \delta W) = 0.$$

Considering in Fig. 1.10 the liquid inside the vessel as a thermodynamic system, this equation summarizes Joule's ideas for the conversion of work into heat and vice-versa. Indeed, Joule's principle of the mechanical equivalent of heat can be written as

$$\oint \delta W = -\oint \delta Q,$$

where the l.h.s. of the above equation describes the friction work performed on the liquid by the paddles, with the weight fall, and the r.h.s. the heat that must be removed from the liquid for recovering its initial temperature.

In addition to its kinetic (E_c) and potential energy (E_p) a thermodynamic system has the internal energy, U (or the mechanical energy of its molecules) as a third kind of energy. The total energy of this system can be exchanged with its environment by heat or work. This is the *principle of energy conservation*

$$\delta Q + \delta W = dU + dE_c + dE_p.$$

1.4.5 The Joule Expansion

In 1843, Joule did the experiment shown in Fig. 1.11. In the left glass sphere, he introduced an *elastic* gas (*as the ideal gases were known at the time*), while the right sphere was under vacuum. The set was insulated from the ambient. When the tap was suddenly opened, the gas in the left sphere flowed without resistance into the right sphere. No work was performed, no heat was transferred and, in spite of the abrupt gas expansion, the gas temperature remained constant.

This was a surprising result since we would expect that the gas would cool after an adiabatic expansion. Nevertheless, by considering the first principle of thermodynamics for this problem

$$\delta Q + \delta W = dU, \tag{1.4}$$

it is easily seen that under an adiabatic free expansion both δQ and δW are null, so, the internal energy remains unchanged in this expansion.

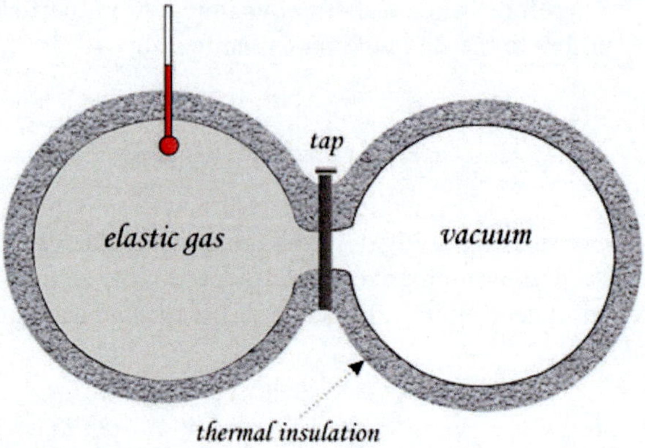

Fig. 1.11 Joule's experiment on gas expansion

The conclusion is that the internal energy of ideal gases depends only on the temperature $U = U(T)$.

1.4.6 Specific or Molar Heats

Fig. 1.12 shows an ice calorimeter developed by Lavoisier and Laplace to measure the heat that is released by chemical reactions. Ice calorimeters were based on Black's idea of latent heat.

The addition of a certain amount of heat δQ to a constant volume of a gas will elevate its temperature by ΔT (Fig. 1.13). The specific heat capacity at constant volume is defined as

$$c_v = \lim_{\Delta T \to 0} \left(\frac{\delta Q}{m \Delta T} \right)_{V=const},$$

whereas the molar heat capacity at constant volume as,

$$\overline{c}_v = \lim_{\Delta T \to 0} \left(\frac{\delta Q}{N \Delta T} \right)_{V=const},$$

where m and N are, respectively, the mass and the molar amount of the gas in the chamber.

The same definitions apply to the specific heat capacity at a constant pressure, but, in this case, is the pressure that is maintained constant during the heating process (Fig. 1.13b).

From the first principle of thermodynamics, Eq. (1.4), when heat is added at constant volume the amount $\delta Q = dU$, because the volume is constant and there is

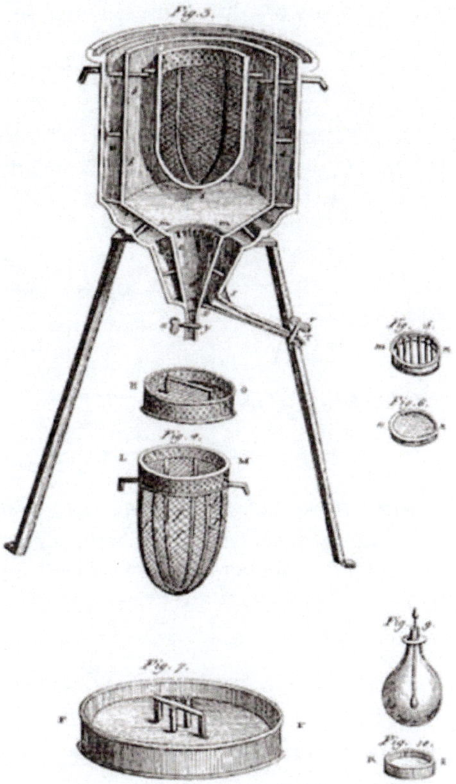

Fig. 1.12 Ice calorimeter created by Lavoisier and Laplace based on Black's idea of latent heat

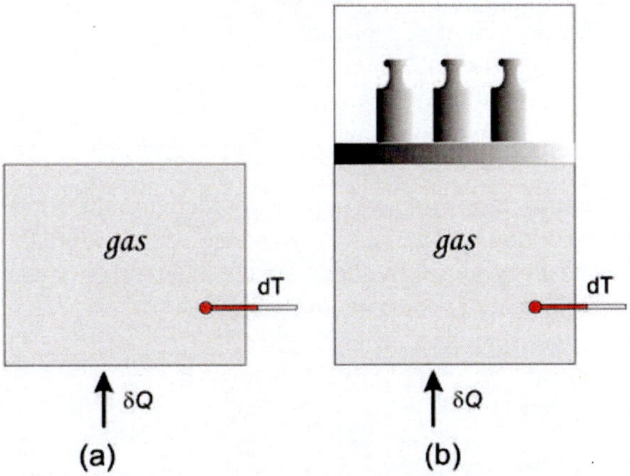

Fig. 1.13 Sketch for the specific or molar heat capacity measurement **a** at constant volume and **b** at constant pressure

no expansion work during the heating. So,

$$dU = N\bar{c}_v dT \quad \text{or} \quad dU = mc_v dT.$$

On the other hand, by recalling that the enthalpy is defined as $H = U + PV$ the amount $\delta Q = dH$ when this heat is added at constant pressure (Fig. 1.13b), because the related expansion work is $-PdV$. So

$$dH = N\bar{c}_p dT \quad \text{or} \quad dH = mc_p dT.$$

1.4.7 Ideal Gases

The molar, $\bar{c}_v$, and specific, c_v, heat capacities depend on both the temperature and pressure. Nevertheless, for ideal gases Joule's expansion experiments showed that they are only dependent on the temperature $\bar{c}_v = \bar{c}_v\,(T)$ and $c_v = c_v\,(T)$.

In addition, since for ideal gases $H = U + N\mathscr{R}T$, we are led to the conclusion that

$$c_p = c_v + R \quad \text{or} \quad \bar{c}_p = \bar{c}_v + \mathscr{R}$$

where $R = \mathscr{R}/M$, M being the molecular mass of the gas.

Specific or molar heat capacities measure the ability of the molecules to store thermal energy. It is thus an intrinsic molecular property and is related to the degrees of freedom of each molecule,

$$\bar{c}_v = \bar{c}_{v,trans} + \bar{c}_{v,rot} + \bar{c}_{v,vib},$$

where $\bar{c}_{v,trans}$, $\bar{c}_{v,rot}$, $\bar{c}_{v,vib}$, are respectively the translational, rotational, and vibrational molar heat capacities of the molecules. Monatomic molecules have only translational degrees of freedom and the kinetic theory shows that the translational molar heat capacity is constant and given by

$$\bar{c}_{v,trans} = \frac{3}{2}\mathscr{R}.$$

On the other hand, polyatomic molecules have also rotational an vibrational degrees of freedom and these are both dependent on the temperature. Many textbooks and thermodynamic tables show the specific or molar heat capacities of ideal gases as the result of numerical interpolation on experimental data. A frequently used expression is

$$\bar{c}_v = A + BT + C/T^2,$$

where A, B, C are interpolation constants.

1.5 Equilibrium and Non-equilibrium States

We can consider a thermodynamic system as a portion of matter with a well-defined frontier. This frontier can be deformed and the total volume of this system can change with time. When the system does not exchange mass with the external environment, the system is said to be *closed*, otherwise it is called an *open* system.

For the moment, let us limit ourselves to closed systems in *equilibrium* and we understand an *equilibrium state* as a state where all the *intensive variables* are the same everywhere, in all points of this system. We designate intensive variables as the ones which do not depend on the *mass* of the system such as the temperature, the pressure, and the density. In other words, intensive variables are the ones that do not change when you move from one point to another inside a system in equilibrium.

For closed systems that change their equilibrium state by exchanging heat and/or work with its external environment, the first principle of thermodynamics reads as,

$$Q_{1-2} + W_{1-2} = U_2 - U_1,$$

where U is the thermal or internal energy of the thermodynamic system, Q_{1-2} is the heat exchanged between the equilibrium states 1 and 2, and W_{1-2} is the corresponding work exchanged between these two equilibrium states.

Nevertheless, some important points must be cleared before going to the Second Principle:

(a) the first principle as above written gives an energy balance between two *equilibrium states*; both states 1 and 2 are equilibrium states and the heat and work exchanged are the result of an energy balance between these two equilibrium states;
(b) when a system moves from an equilibrium state to another it will necessarily experience *non-equilibrium* in its intensive variables; when, e.g., heat is added to a cold system, this addition will first heat the frontier and the equilibrium state will be only reached after the heat is transferred from the hot frontier to the inner parts of the system.

Therefore, in real processes, thermodynamic systems always move through non-equilibrium states. These processes are called *irreversible* because they always involve the transformation of a certain amount of mechanical energy into heat. In other words, real processes are always accompanied by dissipation.

In the following chapter, we will try to better understand the role of energy dissipation in real processes of a thermodynamic system.

Chapter 2
The Second Principle

Abstract For producing motive power in a cyclic operation, a heat engine is required to operate between two thermal reservoirs at different temperatures. What is nowadays known as Planck's statement of the second principle of thermodynamics was first perceived by Carnot and was the main difficulty for the acceptance of the equivalence between heat and work and of the principle of energy conservation. In the reverse sense, as stated by Clausius, heat cannot be transferred from a cold to a hot body by itself. Although with different words and worries, the underlying meaning behind both Plank and Clausius's statements is the same: *nature has an arrow*. We start this chapter by presenting the role of dissipation when reversible and irreversible processes are compared. This subject is followed by the presentation of the statements of the second principle and the demonstration of their equivalence and by an analysis of the works of Carnot and Clapeyron and of reversible cycles other than Carnot's cycles. Real processes are irreversible and irreversible processes are always dissipative and sources of entropy, as presented in Sect. 2.5. The chapter ends with a section on the meaning of entropy following the outstanding findings of Clausius, Maxwell, and Boltzmann.

2.1 Reversible and Irreversible Processes

2.1.1 An Abrupt Isothermal Expansion

Consider Fig. 2.1. One mole of an ideal gas is inside a cylinder with a 10 cm^2 straight section, under the action of the atmospheric pressure and three 50 kg balance weights and in thermal contact with a heat reservoir at 300 K. In these conditions, considering the atmospheric pressure to be $P_{atm} = 101325$ Pa, the initial equilibrium state of the gas is given in Table 2.1.

Now, we are interested in knowing what happens with this thermodynamic system, when a single balance weight is removed from the top of the piston, here supposed to have negligible weight. Since the gas is in thermal contact with a bath at a constant temperature and under the action of the two remaining balance weights, the final equilibrium state will be established at $T = 300$ K and $P = 1.0823$ MPa (Table 2.1).

P. C. Philippi, *Thermodynamics*, https://doi.org/10.1007/978-3-031-49357-7_2

Fig. 2.1 Isothermal expansion of an ideal gas

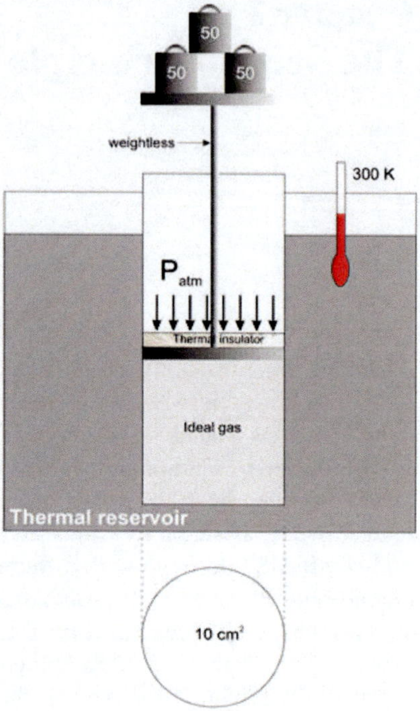

Table 2.1 Initial and final equilibrium states

Thermodynamic variables	T (K)	P (MPa)	V (ℓ)
Initial equilibrium state (1)	300	1.5728	1.5858
Final equilibrium state (2)	300	1.0823	2.3045

Since the gas is ideal, it follows the Clapeyron equation of state and the final volume is, then,

$$V = \frac{N\mathcal{R}T}{P} = 2.3045\ell.$$

Nevertheless, the sudden removal of a 50 kg weight will produce an appreciable non-equilibrium in the internal fields of temperature, pressure and density, with a rarefaction wave moving downward from the top of the gas chamber, with the speed of sound.

Consider, for instance, the temperature field produced by gas expansion. During the sudden expansion, the gas will cool below the temperature $T = 300$ K that was imposed on it in the beginning. In fact, the immediate neighboring of the cylinder walls is the only place where it is possible to assert that the temperature remains at 300 K in the course of the gas expansion.

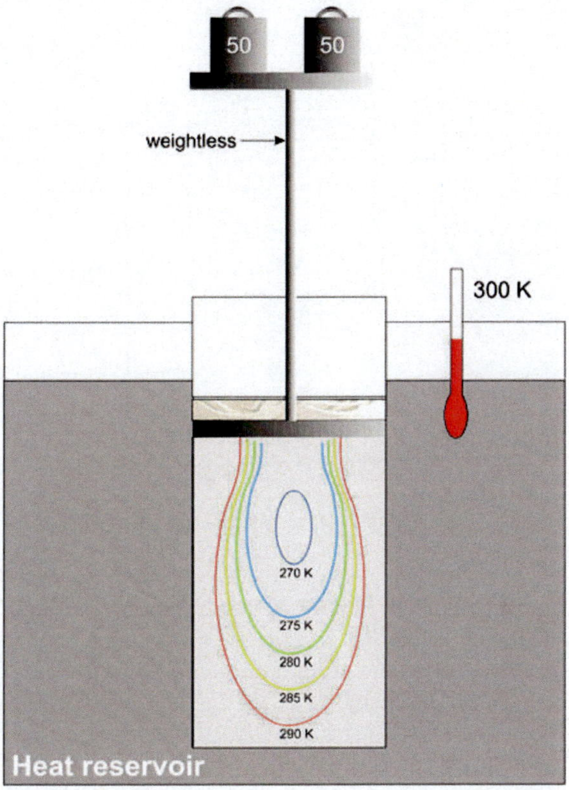

Fig. 2.2 Isotherms in an intermediate time of the abrupt expansion

Figure 2.2 shows a sketch of the isotherms in an intermediate time of the abrupt expansion. Temperature can reach very low values near the center of the gas chamber and the transfer of heat from the wall surface to the inner core is dependent on the gas thermal conductivity and on the advection currents that are formed with the piston displacement. Since the thermal conductivity of gases is very small, the attainment of thermal equilibrium can be a very slow process.

Let us try to get a deeper understanding of what happens between the initial and the final equilibrium states, by sketching the time evolution of the pressure that would be measured in two points, just above and below the piston base (Fig. 2.3). So, let P_{ext} be the external pressure just above the piston base and P_{int} be the internal pressure just below this base.

The external pressure P_{ext} falls, instantaneously, from P_1, the initial pressure in the gas chamber to P_2, the final pressure. Nevertheless, the internal pressure P_{int} is not able to follow P_{ext}, because when the pressure falls just below the piston base, the remaining gas in the chamber will move upward, trying to restore the pressure. The time evolution of P_{ext} and P_{int} are pictured in Fig. 2.3. This pressure delay creates a pressure difference between the outer and inner faces of the piston base. So, the

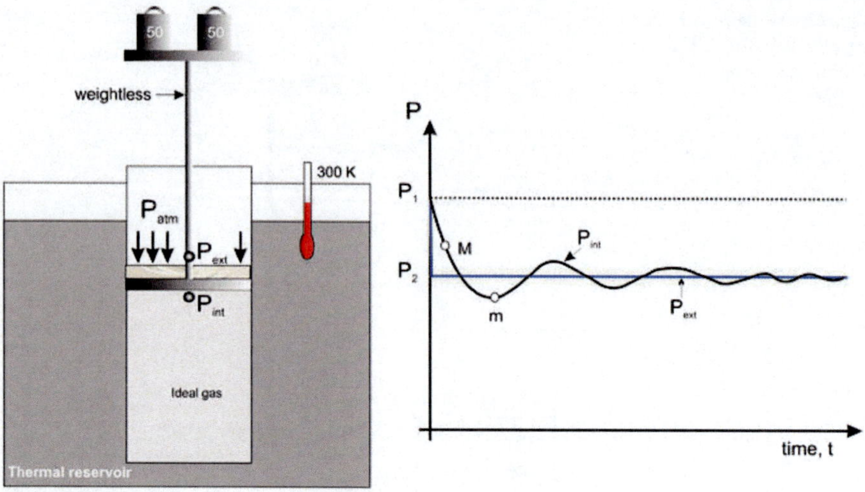

Fig. 2.3 Time evolution of the external and internal pressure just above and below the base plate of the piston

piston will be accelerated upward when $P_{int} > P_{ext}$. This is the case of point M in the figure. When $P_{int} = P_{ext}$, the net force on the piston is null, but the piston has a finite velocity and will continue to move upward due to its inertia (the Newton's First principle of Mechanics). After that, the piston is braked, because $P_{int} < P_{ext}$ until it has a null displacement velocity at point m, returning back to its final equilibrium position with an accelerated motion.

It is thus easy to conclude that the piston will have an oscillatory motion around its final equilibrium position. These oscillations will be damped by two kinds of physical mechanisms: (a) friction between the piston and the cylinder wall and (b) viscous forces.

The final equilibrium state is reached when: (a) the heat transferred from the liquid bath restores the initial temperature $T = 300$ K in the whole gas phase; (b) the viscous forces inside the gas dissipate the advection flows that appear during the expansion, transforming kinetic energy into internal energy. So, the time that the gas needs to reach the final equilibrium state is dependent on both its thermal conductivity and viscosity.

2.1.2 Expansion Work

In classical mechanics, the work δW that is required for displacing a particle from a point $\mathbf{x}$ to a point $\mathbf{x} + d\boldsymbol{\ell}$ is defined as

$$\delta W = \mathbf{F} \cdot d\boldsymbol{\ell},$$

where **F** is the force that is required for this displacement.

In the present case, this force F is given by the internal pressure P_{int} depicted in Fig. 2.3 multiplied by the area A of the cylinder cross-section. On the other hand the displacement $d\ell$ is given by $d\ell = \mathsf{v}dt$, where v is the piston velocity. Both P_{int} and v are dependent on the time, t. So, the total work that is released by the gas between its initial and final equilibrium states will be given by

$$W_{1-2} = -A \int_0^\infty P_{int}(t)\,\mathsf{v}(t)\,dt. \tag{2.1}$$

The minus sign means that this work is being released by the thermodynamic system in the expansion process. In compression processes, when work is added to the system, it would be positive.

The above expression is a simplified form of the general expression for the transfer rate of mechanical work, $\dot{W}$, through a surface A, moving with a velocity $\mathbf{v}(\mathbf{x}, t)$

$$\dot{W} = -\int_A P\mathbf{v} \cdot \mathbf{n}dA,$$

where, in this case, the pressure P is supposed to vary along the boundary A and $\mathbf{n}(\mathbf{x}, t)$ is the unitary vector normal to the surface at point $\mathbf{x} \in A$. In present conditions, the velocity $\mathsf{v}(t)$ in Eq. (2.1) can be found in terms of $P_{int}(t)$ by solving the initial value problem,

$$m\frac{d\mathsf{v}}{dt} = (P_{int} - P_{ext})\,A, \quad \mathsf{v} = 0 \ \text{ for } t = 0,$$

where m is the mass of the two balance weights placed on the upper part of the weightless piston (Fig. 2.3).

Finding the time evolution of the internal pressure $P_{int}(t)$ is, however, a difficult task since it requires solving the fundamental thermo-hydrodynamic equations in their differential form, for a compressible flow problem with a moving boundary.

Nevertheless, it can be argued that what is important in this problem is to take into account the *effective work* W_{1-2} released by the gas, when elevating the two balance weights to a level $h = (V_2 - V_1)/A$. For displacing the two balance weights, the thermodynamic system must overcome a resistance force that is given by P_2A—the external pressure during the whole expansion multiplied by the area of the cylinder cross section—and by the friction force between the piston and the cylinder wall. Friction work is transformed into heat and, in this case, this heat is absorbed by the heat reservoir without any consequence for the problem that is here being treated.

So we are led to the conclusion that the effective work released by the thermodynamic system is

$$W_{1-2} = -P_2 \times (V_2 - V_1) = -778\,\text{J}$$

and this is the work needed for elevating the two balance weights and the atmosphere by a height h. After the final equilibrium state is reached, the change in the potential energy of the two balance weights and the atmosphere will be given by

$$\Delta E_p = (2 \times 50 \times g + P_{atm} \times A)\, h = \underbrace{\left(\frac{2 \times 50 \times g}{A} + P_{atm}\right)}_{=P_2} (V_2 - V_1)$$

So, if we ignore all the complex physical phenomena that happen in the course of the gas expansion and consider only the net balance of energy— from the initial to the final equilibrium states of the gas— what it is important to take into account in this expansion process is the change of the potential energy of the external system (the two balance weights and the atmosphere) due to the work released by the thermodynamic system.

2.1.3 *Quasi-static Isothermal Expansion*

Consider now that the three balance weights of Fig. 2.1 are replaced by a great number of plumb spheres with mass Δm as shown in Fig. 2.4. The total mass of these spheres is 150 kg and the gas expansion is produced by removing these plumb spheres one after the other. The final equilibrium state is reached when the residual mass on the top of the piston is 100 kg.

The initial and final equilibrium states are the same as before and given in Table 2.1 independently of the mass Δm of each plumb sphere. In the same way, the total expansion work delivered by the gas can be found by summing all the elementary work δW that is delivered when each sphere is removed from the top plate. This elementary work can be found using the same procedure of Sect. 2.1.2.

It is easy to conclude that, although the initial and final equilibrium states remain the same as before, the delivered work will be dependent on the mass Δm and that it increases in absolute value, approaching an asymptotic value when Δm becomes smaller (Fig. 2.5).

Consider now the theoretical limiting case when $\Delta m \rightarrow 0$. In this case, the removal of a single plumb sphere reduces the external pressure P_{ext}, but the internal pressure P_{int} will follow this reduction without any meaningful time delay. In addition, it is possible to assert that this removal will not unbalance the pressure field in the gas phase in such a manner that the pressure P in the whole gas phase can be considered equal to P_{int}. So, each elementary work δW can be calculated as

$$\delta W = -\, P dV,$$

and the total work W_{1-2} as the integral

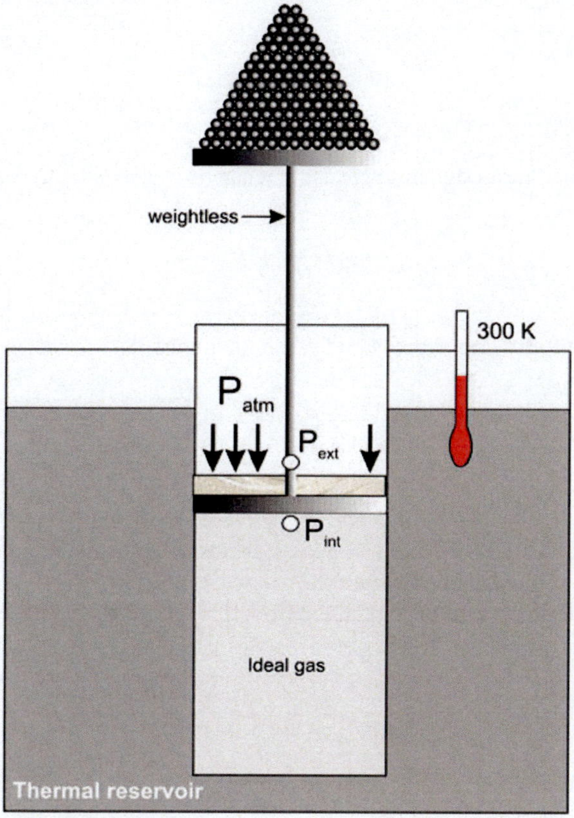

Fig. 2.4 Quasi-static isothermal expansion

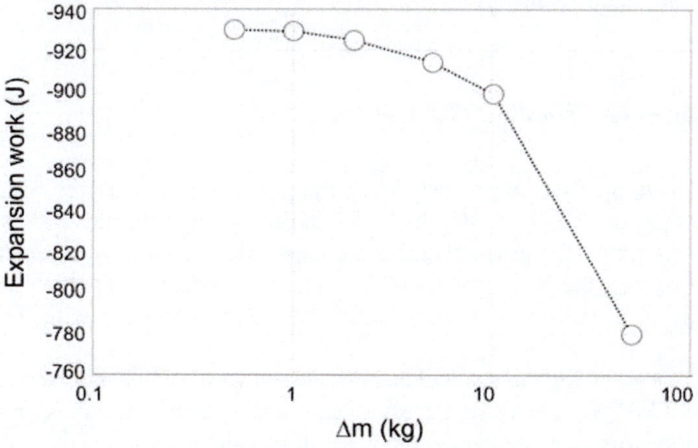

Fig. 2.5 Work delivered by the isothermal gas expansion when a certain number of plumb spheres with mass Δm are removed, one after the other, from the top of the piston in Fig. 2.4

$$W_{1-2} = -\int_{V_1}^{V_2} P dV.$$

Since P is the thermodynamic pressure, it can be replaced by the equation of state of an ideal gas

$$P = \frac{N\mathscr{R}T}{V},$$

resulting in the following expression for the total work

$$W_{1-2} = -N\mathscr{R}T \ln \frac{V_2}{V_1}.$$

In the present case, using the data given in Table 2.1, the total work can be calculated as $W_{1-2} = -932$ J. So, since the delivered work $W_{1-2} = -930$ J when the mass Δm of each plumb sphere is 0.5 kg, it is possible to say that with this Δm, we are very close to the asymptotic case $\Delta m \to 0$.

What is more important to be retained from this section is that the expansion work can be only calculated as

$$W_{1-2} = -\int_{V_1}^{V_2} P dV.$$

when *all* the states between V_1 and V_2 are *equilibrium states*, because we are only authorized to replace the external pressure P_{ext} by the gas pressure P when the gas is in thermodynamic equilibrium.

2.1.4 Inverse Process. The Cycle

Consider, now, the inverse process, when the plumb spheres are repositioned, one after the other, on the top of the piston till the initial equilibrium state is restored. When Δm is finite, the gas will suffer a compression process with an appreciable unbalance in their fields of temperature, pressure and density and we will now have a compression wave traveling from the top of the gas chamber, with the speed of sound.

When $\Delta m = 50$ kg, the gas will be compressed with a constant external pressure P_1 =1.5728 MPa and, by using the same reasoning of Sect. 2.1.2, the corresponding compression work that is added to the gas will be calculated as

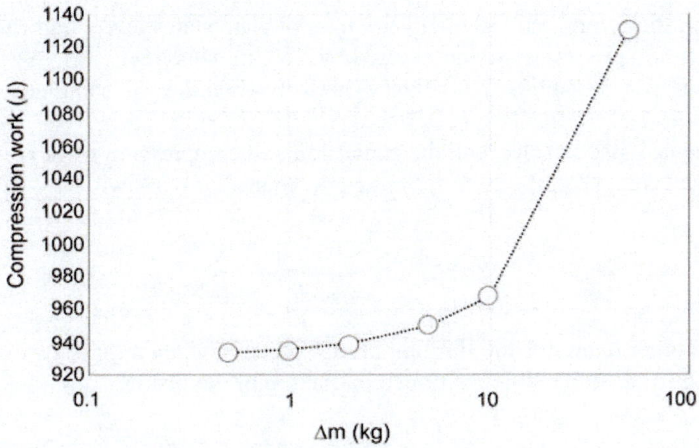

Fig. 2.6 Compression work when sequentially adding plumb spheres of mass Δm to the top of the piston

$$W_{2-1} = -P_1 \times (V_1 - V_2) = 1130\,\text{J}.$$

The compression work W_{2-1} can be found for different Δm in the same way as it was done for the expansion work. Figure 2.6 shows that W_{2-1} tends asymptotically to the quasi-static work $W_{2-1} = 932$ J when $\Delta m \to 0$.

Consider now the entire cycle 1-2-1 when the two limiting cases are compared. Figure 2.7 shows a P-V equilibrium diagram representing the two processes along the 300 K isotherm. The abrupt process is represented by a dashed line because only the states 1 and 2 are equilibrium states. All the intermediate states are non-equilibrium states and cannot be represented in this equilibrium diagram. When $\Delta m \to 0$, in

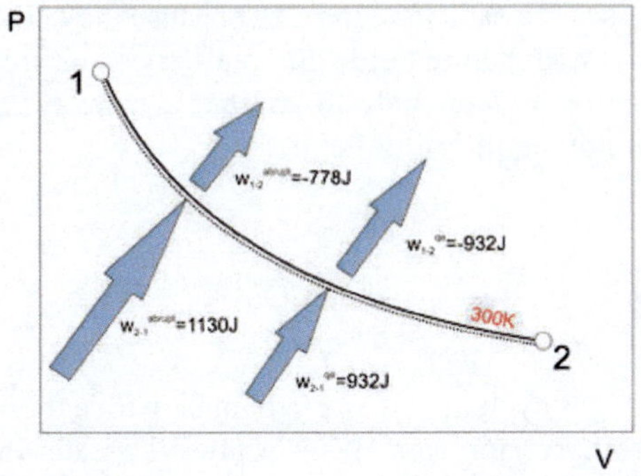

Fig. 2.7 P-V diagram of the entire cycle 1-2-1. For the abrupt process, only the states 1 and 2 are equilibrium states. For the quasi-static process, when $\Delta m \to 0$, in addition to these terminal states, all the intermediate states are equilibrium states

addition to these terminal states, all the intermediate states are equilibrium states. So, quasi-static processes are represented by a continuous line.

Let us now calculate the net balance of work when the gas completes each full cycle 1-2-1.

In the quasi-static cycle, both the expansion and compression work are the same in absolute value, $|W_{1-2}| = 932$ J. So the net balance of work will be

$$\oint \delta W_{quasi-static} = \underbrace{-932}_{\text{expansion}} + \underbrace{932}_{\text{compression}} = 0.$$

On the other hand, for the limiting abrupt process, when a mass $\Delta m = 50kg$ is removed from the top and repositioned on the top of the piston,

$$\oint \delta W_{abrupt} = \underbrace{-778}_{\text{expansion}} + \underbrace{1130}_{\text{compression}} = 352\,\text{J}.$$

Therefore, while each quasi-static cycle can be completed without any expense of work, the abrupt cycle requires an expense of 352 J of work to be completed. This is a general rule for thermodynamic cycles that are run in thermal contact with a constant temperature heat reservoir,

$$\oint \delta W \geq 0,$$

meaning that, independently of the mass Δm that is used for performing the cycle, a certain amount of work must always be added to the system to complete the cycle. This added work will be only null in the limit $\Delta m \to 0$.

So, while quasi-static processes are *reversible*, abrupt processes are *irreversible* and by *irreversible* we mean a process that requires a certain amount of work to be reversed—*and not that such a process cannot be reversed.*

Irreversible processes are easy to recognize because they always present a spatial imbalance in their thermodynamic fields. So, real processes are always irreversible, because the transition of a given state to another are always accompanied by the formation of internal non-equilibrium fields.

2.1.5 Exercises

1. Show that in every gas cycle, 1-2-1 ran in contact with a thermal pond at constant temperature the compression work W_{2-1} is always greater in absolute value than the expansion work W_{1-2}.
2. Considering that the liquid phase will remain at 90 °C, due to its high thermal capacity, explain the physical mechanisms that will heat the gas from 20–90 °C

in Fig. 2.8, after the insulation plate separating the liquid from the gas phase is removed. Is this heating process a reversible or an irreversible process? Why?

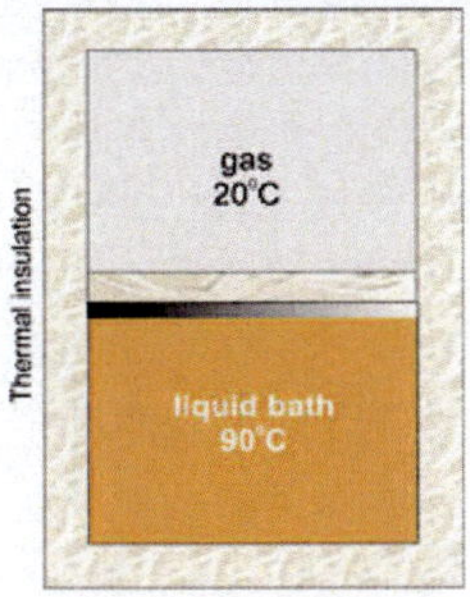

Fig. 2.8 Isovolumetric heating

Further reading: Rayleigh–Bénard convection is a type of natural convection, occurring in a plane horizontal layer of fluid heated from below, in which the fluid develops a regular pattern of convection cells known as Bénard cells. Rayleigh–Bénard convection is one of the most commonly studied convection phenomena because of its analytical and experimental accessibility. The convection patterns are the most carefully examined example of self-organizing nonlinear systems. There are several papers and internet articles written about the Rayleigh–Bénard convection. The Landsat image in Fig. 2.9 shows a pattern of stratocumulus cloud cells, each about 10–15 km over the Pacific Ocean. As warmer moist air rises in convection cells over the ocean and cools, condensing the water vapor into cloud droplets—*which usually coalesce to form clouds*—cold air then sinks around the sides of the cells (from https://earthobservatory.nasa.gov/images/2387/closed-small-cell-clouds-in-the-south-pacific).

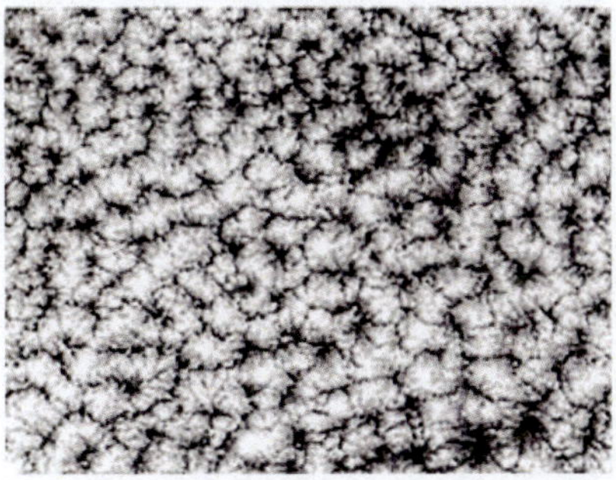

Fig. 2.9 Cloud cells formation by Rayleigh-Bénard convection over the Pacific Ocean. Each cell has 10–15 km characteristic length (from https://earthobservatory.nasa.gov/images/2387/closed-small-cell-clouds-in-the-south-pacific). Image courtesy NASA/GSFC/LaRC/JPL, MISR Team

3. One mole of an ideal gas is in equilibrium inside a cylinder with a 10 cm^2 straight section (Fig. 2.10). The system is under the action of atmospheric pressure and one 50 kg balance weight and in thermal contact with a heat reservoir at 300 K. Consider now that we place two others 50 kg balance weights on the top plate of the piston. (a) Find the final equilibrium state of the gas; (b) try to explain what happens with P_{ext} and P_{int} in the course of time and detail the physical mechanisms that lead the gas to a final equilibrium state; (c) calculate the effective compression work between the initial and final equilibrium states.

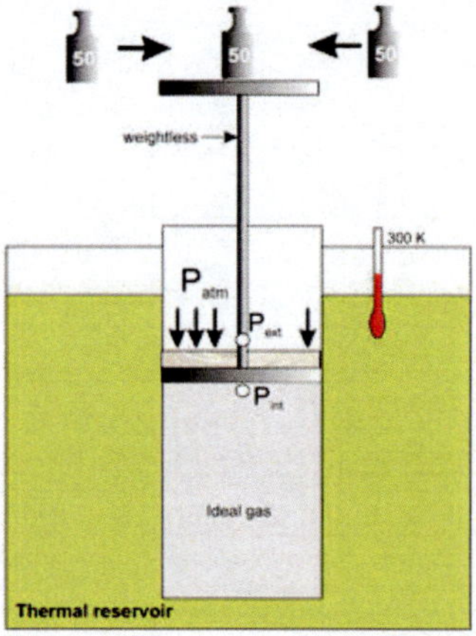

Fig. 2.10 Isothermal abrupt compression

2.1.6 *Work and Heat Conversion. The Role of Dissipation*

It is now time to go back to Sect. 1.4 and to the first principle of thermodynamics. When a gas expands it cools, but if it is in thermal contact with a heat reservoir at a constant temperature, this reservoir will supply the heat that is needed to restore its equilibrium temperature.

If the gas is an ideal one, its internal energy is only dependent on the temperature and so, by integrating the first principle of thermodynamics, Eq. (1.4), between the equilibrium states 1 and 2,

$$Q_{1-2} + W_{1-2} = U_2 - U_1 = 0.$$

It was found in Sect. 2.1.2 that, when a weight of 50 kg is retired from the top plate of the piston, the work in the irreversible expansion was $W_{1-2} = -778$ J, the heat absorbed by the gas will be $Q_{1-2} = 778$ J. The conclusion is that, in this expansion process, heat is entirely converted into work, i.e., that the work delivered by the gas results from the withdraw of the equivalent amount of heat from the reservoir. On the other hand, when the mass of the balance weight $\Delta m \rightarrow 0$ and the gas expansion is reversible the heat absorbed by the gas and transformed into work is $Q_{1-2} = 932\,\mathrm{J}$.

The question we must answer is why the amount of heat that is absorbed by the gas in an irreversible expansion is $158J$ smaller than the heat that is absorbed in a reversible expansion? In other words, why in irreversible gas expansions in contact with a thermal pond the gas requires less amount of heat to reestablish its equilibrium temperature?

An irreversible expansion such as the one of Sect. 2.1 is always accompanied by advection currents with the formation of a velocity field, in addition to the temperature field sketched in Figure 2.2, picturing a complex non-equilibrium state. For restoring its final equilibrium state these velocity and temperature fields must be destroyed. This is achieved by the heat transfer from the wall and, also, by viscous dissipation which transforms the kinetic energy of the advection currents into internal energy. Viscous dissipation is an intrinsically irreversible process and is always responsible for the transformation of mechanical energy into heat (or internal energy).

The result is that in irreversible expansions the thermodynamic system does not require the same amount of heat from the thermal pond to attain its final equilibrium state when compared to a reversible expansion. In this example, this also means that the 158 J above mentioned corresponds to the mechanical energy of the advection currents that is dissipated in the course of the irreversible expansion.

On the other hand, considering the inverse compression process $2-1$, the compression work is now $W_{2-1} = 1130$ J, when the compression is irreversible. So the gas needs to deliver $Q_{2-1} = -1130$ J to the reservoir to restore its equilibrium temperature, an amount that is $198J$ greater in absolute values than the one corresponding to the reversible compression. In compression processes, work is converted into heat, and the surplus of $198J$ corresponds to the mechanical energy of the advection currents that is dissipated, i.e., transformed into heat and delivered to the thermal pond.

Figures 2.11 and 2.12 summarizes the energy balance for this isothermal cycle when a weight of 50 kg is retired and put it back again on the top plate of the piston. From Figs. 2.1 and 2.10 and related comments in Sect. 2.1, an important conclusion is that, in isothermal cycles,

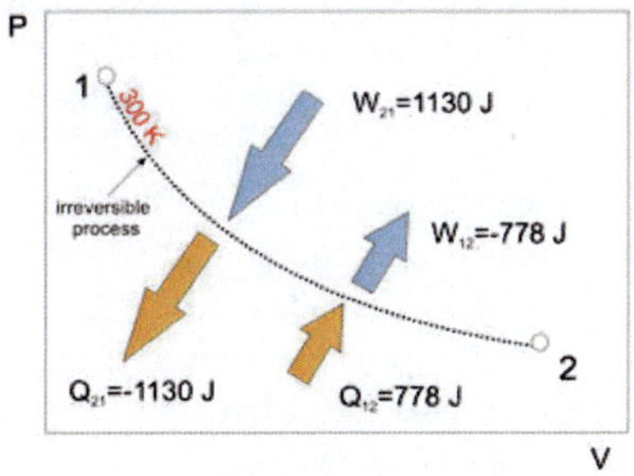

Fig. 2.11 Energy balance for the isothermal expansion of section

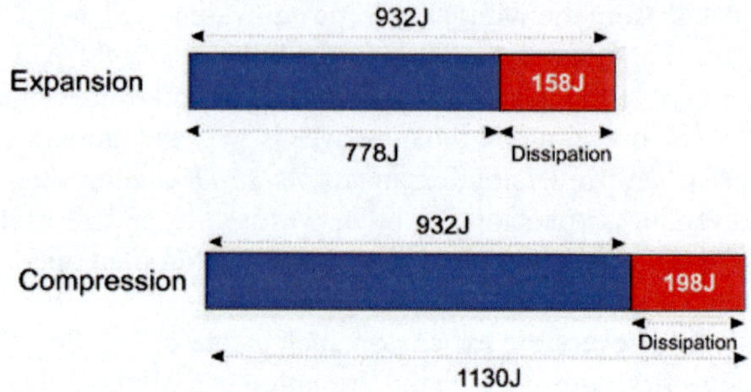

Fig. 2.12 The role of dissipation

$$\oint \delta W \geq 0 \text{ and } \oint \delta Q \leq 0,$$

independently of the way the gas expansion and compression are performed, the equality being restricted to reversible isothermal cycles.

Therefore, when a thermodynamic system follows an irreversible cycle and returns to its initial state at the same temperature, *we will always have the conversion of work into heat, and not the inverse.*

Indeed, the natural tendency of physical phenomena to transform mechanical energy into heat (*a dissipation process*) was a major objection to the equivalence principle of work and heat. This question will be seen in more detail in the following sections.

In isothermal cycles, there is, always, a conversion of work into heat and, never, the inverse.

2.1.7 Exercises

1. One mole of an *ideal monatomic* gas is in equilibrium inside a cylinder with a 10 cm^2 straight section (Fig. 2.13). The system is under the action of the atmospheric pressure and three 50 kg balance weights. Thermal insulation separates the gas from the outer ambient. The initial temperature is 300 K. Consider now that we remove one of the three 50 kg balance weights from the top plate of the piston. (a) Considering the friction work between the piston and the cylinder wall, is it possible to find the final equilibrium state of the gas? (b) try to explain what happens with P_{ext} and P_{int} in the course of time and detail the physical mechanisms that lead the gas to a final equilibrium state; (c) calculate the final temperature of the gas, when the friction work can be neglected; (d) what would be the influence of the friction work on the final temperature? (e) what would be the maximum temperature at the final equilibrium state? (f) calculate the expansion work between the initial and final equilibrium states, when the friction work is neglected.

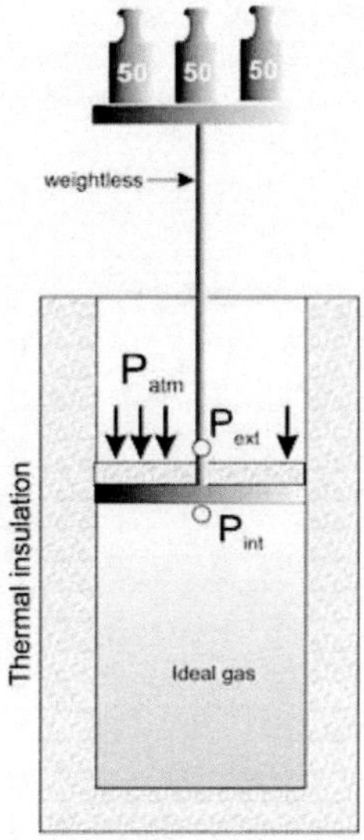

Fig. 2.13 An irreversible adiabatic expansion

2. In the previous exercise, consider now that we have replaced the three 50 kg balance weights by 150 kg of plumb spheres (Fig. 2.14). Consider also that the mass of each sphere Δm is, alternatively, 10, 5, 1, and 0.5 kg. By neglecting the friction work between the piston and the cylinder wall, calculate the final equilibrium temperature of the gas and the delivered work when 50 kg are removed from the piston top plate. (a) Plot the expansion work in terms of the mass of the plumb spheres and calculate the expansion work when $\Delta m \to 0$. What would be the friction work in this last case?
3. In Exercise 2 what is the minimum temperature that is attained by the gas after the adiabatic expansion? Why? In adiabatic expansions, the delivered work is given by the internal energy variation between the initial and final equilibrium states. Try to explain why the internal energy is not entirely converted into useful work in irreversible gas expansions. The internal energy is often called the *free—or available—energy in adiabatic expansions*. Try to explain the reasons for this designation.
4. Calculate the temperature and the expansion work for $\Delta m \to 0$ in Exercise 2, when the gas is pure nitrogen. The molar heat capacity of Nitrogen was found as

$$\bar{c}_v\left(\frac{J}{\text{mol K}}\right) = 20.586 - 0.1571 \times 10^{-2}T + 0.8081 \times 10^{-5}T^2 - 2.8730 \times 10^{-9}T^3,$$

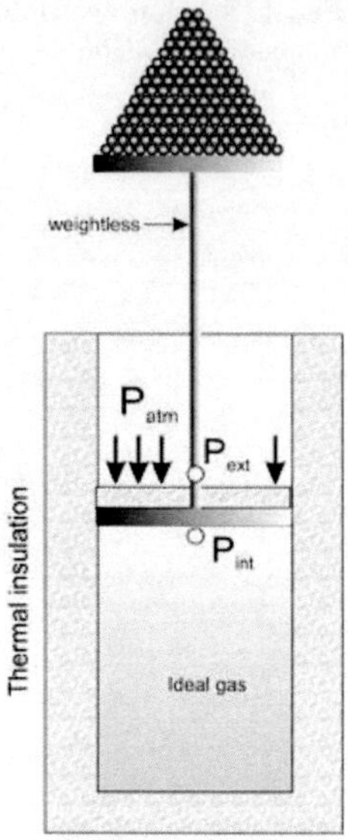

Fig. 2.14 Adiabatic expansion produced by the removal of balance weights of mass Δm

in the range 273–1800 K,[1] with an average deviation with respect to experimental data of 0.34%. What would be these values if the molar heat capacity of nitrogen at 273 K, $\bar{c}_v = 20.701\ J/\,(\text{mol K})$, was used in the calculations?

5. Consider that the gas in Exercise 2 is in its final equilibrium state, after a reversible adiabatic expansion ($\Delta m \to 0$). By neglecting the friction work, calculate the final temperature and the compression work when its initial pressure is restored by successively adding balance weights with 10, 5, 1, and 0.5 kg.
6. The first principle is usually written as

$$\oint (\delta Q + \delta W) = 0.$$

Consider the two adiabatic expansions of Fig. 2.15. Both expansions start from the same initial equilibrium state and finish on the same final volume V_2. Expansion 1-2 is irreversible while expansion 1-3 is reversible. Use the first principle as given above to show that the work delivered in the reversible expansion is necessarily greater than in the irreversible expansion when their absolute values are compared, $|W_{1-3}| \geq |W_{1-2}|$.

7. Show that, when the molar heat capacities $\bar{c}_p$ and $\bar{c}_v$ are considered constants,

$$P\mathrm{v}^{\gamma} = const.,$$

[1] B. G. Kyle, Chemical and Process Thermodynamics (Englewood Cliffs, NJ: Prentice-Hall, 1984).

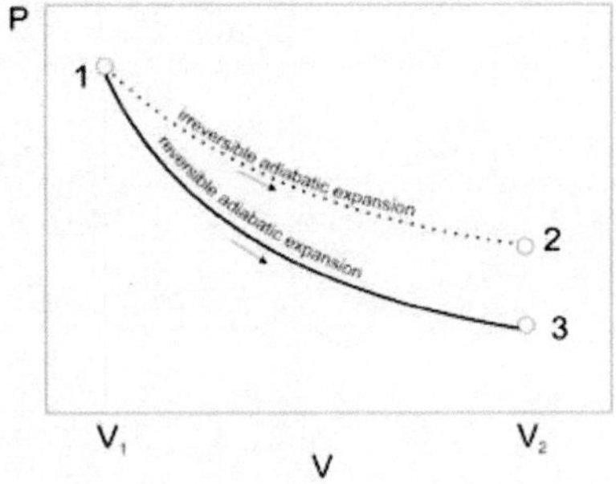

Fig. 2.15 Using the first principle of thermodynamics to show that the work W_{1-3} delivered in a reversible adiabatic expansion is greater (in absolute value) than the work W_{1-2}

in adiabatic and reversible expansions of ideal gases. Parameter γ is given by $\gamma = c_p/c_v$. In the sequence, show that the adiabatic reversible lines and the isotherms are both hyperboles, but the adiabatic lines are more inclined than the isotherms.

8. A gas is compressed adiabatically from an initial equilibrium state 1 to a final equilibrium state 2 (Fig. 2.17). Show that, independently of the way the compression is performed, the compression work $|W_{1-2}|$, in absolute value, will be always greater than the compression work $|W_{1-3}|$ in a reversible process, with the same compression rate. (Hint: *consider an isotherm starting from state 2 and crossing the extension of the reversible line* $Pv^\gamma = const.$ *that joins states 1 and 3 and use the first principle*).
9. A cylinder is divided into two equal-volume chambers A and B. Each chamber contains 1 mole of Helium initially at 20 °C (Fig. 2.16). Chambers A and B are separated by a thermally insulated plate, free to move inside the cylinder, in such a manner as to equalize the pressures in the two chambers. The base plate of the cylinder is a thin thermal contact plate. Consider now that this base plate is put in contact with a thermal reservoir at 200 °C. The Helium in chamber B will expand moving the dividing plate upward.

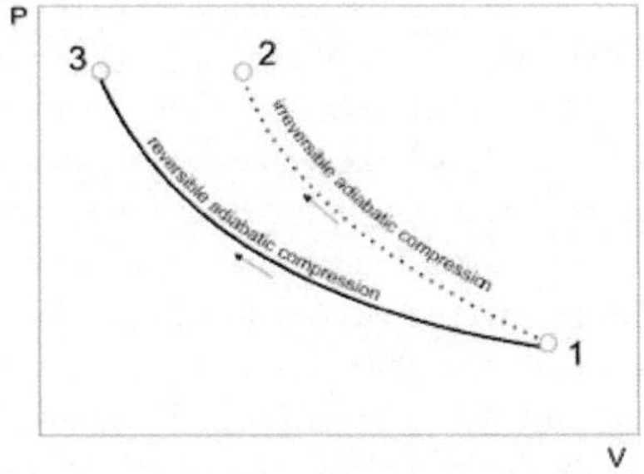

Fig. 2.16 A Rayleigh-Bénard cellular convection heats the Helium in the chamber B. Nevertheless, the adiabatic compression of the Helium in the chamber A is a very slow process and supposing this compression to be reversible can be a fair hypothesis

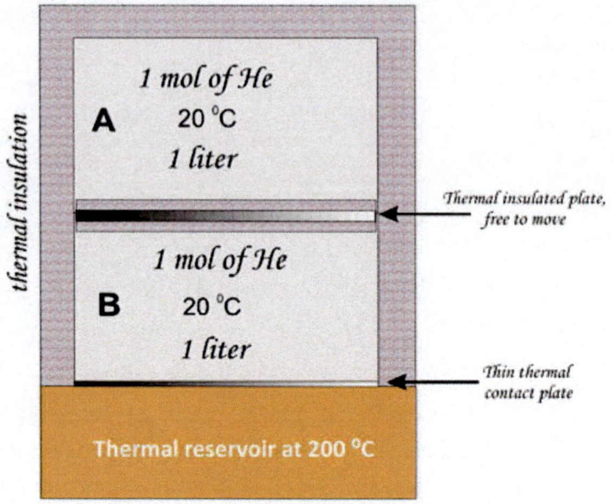

Fig. 2.17 Using the first principle of thermodynamics to show that the work W_{1-3} delivered in a reversible adiabatic compression is smaller (in absolute value) than the work W_{1-2}

Try to picture the heating process of the Helium in chamber B. A Rayleigh-Bénard cellular convection heats the Helium in chamber B and this is clearly an irreversible process. Nevertheless, the adiabatic compression of the Helium in chamber A is a very slow process in such a way that supposing this compression to be reversible can be a fair hypothesis. With this assumption: (a) Find the final equilibrium state of the system; (b) calculate the work released by the Helium in chamber B to compress the Helium in chamber A; (c) calculate the heat that is withdrawn from the thermal reservoir at 200 °C between the initial and final equilibrium states of the Helium in the two chambers.

10. A thermodynamic system is composed of two chambers A and B in thermal contact with a thermal reservoir at 300 K. The two chambers are separated by a metallic plate free to move. Each one of these chambers is filled with 1 mole of an ideal gas, but the pressure in A is, initially, greater than the pressure in B due to two pins that maintain the separating plate at the position shown in Fig. 2.18. Consider now that the pins are removed: (a) Find the final equilibrium state of the gas in chambers A and B; (b) try to describe the physical phenomena that happen in this process; in particular, try to picture the movement of the separating plate with the time; (c) since the gas in chamber A suffers an expansion process, it will receive heat from the heat reservoir, Q_{A-R} and from chamber B, Q_{A-B}; in the inverse sense the gas in chamber B will be compressed and will release heat to the reservoir, Q_{B-R} and chamber A, Q_{B-A}. We know that $Q_{A-B} = -Q_{B-A}$, but what can we say about Q_{A-R} and Q_{B-R}? d) in this process, is it possible to calculate the expansion work W_{A-B} that is delivered to compress the gas in chamber B?

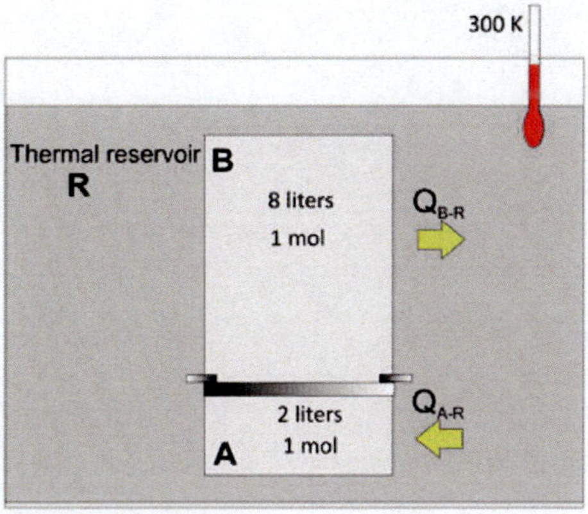

Fig. 2.18 A thermodynamic system composed of two chambers A and B in thermal contact with a thermal reservoir

2.2 The Second Principle of Thermodynamics

> La production de la puissance motrice est donc due, dans les machines à vapeur, non à une consommation réelle du calorique, mais à son transport d'un corps chaud à un corps froid (....) D'après ce principe, il ne suffit pas, pour donner naissance à la puissance motrice, de produire de la chaleur: il faut encore se procurer du froid; sans lui la chaleur serait inutile. Et en effet, si l'on ne rencontrait autour de soi que des corps aussi chauds que nos foyers, comment parviendrait-on à condenser la vapeur?.... (Sadi Carnot, 1824 [27])

Denis Papin (1647—1712) was a French physicist, mathematician and inventor, best known for his pioneering invention of the first prototype of a steam engine in 1691. Papin's engine, Fig. 2.19, used the vapor, produced by heating water, to move a piston upward inside a cylindrical chamber. When the piston was in the top of this chamber, cold water was admitted into the jacket around the chamber condensing the vapor. The vacuum created by vapor condensation pulled the piston back to its lower position. The refrigerating water was then stopped from entering the jacket, allowing the cycle to be restarted.

Thermal machines such as Papin's engine, working in cycles can be represented by using a simplified scheme like the one shown in Fig. 2.20.

Although a very simplified sketch of a real thermal engine, this sketch represents the main processes that may concern us when performing a thermodynamic analysis of such engines. The first thing that it is important to observe is that a source of heat—the heat from the combustion of fuel—and a heat sink—the cold water—are both needed if one wants the engine to work in a cycle. In this sketch, the heat source is represented by a heat reservoir at a temperature T_H, related to the combustion temperature where the combustion gases are sensed to be after the fuel reacts with the atmospheric air. The heat sink is, here, represented by a heat reservoir at a temperature T_L, related to the temperature of the cold water that was used to condense the vapor.

In each cycle, the combustion process releases an amount of energy Q_H to the cycle, which is the amount of energy that is needed to produce vapor and to heat this

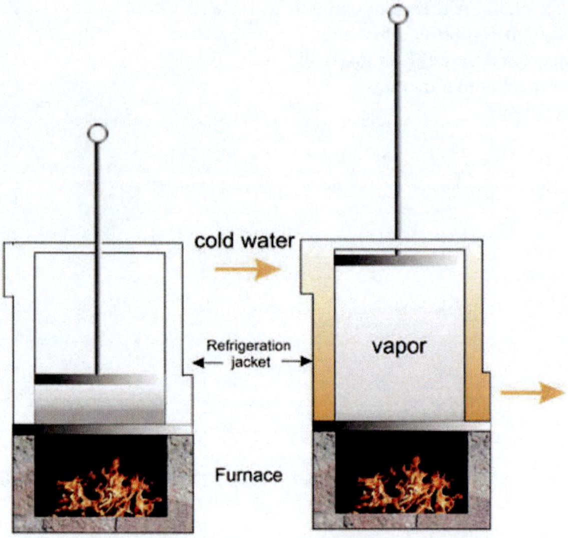

Fig. 2.19 Papin steam engine

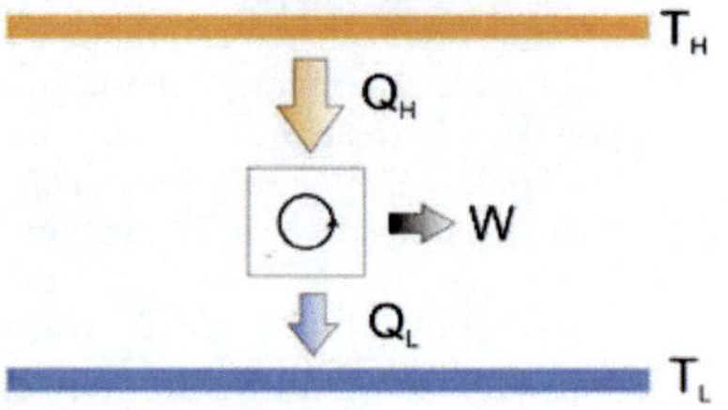

Fig. 2.20 A simplified scheme for a thermal engine

vapor until the vapor pressure inside the chamber is high enough to move the piston upward.

After the principle of energy conservation was established in the first half of the nineteenth century and heat was recognized as a kind of energy,[2] what was considered to happen was the partial conversion of Q_H into mechanical work W, a part of Q_H being used to cool the vapor bringing it to its initial state. This amount Q_L represents the heat that must, necessarily, be lost if one wants the engine to work in a cycle.

The total amount of energy that is exchanged by the thermodynamic system, i.e., by the water inside the chamber, must be null after each cycle. This is the principle of energy conservation—the first principle of thermodynamics—and means that the output work W must be, in absolute value, smaller than Q_H. So the thermodynamic efficiency η of the cycle must, necessarily, be smaller than 1,

$$\eta = \frac{|W|}{|Q_H|} = -\frac{W}{Q_H} < 1, \tag{2.2}$$

[2] In the early years of thermodynamics it was thought that a heat engine was like a hydraulic turbine, the flow of caloric from the heat source to the heat sink being responsible to put the engine to work as water makes a hydraulic turbine to work when it flows from a higher to a lower level.

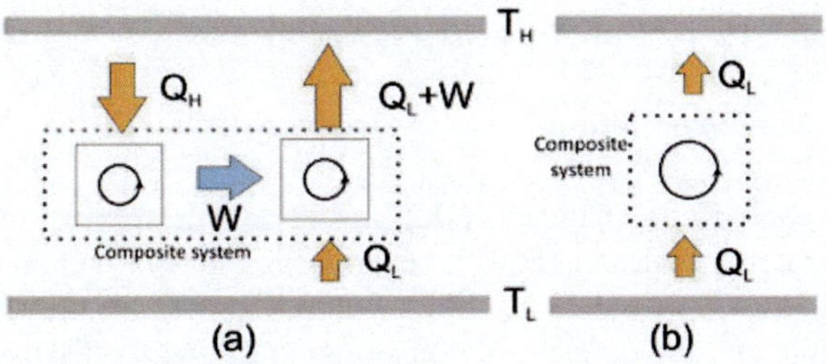

Fig. 2.21 Demonstration that it is impossible to conceive a heat engine that entirely converts heat into mechanical work. In **a** it is supposed that a thermal engine with $\eta = 1$ can be used to drive a heat pump operating between the same thermal reservoirs. In **b** an energy balance is performed for the composite system

because, from the first principle,

$$\oint (\delta Q + \delta W) = \underbrace{Q_H}_{>0} + \underbrace{Q_L}_{<0} + \underbrace{W}_{<0} = 0,$$

and the heat loss Q_L is, necessarily, different from zero.

Is there any thermal engine with an energy conversion efficiency $\eta = 1$?

The answer was given by Rudolf Clausius in 1850 [34]: *If it is possible to conceive a thermal engine, operating in a cycle, that entirely converts, heat into mechanical power, without any loss, it would, also, be possible to transfer heat from a cold to a hot body without any supply of mechanical work.*

Indeed, if it is possible to conceive a thermal engine with a conversion efficiency equal to 1, this thermal engine can be used to drive a heat pump operating between the same thermal reservoirs, as sketched in Fig. 2.21a. Following the operating principle of *all* heat pumps, this heat pump withdraws the amount Q_L from the cold reservoir and delivers $Q_L + W$ to the hot reservoir.

Consider now the composite system, i.e., the set 'thermal engine + heat pump'. Since the work W is delivered by the heat engine to the heat pump, it becomes an internal transfer of energy of the composite system and consequently does not appear in the energy balance of Fig. 2.21b. Now, the work W has the same absolute value of Q_H, because the conversion efficiency of the power engine was supposed as $\eta = 1$. Therefore the heat that will be delivered to the hot reservoir by the composite system is given by

$$|Q_L| + |W| - |Q_H| = |Q_L|$$

This result is shown in Fig. 2.21b and demonstrates the impossibility of conceiving a thermal engine with unitary conversion efficiency $\eta = 1$, since heat cannot be transferred from a cold to a hot body without any supply of work.

2.2.1 *Statements of the Second Principle*

2.2.1.1 Clausius Formulation

The German scientist Rudolf Clausius (1822–1888) laid the foundation for the second principle of thermodynamics in 1850[3] by examining the relation between heat transfer and work. In this 1850 paper, he showed that there was a contradiction between Carnot's principle and the concept of conservation of energy. Clausius restated the two principles of thermodynamics to overcome this contradiction. His version of the second principle, which was published in German in 1854[4] reads as follows:

> Heat can never pass from a colder to a warmer body without some other change, connected therewith, occurring at the same time.

He explains this principle: "*Everything we know concerning the interchange of heat between two bodies of different temperature confirms this; for heat everywhere manifests a tendency to equalize differences of temperature, and therefore to pass, from warm to cold bodies. Without further explanation, therefore, the truth of this principle will be granted. The principle may be more briefly expressed thus: Heat cannot by itself pass from a colder to a warmer body; the words by itself, however, here requires explanation. Their meaning will, it is true, be rendered sufficiently clear by the exposition contained in the present memoir, nevertheless, it appears desirable to add a few words here in order to leave no doubt as to the signification and comprehensiveness of the principle.*

The Clausius principle can be stated in a graphical form as shown in Fig. 2.22.

The Clausius principle can be better understood from the sketch given in Fig. 2.23. The natural *arrow* of heat is from the hot to the cold bodies and one wants to pump heat from a cold to a hot body, using a thermodynamic system working in a cycle. For this purpose, one needs to add the work W taken, for instance, from a heat engine that works, e.g., using the hot body as its source of heat Q'_H and the cold body as its heat sink to which it will release the heat Q'_L. We do not know what the values Q'_H and Q'_L are to be, individually, but we must assure $Q'_H + Q'_L = -W$. Consider now a composite system, i.e., a system composed of the heat pump and the power engine. Since the mechanical work that is released by the power engine will be used as a shaft

---

[3] Clausius R (1850) *Ueber die bewegende Kraft der Wärme und die Gesetze, welche sich daraus für die Wärmelehre selbst ableiten lassen.* Ann der Phys und Chemie 155:500–524. https://doi.org/10.1002/andp.18501550403 [34]Translated into English: Clausius, R. (July 1851). *On the Moving Force of Heat, and the principles regarding the Nature of Heat itself which are deducible therefrom.* London, Edinburgh and Dublin Philosophical Magazine and Journal of Science. 4th 2 (VIII): 1–21; 102–119.

[4] Clausius R (1854) *Ueber eine veränderte Form des zweiten Hauptsatzes der mechanischen Wärmetheorie.* Ann der Phys und Chemie 169:481–506. https://doi.org/10.1002/andp.18541691202 [35]. Translated into English: Clausius, R. (July 1856). *On a Modified Form of the Second Fundamental Theorem in the Mechanical Theory of Heat.* London, Edinburgh and Dublin Philosophical Magazine and Journal of Science. 4th 2: 86. Reprinted in: Clausius, R. (1867). *The Mechanical Theory of Heat—with its Applications to the Steam Engine and to Physical Properties of Bodies.* London: John van Voorst [36].

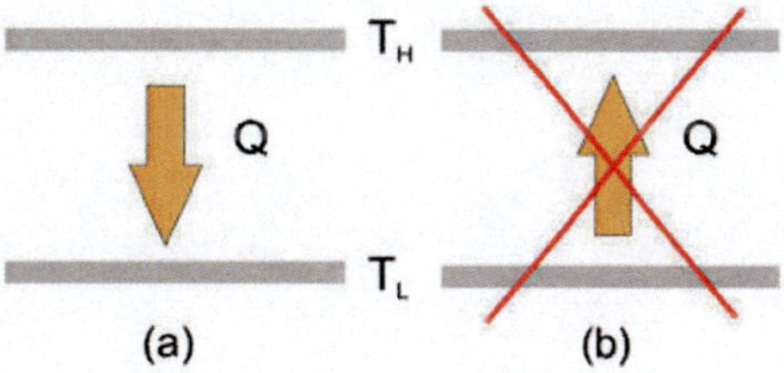

Fig. 2.22 Clausius principle in graphical form. Heat can pass from a warmer to a colder body (**a**), but, in the words of Clausius, cannot *by itself* pass from a colder to a warmer body (**b**)

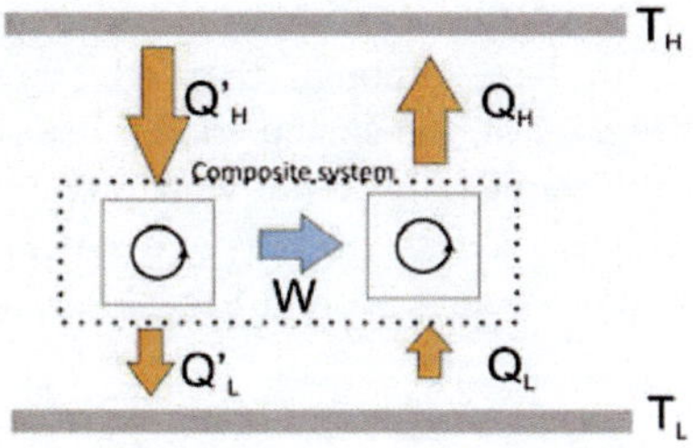

Fig. 2.23 The natural arrow of heat

work to the heat pump compressor, the net effect will be the flow of heat following its natural arrow from the hot to the cold body, when $|Q'_H| \geq |Q_H|$, otherwise, we would break a fundamental principle, since heat must flow from a hot to a cold body. The limiting case $|Q'_H| = |Q_H|$ means that the amount of heat that is to be delivered to the heat engine can be furnished by the heat pump itself, This is only possible for theoretical engines operating under reversible cycles.

Therefore, when a heat pump is built to transfer heat from a colder to a warmer body, what can be concluded from Fig. 2.23 is that the net effect will be a transfer of heat from the warmer to the colder body.

2.2.1.2 Planck Formulation

Max Karl Ernst Ludwig Planck (1858–1947) was a German theoretical physicist whose work on quantum theory won him the Nobel Prize in Physics in 1918. Planck stated the following proposition as derived directly from experience. This is often considered as his statement of the second principle, but he regarded it as a starting point for the derivation of the second principle.[5]

> It is impossible to construct an engine that will work in a complete cycle, and produce no effect except the raising of a weight and cooling of a heat reservoir.

For a better understanding of the Planck statement of the second principle, consider Fig. 2.24 which represents a thermal engine which works in a cycle in contact with a single heat reservoir.

[5] Planck, M. (1903). *Treatise on Thermodynamics*. Published in German in 1897. Translated to English by Alexander Ogg. London: Longmans, Green & Co. This book was also edited by Dover [136].

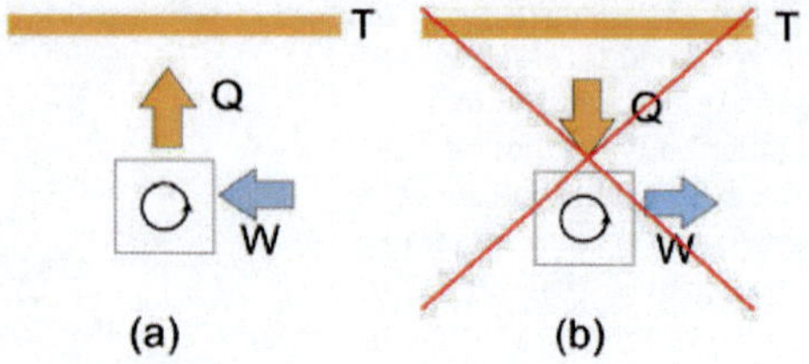

Fig. 2.24 Graphical formulation of the Planck statement

In fact, although the full conversion of work into heat is always possible (Fig. 2.24a, the sketch shown in Fig. 2.24b represents a thermal engine with a full conversion of heat into work, i.e., with a thermal efficiency $\eta = 1$, without any other effect *except the raising of a weight (work) and cooling of a (single) heat reservoir.*

As we have seen in Sect. 2.1.6, this statement can also be written as

$$\oint \delta W \geq 0 \text{ and } \oint \delta Q \leq 0,$$

for a thermodynamic system that performs a cycle in contact with a *single heat reservoir.*

2.2.1.3 Kelvin Formulation

Before talking about Kelvin's statement for the second principle, it is perhaps interesting to read a text from William Thomson, the Lord Kelvin, extracted from its 1851 famous paper '*On the dynamical theory of heat; with numerical results deduced from Mr. Joule's equivalent of a thermal unit and M. Regnault's observations on steam*'.[6]

In his words: *I believe the tendency in the material world is for the motion to become diffused, and that as a whole the reverse of concentration is gradually going on—I believe that no physical action can ever restore the heat emitted from the sun, and this source is not inexhaustible; also that the motions of the earth and other planets are losing vis viva which is converted into heat; and that although some vis viva may be restored for instance to the earth by heat received from the sun, or by other means, that the loss cannot be precisely compensated and I think it probable that it is under compensated.*

---

[6] Thomson, W. (1851). *On the Dynamical Theory of Heat, with numerical results deduced from Mr Joule's equivalent of a Thermal Unit, and M. Regnault's Observations on Steam*. Transactions of the Royal Society of Edinburgh XX (part II): 261–288 [172].

Thomson was reluctant to accept Joule's ideas of the equivalence between heat and work. In an 1848 paper[7] he wrote that *the conversion of heat (or caloric) into mechanical effect is probably impossible, certainly undiscovered.*

In his 1851 publication, Thomson retreated from a radical departure and declared *the whole theory of the motive power of heat is founded on two propositions, due respectively to Joule, and to Carnot and Clausius*. Thomson went on to estate a form of the second principle:

> It is impossible, by means of inanimate material agency, to derive mechanical effect from any portion of matter by cooling it below the temperature of the coldest of the surrounding objects.

This statement is followed by a footnote at the bottom of page 265 of his paper:

> If this axiom be denied for all temperatures, it would have to be admitted that a self-acting machine might be set to work and produce a mechanical effect by cooling the sea or earth, with no limit but the total loss of heat from the earth and sea or, in reality, from the whole material world.

Kelvin's statement is considered to be analogous to the one enunciated by Planck and, in some textbooks, the Planck statement is called as '*Kelvin-Planck statement*'. Figure 2.25 helps to understand this analogy, based on the footnote of Kelvin's principle: *if it would be possible for a boat to use the heat pumped from the ocean to generate the vapor for powering the boat steam engine, the composite system 'steam engine + heat pump' would be able to fully convert heat into mechanical work as shown in the figure and this is in contradiction with Planck's statement of the second principle*.

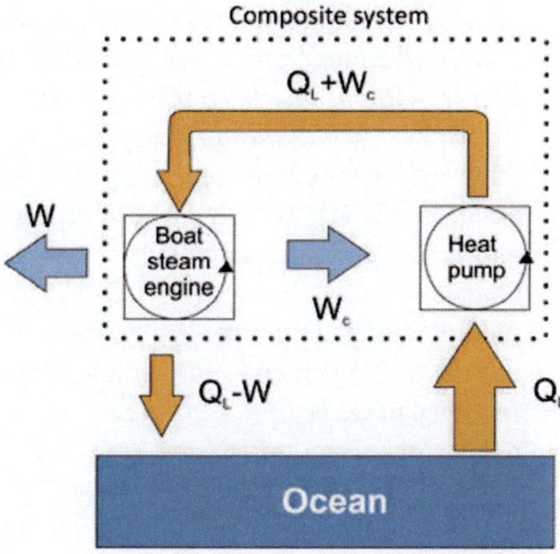

Fig. 2.25 Kelvin statement of the second principle

[7] Thomson W (1848) On an Absolute Thermometric Scale founded on Carnot's Theory of the Motive Power of Heat, and calculated from Regnault's Observations. Proc. Camb. Philol. Soc. [171]. Reprinted in Thomson, William (1882). Mathematical and Physical Papers. Cambridge, England: Cambridge University Press. pp. 100–106 [173].

2.2.1.4 The Arrow of Natural Phenomena

The Clausius and Plank statements can be shown to be equivalent (see Exercise 1 in Sect. 2.2.2). Two assertions are equivalent when the second can be proved from the first and vice-versa. Nevertheless, in Clausius statement the focus is the arrow of heat, from warmer to colder bodies and not from colder to warmer bodies, and the Planck statement is focused on the direction of the energy conversion: work to heat and not heat to work.

What do Clausius and Planck's statements have in common?

Figures 2.22 and 2.24 try to clear this question. Without any supply of mechanical work, heat will flow from warmer to colder bodies and never from colder to warmer bodies. When a thermodynamic system performs a cycle in contact with a single heat reservoir it will convert work into heat and never heat into work.

Nature has an *arrow* and this *arrow* is what the two statements have in common.

2.2.2 Exercises

1. Show that "*If it is possible to transfer heat from a colder to a warm body without any supply of work, then it is also possible to conceive a heat engine working in a cycle that fully converts heat into work*". Showing this assertion is the same as showing that when the Clausius statement of the second principle is denied, then the Planck statement is also denied or, in other words, that the Planck statement is a particular case of the Clausius statement.
   In Fig. 2.21 it is shown that when the Planck statement of the second principle is denied, then the Clausius statement is also denied or that *If it is possible to conceive a thermal engine, operating in a cycle, that entirely convert, heat into mechanical power, without any loss, it would, also, be possible to transfer heat from a cold to a hot body without any supply of mechanical work.* This last assertion means that the Clausius statement is a particular case of the Planck statement. When two assertions A and B are shown to be particular cases one of each other, i.e., $A \subseteq B$ and $B \subseteq A$, they are equivalent or $A \equiv B$.
2. Constantin Carathéodory (1873–1950) formulated thermodynamics on a purely mathematical axiomatic foundation. His statement of the second principle is known as the Principle of Carathéodory, which may be formulated as follows[8]: *In every neighborhood of any state S of an adiabatically enclosed system there are states inaccessible from S.*
   In Fig. 2.26, show that if the state I is accessible from state S by an *adiabatic expansion*, then the second principle will be contradicted.

[8] Carathéodory, C. Untersuchungen Über die Grundlagen der Thermodynamik. Mathematische Annalen 67 (3): 355–386. [26].

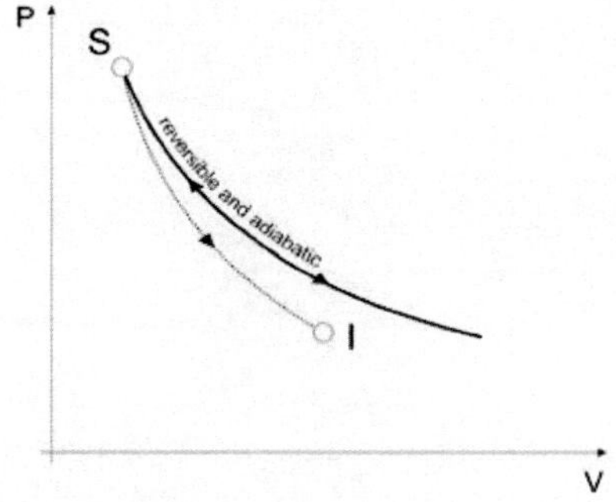

Fig. 2.26 Caratheodory principle

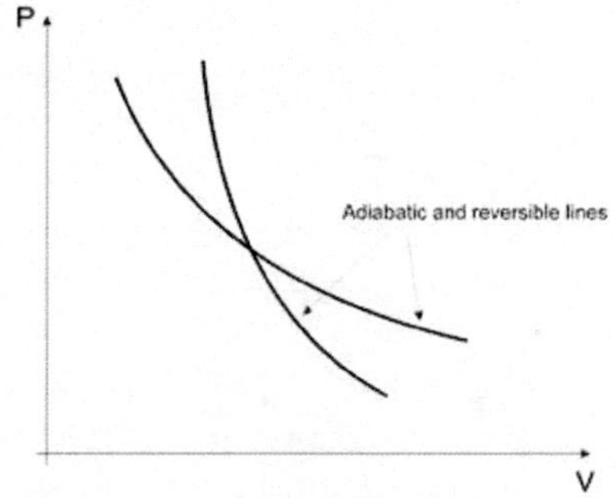

Fig. 2.27 Adiabatic and reversible lines do not intercept

3. In Fig. 2.27, show that two different lines representing an adiabatic reversible process in a P-V equilibrium diagram will never have an equilibrium state in common (Hint: *Suppose that these two lines have a common point and show that this assumption contradicts the second principle*).
4. A *perpetual motion engine of the second kind* is an engine that although does not contradict the principle of energy conservation, it denies the second principle of thermodynamics. Is the thermodynamic system shown in Fig. 2.28 a '*perpetual motion engine of the second kind*'?

2.3 Sadi Carnot and Émile Clapeyron

Sadi Carnot (1796–1832) was a French engineer who lived in a time when the steam engine was founding rapid growth in Europe, after the revolutionary steam engine of James Watt, patented in 1769. Carnot's ideas about the maximum motive power that is possible to extract from a heat engine were described in his book *Reflexions sur la puissance motrice du feu et sur les machine propes á développer cette puissance* published in 1824. Carnot died of cholera eight years later and, one year after his death, his book attracted the attention of Émile Clapeyron who made it known to

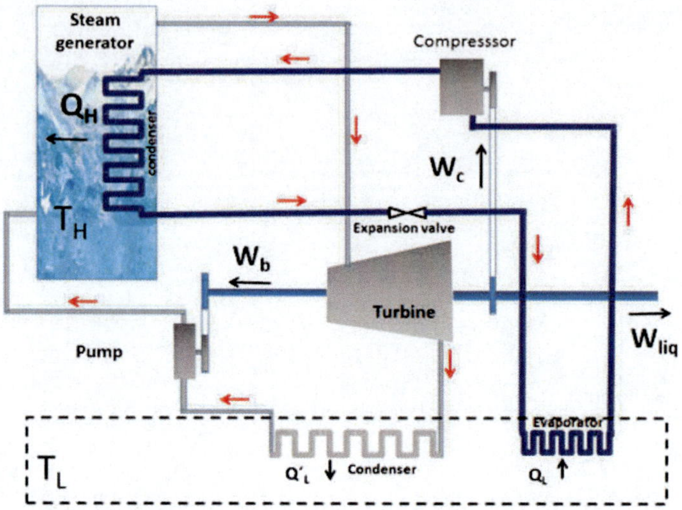

Fig. 2.28 Perpetual motion engine

the scientific community. In 1834, Émile Clapeyron reproduced the main ideas of Carnot in the Journal of the l'École Polytechnique [33].

In the page 154 of his '*Mémoire*' Clapeyron writes:

> Je citerai enfin parmi les travaux qui ont paru sur la théorie de la chaleur, un ouvrage de M. S. Carnot , publié en 1824, sous le titre de Réflexions sur la puissance motrice du feu. L'idée qui sert de base à ses recherches me paraît féconde et incontestable, ses démonstrations sont fondées sur l'absurdité qu'ily aurait à admettre la possibilité de créer de toutes pièces de la force motrice ou de la chaleur.
>
> Voici l'énoncé de plusieurs théorèmes auxquels le conduit cette méthode nouvelle de raisonnement.
>
> 1- Lorsqu'un gaz passe, sans changer de température, d'un volume et d 'une pression déterminés, á un autre volume et á une autre pression également déterminés, la quantité de calorique absorbée ou abandonnée est toujours la même, quelle que soit la nature du gaz choisi comme sujet d'expérience.
>
> 2- La diférence entre la chaleur spécifique sous pression constante, et la chaleur spécifique sous'volume constant, est la même pour tous les gaz.
>
> 3- Lorsqu'un gaz varie de volume sans changer de température les quantités de chaleur absorbées ou dégagées par ce gaz sont en progression arithmétique, si les accroissements ou réductions de volume sont en progression géométrique.

In his paper, Clapeyron gave a graphical representation of Carnot's reversible cycle in a PV diagram, shown in Fig. 2.29.

Carnot reversible cycle deals with a gas—an ideal gas that follows the Clapeyron equation of state—and is composed of two isothermal and two adiabatic processes (Fig. 2.30a). For an isothermal expansion to be reversible it must be quasi-static and the gas and the heat source must have the same temperature, say T_H. Otherwise, it

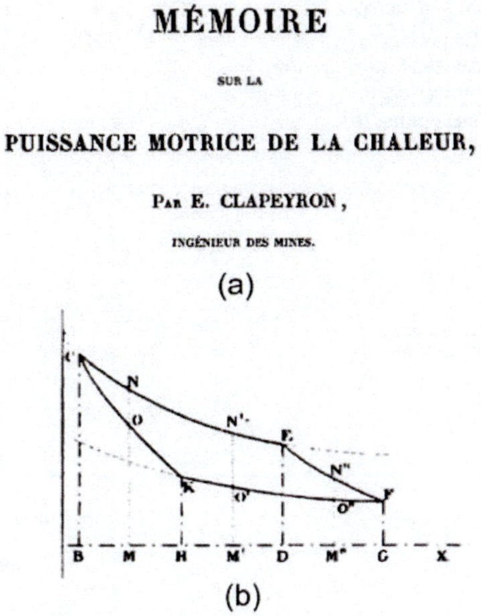

Fig. 2.29 **a** Front page of the Clapeyron 1834 article *Mémoire sur la puissance motrice de la chaleur*; **b** graphical representation by Clapeyron of the Carnot reversible cycle at page 190

will be irreversible (see Sect. 2.1). During this expansion, the gas absorbs the amount Q_H from the heat source. The next process is a quasi-static adiabatic expansion that will cool the gas from T_H to the temperature of the cold reservoir T_L. The heat Q_L is then delivered to the cold reservoir during the quasi-static isothermal compression of the gas, in thermal contact with this reservoir at T_L. Finally, the gas is compressed in a quasi-static adiabatic process and is heated from T_L to T_H.

The cycle of reversible engines can be reverted without changing the the absolute value of the heat and work that are exchanged.

In Carnot's ideas:

> Motive power can be produced wherever there exists a difference of temperature and the maximum motive power can be extracted from the heat (caloric) that flows from the hot to the cold reservoir, when the power cycle used for that is reversible.

This is the statement of the Carnot's first theorem.

A reversible engine is one that can be reverted to work as a heat pump without affecting the amounts of heat and work that are exchanged (Fig. 2.30b).

Carnot used the caloric theory to explain that a reversible engine must produce the maximum amount of work.

Consider that there is an engine operating in accordance with a power cycle that is more efficient than another engine that we suppose is capable of following a Carnot reversible engine, the two engines operating between the same heat reservoirs at T_H and T_L. Since we want to compare efficiencies and not power, we will suppose that the heat delivered to both engines is the same Q_H (2.31a).

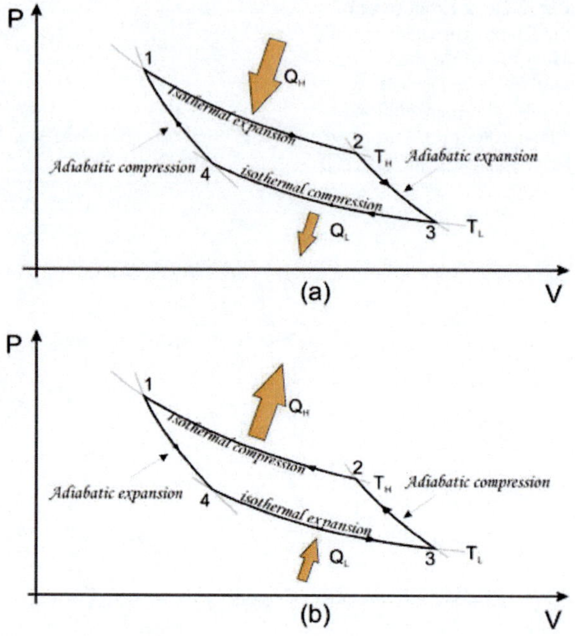

Fig. 2.30 The Carnot cycle as presented by Clapeyron in its 1834 *Mémoire* for a power engine (**a**) and for a heat pump (**b**)

The Carnot engine was supposed to be less efficient, so it produces a smaller shaft work $|W_c| < |W|$ and has a greater heat loss $\left|Q_{L,c}\right| > |Q_L|$.

Nevertheless, Carnot's engine is reversible, so the cycle can be reverted to make the engine operate as a heat pump delivering the same amount of heat Q_H to the heat reservoir T_H. Remember that reversible cycles can be reverted without changing the absolute values of the heat and work that are exchanged. This amount Q_H can be used to supply the heat that is required by the more efficient power engine, and the heat reservoir T_H can, thus, be discarded (Fig. 2.31b).

Finally, under the light of the first principle of thermodynamics and by making an energy balance for the composite system (Fig. 2.31b), we see that this composite system contradicts the Planck statement of the second principle of thermodynamics because it entirely converts the delivered heat $\left|Q_{L,c}\right| - |Q_L|$ into work $|W| - |W_c|$.

In Carnot's words: '*This would be not only perpetual motion, but an unlimited creation of motive power, without the consumption of either caloric or any other agent whatever.*'

Figure 2.31b shows what is now know as Carnot's first theorem.

Carnot's second theorem

Carnot's second theorem may be enunciated as follows:

> Two reversible engines operating in accordance with Carnot's cycle between the same heat reservoirs, will have the same efficiency.

The demonstration of this assertion follows the same reasoning as given in (Fig. 2.31) and is left as an exercise (see Exercise 2.3.3)

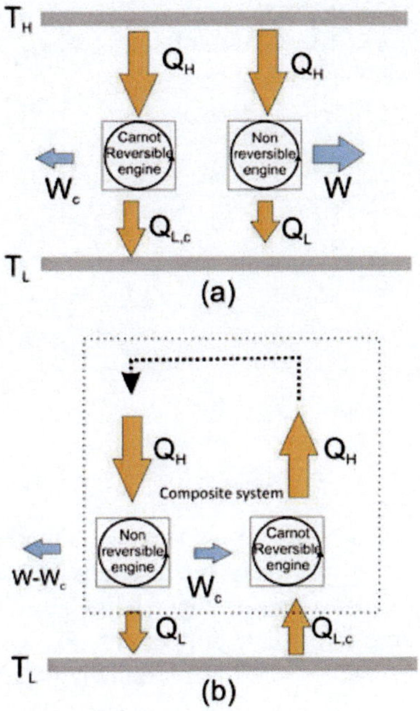

Fig. 2.31 If a non-reversible power engine is more efficient than a Carnot reversible engine when both engines are operating between the same heat reservoirs as in (**a**), then the reversible engine can be reverted and pump the rejected heat Q_H to the non-reversible engine (**b**)

2.3.1 Thermodynamic Scale of Temperature

> The relation between motive power and heat, as established by Carnot, is such that quantities of heat, and intervals of temperature, are involved as the sole elements in the expression for the amount of mechanical effect to be obtained through the agency of heat; and since we have, independently, a definite system for the measurement of quantities of heat, we are thus furnished with a measure for intervals according to which absolute differences in temperature may be estimated. (William Thomson, 1848[9])

Following Carnot' s second theorem, the thermodynamic efficiency η of a reversible engine operating in accordance with a Carnot' s cycle will be only dependent on the temperatures of the heat source and sink.

Since

$$\eta = 1 + \frac{Q_L}{Q_H},$$

and supposing that we are using the symbol θ to represent, e.g., the Celsius scale, this would mean that,

[9] William Thomson. *On an Absolute Thermometric Scale founded on Carnot's Theory of the Motive Power of Heat, and calculated from Regnault's Observations*. Philosophical Magazine October 1848 [from Sir William Thomson, Mathematical and Physical Papers, vol. 1 (Cambridge University Press, 1882), pp. 100–106.].

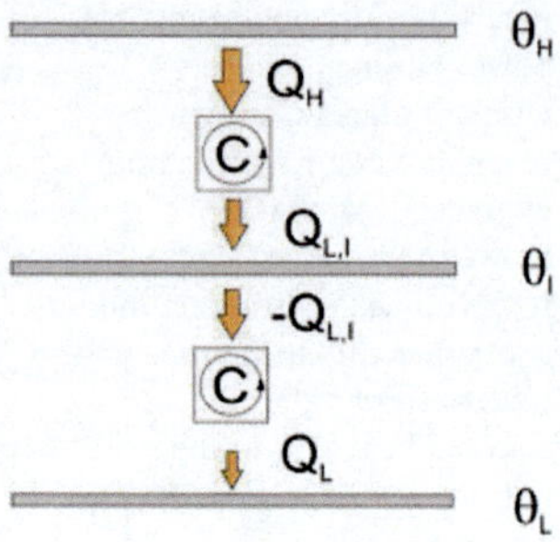

Fig. 2.32 The thermodynamic scale of temperature

$$\frac{Q_L}{Q_H} = f\,(\theta_L, \theta_H)\,,$$

where θ_L, θ_H designate the temperatures that a Celsius thermometer would measure for, respectively, the heat sink and source.

Consider now that a first Carnot' s engine operates between temperatures θ_H and an intermediate temperature θ_I, while a second Carnot' s engine operates between θ_I and θ_L(Fig. 2.32). Following Carnot' s second theorem

$$\frac{Q_{L,I}}{Q_H} = f\,(\theta_I, \theta_H)\,, \quad \frac{Q_L}{\left(-Q_{L,I}\right)} = f\,(\theta_L, \theta_I)\,. \tag{2.3}$$

From Carnot' s second theorem, remark that, for a given Q_H, the heat loss Q_L is only dependent on θ_H and $\theta_L$and that when these temperatures are also given, the amount $Q_{L,I}$ is only dependent on the intermediate temperature θ_I: $Q_{L,I}$ will approach $Q_L$when θ_I becomes close to θ_L and $\left(-Q_{L,I}\right)$ will approach $Q_H$when θ_I becomes close to θ_H This was the main idea of Kelvin: *'since we have, independently, a definite system for the measurement of quantities of heat, we are thus furnished with a measure for intervals according to which absolute differences of temperature may be estimated'*.

Therefore, from Eq. (2.3)

$$\frac{Q_L}{Q_H} = \frac{Q_L}{Q_{L,I}}\frac{Q_{L,I}}{Q_H} = -\,f\,(\theta_I, \theta_H)\,f\,(\theta_L, \theta_I)\,,$$

and since

$$\frac{Q_L}{Q_H} = f\,(\theta_L, \theta_H)\,,$$

we see that

$$f\,(\theta_L, \theta_H) = -\,f\,(\theta_I, \theta_H)\,f\,(\theta_L, \theta_I)\,.$$

The above equation can be only satisfied when its r.h.s. becomes independent of θ_I. A possible choice for eliminating this dependence is to write

$$f(\theta_L, \theta_H) = -\frac{T(\theta_I)}{T(\theta_H)}\frac{T(\theta_L)}{T(\theta_I)}.$$

The conclusion is that in a Carnot' s cycle the ratio between the heat loss Q_L and the heat gain $Q_H$is only dependent on the ratio $\frac{T(\theta_L)}{T(\theta_H)}$

$$\frac{Q_L}{Q_H} = -\frac{T(\theta_L)}{T(\theta_H)}.$$

For a given θ_H the thermodynamic efficiency of a Carnot' s cycle approaches 1 when $T(\theta_L) \to 0$ and approaches 0 when $\theta_L \to \theta_H$. Between these two limits, the thermodynamic efficiency will depend on θ_L. Nevertheless, the main idea of Kelvin was to measure the absolute temperature T by measuring the quantity of heat $Q(T)$.

2.3.2 Using the Ideal Gas Properties for Calculating the Conversion Efficiency for the Carnot's Cycle

Following the results of Kelvin, the thermodynamic efficiency of a Carnot' s engine is given by

$$\eta = 1 - \frac{T(\theta_L)}{T(\theta_H)}.$$

In this section, we will show that the Kelvin' s relationship $T(\theta)$ of the previous section can be found by using the thermodynamic properties of an ideal gas.

An important consequence of Carnot's second theorem is that the thermodynamic efficiency of a Carnot engine is independent of the gas that is used as the working fluid. So let us suppose that this working fluid is an ideal and *monatomic* gas, with a constant molar heat $\bar{c}_v = \frac{3}{2}\mathscr{R}$.

The conversion efficiency of *any* power engine is given as in Eq. (2.2) by

$$\eta = -\frac{W}{Q_H} = 1 + \frac{Q_L}{Q_H}. \quad (2.4)$$

In the Carnot's reversible cycle the heat, Q_H is delivered during the isothermal expansion 1-2 of Fig. 2.30a. From the first principle

$$\delta Q + \delta W = dU.$$

Since the internal energy of an ideal gas is only dependent on the temperature, $dU = 0$ and

$$\delta Q = -\delta W = PdV.$$

So

$$Q_H = Q_{1-2} = \int_{V_1}^{V_2} PdV = N\mathscr{R}T_H \ln \frac{V_2}{V_1}.$$

In the same way, the rejected heat can be calculated as

$$Q_L = Q_{3-4} = \int_{V_3}^{V_4} PdV = -N\mathscr{R}T_L \ln \frac{V_3}{V_4}.$$

Therefore, replacing Q_H and Q_L into Eq. (2.4), the conversion efficiency becomes

$$\eta = 1 - \frac{T_L \ln \frac{V_3}{V_4}}{T_H \ln \frac{V_2}{V_1}}. \tag{2.5}$$

When the gas suffers a reversible adiabatic expansion from T_H to T_L, pictured as process 2-3 in Fig. 2.30a, the delivered work will be equal to the variation in its internal energy or

$$\delta W = dU,$$

or, since we are dealing with an ideal monatomic gas

$$-Pd\bar{v} = \frac{3}{2}\mathscr{R}dT.$$

Now, from Clapeyron's equation $P\bar{v} = \mathscr{R}T$ and so

$$\frac{d\bar{v}}{\bar{v}} = -\frac{3}{2}\frac{dT}{T}.$$

By integrating the above equation between states 2 and 3 of Fig. 2.30a it can be concluded that

$$\frac{\bar{v}_3}{\bar{v}_2} = \left(\frac{T_H}{T_L}\right)^{\frac{3}{2}}.$$

Similarly, using the same reasoning for the adiabatic compression 3-4,

$$\frac{\bar{v}_4}{\bar{v}_1} = \left(\frac{T_H}{T_L}\right)^{\frac{3}{2}}.$$

So

$$\frac{\bar{v}_3}{\bar{v}_4} = \frac{\bar{v}_2}{\bar{v}_1},$$

and the conversion efficiency, Eq. (2.5) of the Carnot's cycle becomes

$$\eta = 1 - \frac{T_L}{T_H}. \tag{2.6}$$

In the above equation, T_L and T_H are the absolute ideal gas temperatures, following the discoveries of Gay-Lussac and Clapeyron (see Sect. 1.2.1).

Three important conclusions can be derived from this equation:

- The conversion efficiency is only dependent on the ratio between the low and high absolute temperatures T_L/T_H.
- The conversion efficiency increases when the temperature T_H of the hot source is increased with respect to T_L. In steam power engines this implies increasing the vapor pressure in the steam generator above the pressure of the James Watt, eighteenth century, atmospheric steam engines. Also, to produce, in the condenser, the vacuum necessary to condense the vapor at reduced pressures (*decreasing the temperature* T_L).
- The gas absolute temperature T is never negative, otherwise it would be possible to run a cycle with conversion efficiency η greater than 1.

2.3.3 Exercises

1. Two Carnot's engines A and B operate between the same heat reservoirs at T_H and T_L. (a) Suppose that the conversion efficiency of engine A, η_A, is greater than the conversion efficiency of engine B, η_B. Reverting engine B and using it to pump the rejected heat from engine A, show that this hypothesis leads to denying the second principle of thermodynamics and that, therefore, $\eta_B \geq \eta_A$. (b) Use the same reasoning, but now supposing $\eta_B > \eta_A$ for showing that $\eta_B \leq \eta_A$. (c) What is the single solution of the inequalities $\eta_B \geq \eta_A$ and $\eta_B \leq \eta_A$?
2. Suppose that the isothermal and adiabatic processes in a Carnot cycle are irreversible, the irreversible cycle being represented by 1-2-3′-4′-1 as in Fig. 2.33.

Using the analysis method of Sect. 2.1, show that:

(a) after the irreversible adiabatic expansion from state 2, the final equilibrium state 3′ at T_L will necessarily have a larger volume than V_3;
(b) to complete the cycle reaching state 1 the irreversible adiabatic compression must start from a state 4′ at the left of state 4 in the isotherm T_L;
(c) the heat $|Q'_H|$ absorbed by the gas during the isothermal irreversible expansion is necessarily smaller than the heat $|Q_H|$ absorbed during the reversible expansion;

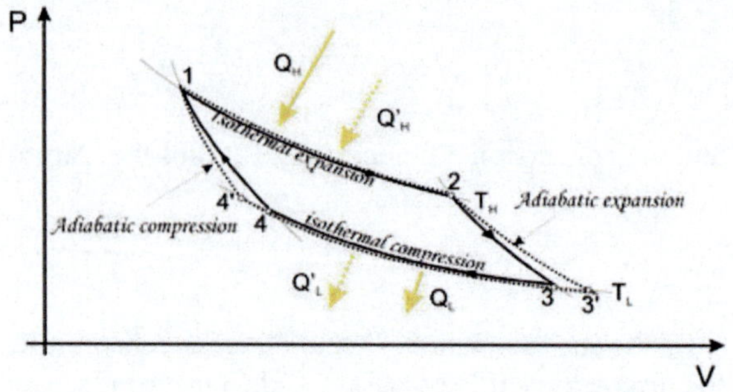

Fig. 2.33 A Carnot cycle where the individual processes fail to be reversible

(d) the heat $|Q'_L|$ rejected by the gas during the isothermal irreversible compression is necessarily larger than the heat $|Q_L|$ rejected during the reversible compression;

(e) the conversion efficiency η_{irr} of the irreversible cycle is necessarily smaller than the one corresponding to the reversible cycle η_{rev}.

2.4 Reversible Cycles Other Than Carnot's Cycle

The main condition for the Carnot first and second theorem to be demonstrated is that a *reversible* power engine can be reverted and work as a heat pump. The next question to be answered is why the Carnot cycle is the one with the highest efficiency and not any other reversible cycle? Since all reversible cycles can be reverted they would satisfy the condition that was imposed when demonstrating these theorems.

Consider the isobaric expansion represented in Fig. 2.34a when a gas at temperature T_1 is put in contact with a hot liquid bath. It is hard to conceive an isobaric expansion of a gas as a reversible process. Indeed, the gas is heated from below giving rise to Rayleigh-Bénard convection cells due to the gas density decrease close to the hot liquid bath. The heated gas will lift upward and cool, transferring heat to the cold gas above it, while the cold gas goes down due to its higher density.

Nevertheless, from a theoretical point of view it is possible to conceive an isobaric expansion as a sequence of very small isothermal expansions followed by adiabatic compressions, as shown in Fig. 2.34b. Each isothermal expansion is produced by removing a small balance weight with mass Δm. After the equilibrium is reached the gas chamber is insulated and this balance weight is put back on the piston head, retrieving the original pressure at a higher temperature $T_1 + \Delta T$. The gas is, then, put in contact with a heat reservoir at the same temperature $T_1 + \Delta T$, and the balance weight is removed a second time. This sequence is repeated till the gas attains temperature T_2. All the states represented as circles along the isobaric line in Fig. 2.34b are equilibrium states and, in the limit, when the mass $\Delta m \to 0$, this isobaric line will represent a reversible isobaric expansion.

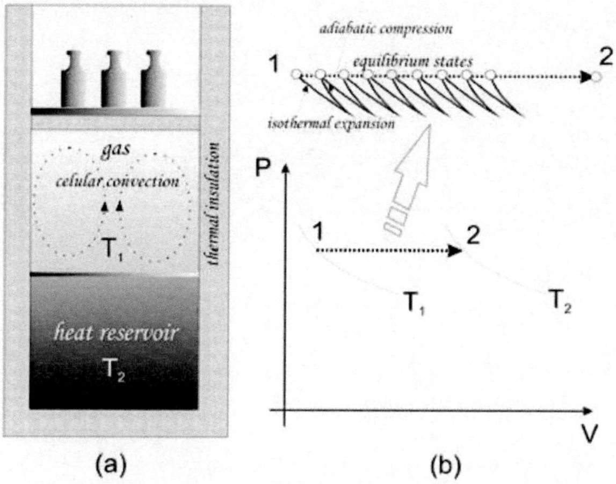

Fig. 2.34 **a** An isobaric expansion of a gas is never reversible: the gas is heated from below giving rise to Rayleigh-Bénard convection cells due to the gas density decrease close to the hot liquid bath. During the heating process, the gas temperature is T_2 close to the heat reservoir, but smaller than T_2 in the remaining space of the gas chamber. **b** From a theoretical perspective, it is possible to conceive an isobaric reversible expansion by considering it as the limit of a sequence of isothermal expansions and adiabatic compressions

Consider now a reversible cycle, composed of two isobaric and two adiabatic process such as cycle 1-2-3-4-1 in Fig. 2.35a. This cycle is known as the Brayton cycle.[10] The isobaric expansion 1-2 is as above: a sequence of reversible isothermal expansions followed by adiabatic compressions in the limit $\Delta m \to 0$. The isobaric compression 3-4 is a sequence of reversible isothermal compressions followed by adiabatic expansions, in the same limit.

We can conceive the Brayton cycle as a set of N Carnot cycles as shown in Fig. 2.35a, with $N \to \infty$. The first Carnot cycle operates between temperatures T_1 and T_4, the lowest temperature of the cycle, $T_4 = T_L$. The last Carnot cycle, cycle N, operates between temperatures T_2 — the highest temperature of the cycle, $T_2 = T_H$ — and T_3.

Since the adiabatic expansion work of each Carnot cycle i will be canceled by the adiabatic compression work of Carnot cycle $i + 1$, the net adiabatic expansion work will be the area under line 2-3 and the net adiabatic compression work will be the area under the line 4-1. Also, the total heat received by the gas during the isobaric expansion 1-2 will be the sum

[10] The cycle is named after George Brayton (1830–1892), the American engineer who developed it originally for use in piston engines. It is also sometimes known as the Joule cycle. There are two types of Brayton cycles, open to the atmosphere and using internal combustion chamber or closed and using a heat exchanger.

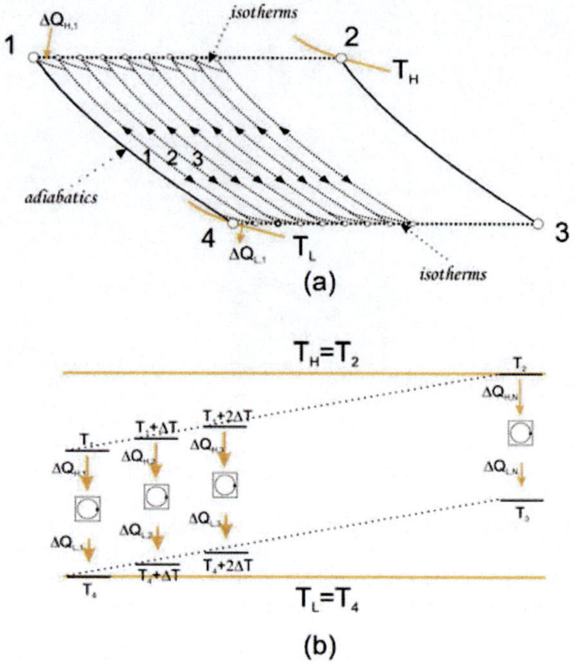

Fig. 2.35 **a** A reversible cycle (Brayton cycle) composed of two isobaric and two adiabatic processes may be decomposed in a set of Carnot cycles. **b** Each Carnot engine operates in a temperature range that is a fraction of the entire temperature range between T_L and T_H

$$Q_H = \sum_{i=1}^{N} \Delta Q_{H,i},$$

and each $\Delta Q_{H,i}$ is the heat absorbed by the gas during the small isothermal expansion at temperature T_i. It can be shown — see Exercise 1 in this section — that in the limit $\Delta T = T_{i+1} - T_i \to 0$, this absorbed heat will be $\Delta Q_{H,i} = c_p(T_i) N \Delta T$ for an ideal gas,[11] and this is the heat that is absorbed by the gas during the isobaric expansion from the equilibrium state i to $i + 1$.

The same reasoning can be made for the total heat that is rejected during the isobaric compression 3-4,

$$Q_L = \sum_{i=1}^{N} \Delta Q_{L,i}.$$

Therefore, the present reversible cycle can be considered as a set of Carnot cycles as shown in Fig. 2.35b, with the same heat and work exchanged as the original cycle.

Let us now calculate the conversion efficiency of this reversible cycle starting from Eq. (2.4)

[11] Since $\Delta T \to 0$ and the molar heats c_p and c_v depend only on the temperature, they can be considered as constants.

$$\eta = 1 + \frac{Q_L}{Q_H} = 1 + \frac{\sum_{i=1}^{N} \Delta Q_{L,i}}{\sum_{i=1}^{N} \Delta Q_{H,i}}. \tag{2.7}$$

For each Carnot engine i,

$$\Delta Q_{L,i} = \Delta Q_{H,i} \left(\eta_{c,i} - 1\right),$$

where $\eta_{c,i}$ is the conversion efficiency of each Carnot engine i. So,

$$\eta = 1 + \frac{\sum_{i=1}^{N} \Delta Q_{H,i} \left(\eta_{c,i} - 1\right)}{\sum_{i=1}^{N} \Delta Q_{H,i}} = \frac{\sum_{i=1}^{N} \eta_{c,i} \Delta Q_{H,i}}{\sum_{i=1}^{N} \Delta Q_{H,i}}. \tag{2.8}$$

Nevertheless, using the concept of the mean value for the conversion efficiency

$$\langle \eta_c \rangle = \frac{\sum_{i=1}^{N} \eta_{c,i} \Delta Q_{H,i}}{\sum_{i=1}^{N} \Delta Q_{H,i}},$$

it results

$$\eta = \langle \eta_c \rangle$$

So, the conversion efficiency of the present reversible cycle is the mean value of the conversion efficiencies of the set of N Carnot cycles in the limit when $N \to \infty$. Since each Carnot cycle operates in a temperature interval which is a fraction of $T_H - T_L$, we conclude that $\eta < 1 - \frac{T_L}{T_H}$.

Indeed, during the (reversible) isobaric expansion heat is supplied to the gas at temperatures that vary from T_1 to $T_2 = T_H$, the highest temperature of the gas, contrary to Carnot's cycle, where heat is supplied at the highest temperature in the cycle; in the reverse direction, during the (reversible) isobaric compression heat is rejected by the gas at temperatures that vary from T_3 to $T_4 = T_L$, the lowest temperature of the gas, contrary to the Carnot's cycle, where heat is rejected at the lowest temperature in the cycle.

In Brayton and other reversible cycles, heat is supplied and rejected at variable temperatures, in contrast with the Carnot' s cycle, where heat is supplied at the highest and rejected at the lowest temperatures of the cycle. This is the reason why their thermodynamic efficiency is smaller than the one of Carnot' s cycle.

2.4.1 Exercises

1. The conversion efficiency of the Brayton cycle shown in Fig. 2.35 can be exactly found as

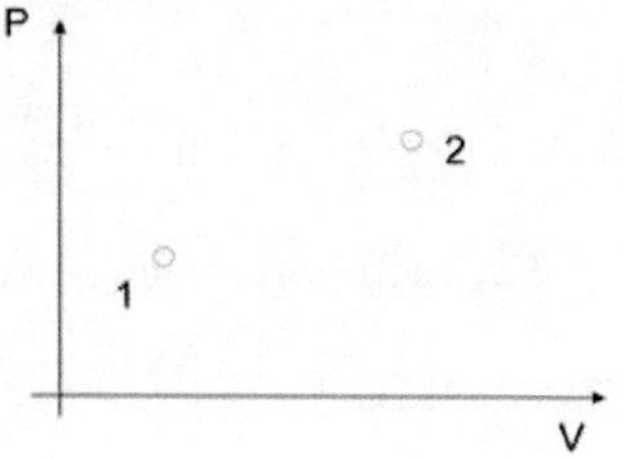

Fig. 2.36 Reversible processes between two arbitrary equilibrium states

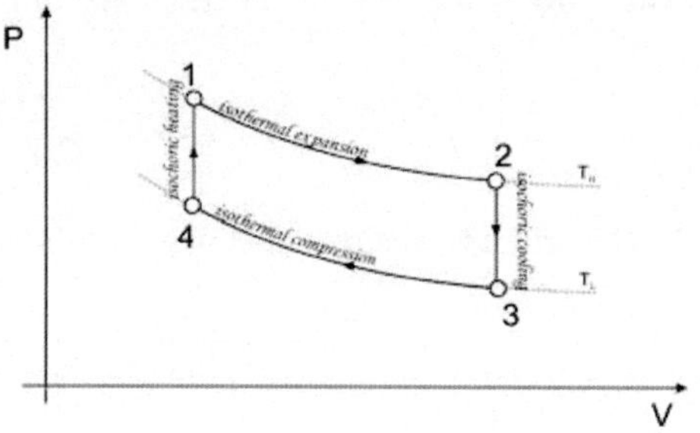

Fig. 2.37 Another reversible cycle

$$\eta = 1 - \zeta \frac{T_L}{T_H}.$$

Find ζ when the gas is an ideal, monatomic[12] gas and show that $\zeta > 1$.

2. How many reversible lines it is possible to draw between the equilibrium states 1 and 2 shown in Fig. 2.36? Is it possible to join points 1 and 2 using a straight line that represents a reversible process?
3. Show that every reversible cycle can be decomposed into Carnot cycles.
4. Show that every reversible cycle can be decomposed into Carnot cycles. Consider, in particular, a reversible Stirling[13] cycle 1-2-3-4-1, composed of two isothermic and two isochoric processes as pictured in Fig. 2.37 and show in detail that this cycle can effectively be decomposed into small Carnot's cycles. Calculate the conversion efficiency of this cycle when the gas is an ideal, monatomic gas. Show that this conversion efficiency is smaller than the one of a Carnot' s cycle operating between T_L and T_H.

[12] For monatomic gases $\bar{c}_v = \frac{3}{2}\mathscr{R}$.

[13] The Stirling cycle is a reversible cycle that theoretically describes the external combustion Stirling engines, invented, developed, and patented in 1816 by Robert Stirling.

2.5 Clausius Inequality and Entropy

Die Energie der Welt ist constant. Die Entropie der Welt strebt einem Maximum zu. (Rudolf Clausius, 1867 [36])

For a Carnot cycle between temperatures T_L and T_H it was seen that the conversion efficiency is given by

$$\eta = 1 + \frac{Q_L}{Q_H} = 1 - \frac{T_L}{T_H}.$$

So, for a Carnot cycle,

$$\frac{Q_H}{T_H} + \frac{Q_L}{T_L} = 0, \tag{2.9}$$

meaning that the absorbed and rejected heat are, respectively, proportional to the temperatures of the hot and cold heat reservoirs.

Consider now an arbitrary reversible cycle such as the one shown in Fig. 2.35. Since, as was shown in the previous section, all reversible cycles can be considered as a set of N Carnot cycles, and so,

$$\sum_{i=1}^{N} \left(\frac{\Delta Q_{H,i}}{T_{H,i}} + \frac{\Delta Q_{L,i}}{T_{L,i}} \right) = 0.$$

which can also be written as

$$\sum_{i=1}^{N} \frac{\Delta Q_i}{T_i} \Bigg)_{rev} = 0,$$

and when $N \rightarrow \infty$,

$$\oint \frac{\delta Q}{T} \Bigg)_{rev} = 0 \tag{2.10}$$

meaning that in reversible cycles, the heat exchanged by a system in thermodynamic equilibrium is proportional to the temperature at which this heat is exchanged.

Since this integral is zero for all reversible cycles, the integrand must be a state variable. In the words of Clausius[14] "*...whenever a body, starting from any initial condition, returns thereto after its passage through any other conditions, then the expression* $\frac{\delta Q}{T}$ *under the sign of integration must be the complete differential of a*

[14] R. Clausius (1867), *The mechanical theory of heat, with its applications to the steam-engine and to the physical properties of bodies*, Edited by T.A. Hirst, John van Voorst, London [36]. Ninth Memoir: *On several convenient forms of the fundamental equations of the mechanical theory of heat*, pp. 327–365.

magnitude that depends only on the present existing condition of the body, and not upon the way by which it reached the latter. Denoting this magnitude by S, we can write $dS = \frac{\delta Q}{T}$."

The name "*entropy* is justified at the page 357 of its Ninth Memoir: "*We might call S the transformational content of the body, just as we termed the magnitude U its thermal and ergonal*[15] *content. But as I hold it to be better to borrow terms for important magnitudes from the ancient languages, so that they may be adopted unchanged in all modern languages, I propose to call the magnitude S the entropy of the body, from the Greek word τροπη, transformation. I have intentionally formed the word entropy so as to be as similar as possible to the word energy; for the two magnitudes to be denoted by these words are so nearly allied in their physical meanings, that a certain similarity in designation appears to be desirable*".

We rewrite the Clausius definition of entropy as

$$dS = \left.\frac{\delta Q}{T}\right)_{rev}, \tag{2.11}$$

meaning that the entropy variation between any two states can be only measured by considering a reversible path joining these states.

Consider, now, a Carnot cycle where heat is supplied and rejected along irreversible isothermal processes such as the ones shown by the dashed lines in Fig. 2.33.

As it was shown in Sect. 2.1 the absorbed heat $|Q'_H|$ will always be smaller than the corresponding absorbed heat $|Q_H|$ if the expansion was reversible; indeed, in irreversible isothermal expansions— the ones due to the removal of a finite balance weight with mass Δm — the delivered work decreases with respect to the work that would be delivered in an isothermal reversible expansion. On the other hand, the rejected heat $|Q'_L|$ during the irreversible compression is always greater than $|Q_L|$, i.e., the heat that would be rejected during an isothermal reversible compression.

Therefore, in this *'irreversible Carnot cycle'*, Eq. (2.9) giving the proportionality between the exchanged heat and the temperature must be replaced by

$$\frac{Q_H}{T_H} + \frac{Q_L}{T_L} \leq 0, \tag{2.12}$$

---

[15] *The term ergon is defined at page 253 of 'The mechanical theory of heat', Sixt Memoir as what would, presently, be designated as 'work': "Now, in the mechanical theory of heat, after admitting that heat can be transformed into work and work into heat, in other words, that either of these may replace the other, it becomes frequently necessary to form a magnitude of which heat and work are constituent parts. (....) Let heat and work continue to be measured each according to its most convenient unit, that is to say, heat according to the thermal unit, and work according to the mechanical one. But besides the work measured according to the mechanical unit, let another magnitude be introduced denoting the work measured according to the thermal unit, that is to say, the numerical value of the work when the unit of work is that which is equivalent to the thermal unit. For the work thus expressed a particular name is requisite. I propose to adopt for it the Greek word (ϵργον) ergon".*

Fig. 2.38 Clausius entropy

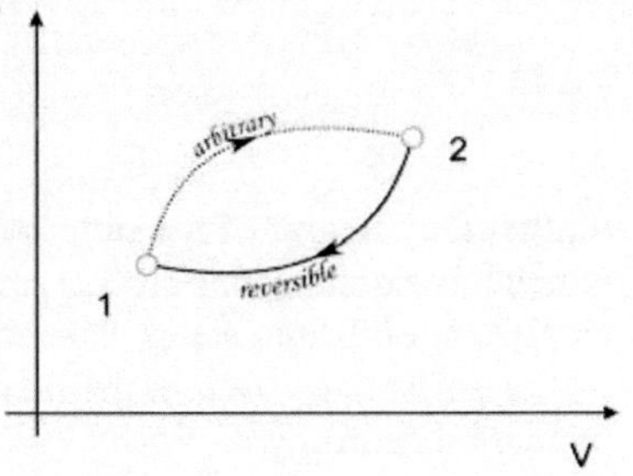

meaning that, in absolute values, the heat gains are smaller, and the heat losses are greater than the ones corresponding to a theoretical Carnot reversible cycle. The equality sign only applies when the mass $\Delta m \to 0$ and the entire cycle is reversible.

For arbitrary cycles, the above inequality can be easily generalized to give

$$\oint \frac{\delta Q}{T} \leq 0.$$

This inequality received the name of '*Clausius inequality*' or '*Clausius theorem*' and is described on page 133 of Clausius Fourth Memoir '*On a modified form of the second fundamental theorem in the mechanical theory of heat*'.[16]

Consider now that a cycle is composed of an arbitrary process (reversible or non-reversible) leading a thermodynamic system from an equilibrium state 1 to another equilibrium state 2. It is always possible to return back from state 2 to the initial equilibrium state 1, by taking one of the infinite reversible processes from 2 to 1 (Fig. 2.38).

Therefore, from the Clausius theorem,

$$\oint \frac{\delta Q}{T} = \int_1^2 \frac{\delta Q}{T} + \int_2^1 \left.\frac{\delta Q}{T}\right)_{rev} \leq 0,$$

and, since the returning process, 2-1 is reversible

$$\int_2^1 \left.\frac{\delta Q}{T}\right)_{rev} = S_1 - S_2.$$

So, we are led to the conclusion that

[16] R. Clausius (1865), *The mechanical theory of heat, with its applications to the steam engine and to the physical properties of bodies*, Edited by T.A. Hirst, John van Voorst, London [36]. Fourth Memoir: *On a modified form of the second fundamental theorem in the mechanical theory of heat*, pp.111–135.

$$S_2 - S_1 \geq \int_1^2 \frac{\delta Q}{T}. \tag{2.13}$$

or, that the integral of the ratio between the heat exchanged and the temperature at which it is exchanged is always smaller than the related variation of entropy between any two equilibrium states. The equality only holds for reversible processes.

For an isolated system the heat exchanged is null, so, any internal process will increase its entropy

$$dS \geq 0. \tag{2.14}$$

This equation can be seen as the mathematical expression of the *arrow of natural processes* since all internal processes that result from a non-equilibrium in the fields of temperature, density, pressure, or local velocity will produce an increase in entropy.

2.5.1 External and Internal Sources of Entropy

Irreversible processes are always escorted by the appearance of non-equilibrium in their local fields. So, when a thermodynamic system is subjected to an irreversible process, its temperature field is unbalanced and it is impossible to access its value at a given point without a local thermodynamic non-equilibrium analysis. Nevertheless, all the heat that is exchanged by a thermodynamic system flows through the boundary that separates it from the external world and Eq. (2.13) is better written as

$$S_2 - S_1 \geq \int_1^2 \left.\frac{\delta Q}{T}\right)_{bound.}. \tag{2.15}$$

In fact, the elementary variation of entropy of a thermodynamic system can be written as[17]

$$dS = d_e S + d_i S,$$

where

$$d_e S = \left.\frac{\delta Q}{T}\right)_{bound.},$$

is the variation of entropy of this system due to the heat that flows through its boundary and the term $d_i S$ is the source of entropy due to internal non-equilibrium. From the

[17] I. Prigogine, Introduction to Thermodynamics of Irreversible Processes (Interscience, 1968).

Fig. 2.39 Irreversible cycle of an ideal gas in contact with a heat reservoir: Heat and work balances for the expansion and compression processes

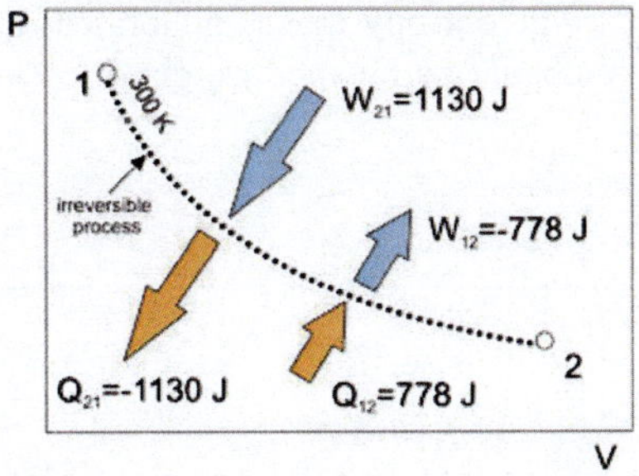

principle of increasing entropy for isolated systems, Eq. (2.14) it is clear that this term must be non-negative,

$$d_i S \geq 0.$$

It is also clear that the elementary variation of entropy $d_e S$ can be positive or negative in accordance with the heat that is flowing through the boundary: when heat is flowing into the system $d_e S > 0$, otherwise it is negative.

Example

Consider the irreversible isothermal cycle of Fig. 2.1, when a balance weight of 50 kg is removed from the piston top plate above a gas chamber, producing an irreversible expansion (Fig. 2.39).

The initial and final equilibrium states are given in Table 2.1.

It is possible to calculate the entropy variation between the equilibrium states 1 and 2, by integrating Eq. (2.11), following a reversible path between states 1 and 2. The most obvious reversible path to follow is the isothermal expansion 1-2. In this case, the heat absorbed by the gas was calculated as 930 J (see Sect. 2.1). So,

$$S_2 - S_1 = \frac{Q_{1-2)_{rev}}}{T} = \frac{930}{300} = 3.10\,\frac{\text{J}}{\text{K}}.$$

On the other hand, the entropy flow from the heat reservoir to the gas will be

$$(S_2 - S_1)_e = \frac{Q_{1-2}}{T} = \frac{778}{300} = 2.5933\,\frac{\text{J}}{\text{K}},$$

This ratio becomes closer to 3.10 $\frac{\text{J}}{\text{K}}$ when the gas expansion is produced by sequentially removing lighter balance weights with, e.g., 10, 5, 2, 1, and 0.5 kg, but in all these expansion processes the entropy variation of the gas is always the same because entropy is a state variable that "*depends only on the present existing condition of the body, and not upon the way by which it reached this state*".

The entropy that is generated due to internal non-equilibrium produced by the removal of a balance weight of 50 kg from the piston head can be calculated as

$$(S_2 - S_1)_i = (S_2 - S_1) - (S_2 - S_1)_e$$
$$= 0.5067\,\frac{\text{J}}{\text{K}}.$$

Gas compression
Let us now repeat the same calculation for the compression process when the 50 kg balance weight is placed back on the piston plate. In this case

$$S_1 - S_2 = -\,3.10\,\frac{\text{J}}{\text{K}},$$

and the heat that is delivered from the gas to the heat reservoir during the compression is $Q_{2-1} = -\,1130$ J. So

$$(S_1 - S_2)_e = -\,\frac{1130}{300} = -\,3.77\,\frac{\text{J}}{\text{K}}.$$

On the other hand, the entropy production will be given by

$$(S_1 - S_2)_i = 0.67\,\frac{\text{J}}{\text{K}}.$$

Cycle
It was seen that the heat and work exchanged by the system in an isothermal cycle must satisfy

$$\oint \delta W = -\oint \delta Q \geq 0,$$

otherwise, the second principle would be denied.

The Clausius inequality means that

$$\oint \frac{\delta Q}{T} = 2.5933 - 3.77 = -\,1.1767\,\frac{\text{J}}{\text{K}},$$

is negative for *all* irreversible cycles or that, in absolute value, the heat gains are smaller and the heat losses are greater than the corresponding heat gains and losses in a reversible cycle.

Since entropy is a state variable,

$$\oint dS = 0.$$

Nevertheless, there is a net production of entropy after each irreversible cycle

$$\oint dS_i = 0.5067 + 0.67 = 1.1767 \, \frac{\mathrm{J}}{\mathrm{K}}.$$

Also,

$$\oint dS_e = 2.5933 - 3.77 = -1.1767 \, \frac{\mathrm{J}}{\mathrm{K}},$$

meaning that all the entropy that is internally produced flows to the heat reservoir after each cycle.

Reservoir

Consider now the heat reservoir as our thermodynamic system. Since the reservoir is a system that, theoretically, is always in equilibrium, its entropy variation will be only due to the flow of heat. So, after the gas expansion

$$(S_2 - S_1)_R = -\frac{778}{300} = -2.5933 \, \frac{\mathrm{J}}{\mathrm{K}},$$

and after the gas compression

$$(S_1 - S_2)_R = \frac{1130}{300} = 3.77 \, \frac{\mathrm{J}}{\mathrm{K}},$$

For the entire cycle

$$\oint dS_R = 1.1767 \, \frac{\mathrm{J}}{\mathrm{K}},$$

as would be expected, since each cycle that is performed by the gas system produces a flow of entropy from the gas system to the reservoir.

Whole system gas + reservoir

The whole system gas + reservoir is an isolated system and the entropy variation must be positive for both the expansion and compression processes. So, after the gas expansion

$$(S_2 - S_1)_{R+G} = 0.5067 \, \frac{\mathrm{J}}{\mathrm{K}},$$

and after the gas compression

$$(S_1 - S_2)_{R+G} = 0.67 \, \frac{\mathrm{J}}{\mathrm{K}}.$$

For the entire cycle

$$\oint dS_{R+G} = 1.1767\,\frac{\text{J}}{\text{K}}.$$

During each cycle, work is converted into heat (*and not the inverse*) and this is a general rule for all isolated systems. The *arrow of nature* is clearly pictured in this rule and is translated by a cumulative increase of entropy,

In concluding this section it is perhaps important to read again William Thomson in his 1851 famous paper 'On the dynamical theory of heat': "*I believe the tendency in the material world is for motion to become diffused, and that as a whole the reverse of concentration is gradually going on - I believe that no physical action can ever restore the heat emitted from the sun, and that this source is not inexhaustible; also that the motions of the earth and other planets are losing vis viva which is converted into heat; and that although some vis viva may be restored for instance to the earth by heat received from the sun, or by other means, that the loss cannot be precisely compensated and I think it probable that it is under-compensated.*

2.5.2 *Exercises*

1. A single mole of Helium initially at T_1 =300 K and P = 1 MPa, is put in contact with a heat reservoir at T_2 =800 K as shown in Fig. 2.40. In consequence, it has an isobaric expansion till equilibrium is reached at 800 K. Helium can be considered as satisfying the Clapeyron equation of state in this range of temperatures. Calculate: (a) the entropy variation $S_2 - S_1$ of the gas ; (b) the heat Q_{1-2} that is transferred to the gas during the expansion; (c) the increase in the gas entropy $(S_2 - S_1)_e$ due to the flow of heat from the reservoir; (d) the entropy that is generated inside the gas phase $(S_2 - S_1)_i$; (e) the entropy variation of the heat

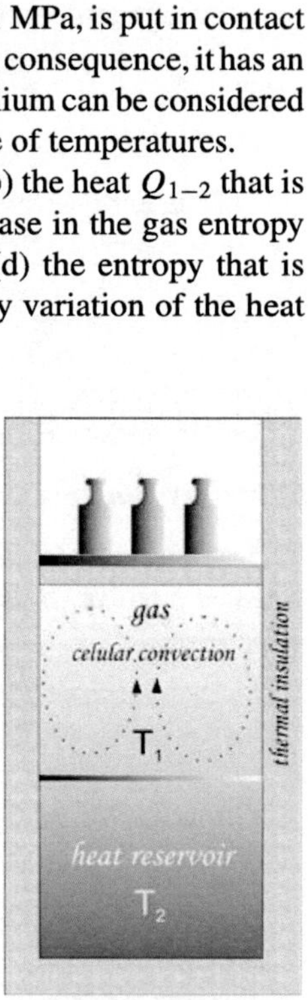

Fig. 2.40 Isobaric expansion of an ideal gas in contact with a heat reservoir

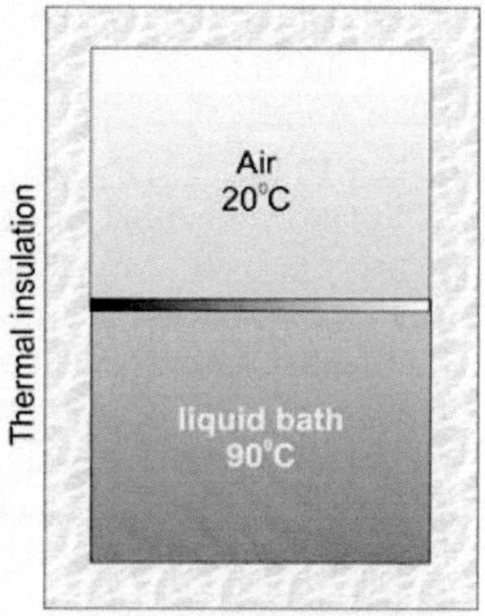

Fig. 2.41 Isochoric heating of an ideal gas in contact with a heat reservoir

reservoir after the gas expansion; (f) the entropy variation of the whole system gas+reservoir $(S_2 - S_1)_{R+G}$; (g) comment on the results that were found.
Repeat the same calculations, considering now that the Helium is replaced by Nitrogen. The molar thermal capacity at a constant volume of Nitrogen is given in terms of the temperature by

$$\bar{c}_v \left(\frac{\text{J}}{\text{mol K}}\right) = 20.586 - 0.1571 \times 10^{-2} T + 0.8081 \times 10^{-5} T^2 - 2.873 \times 10^{-9} T^3.$$

2. One mole of air initially at 20 °C is put in thermal contact with a heat reservoir at T_2 =90 °C as shown in Fig. 2.41 and suffers iso-volumetric heating till it attains the final equilibrium state at 90 °C. For this problem, the specific thermal capacity of the air can be considered as a constant and given by $c_v = 0.718 \frac{kJ}{kg \cdot K}$. Air can be considered as satisfying the Clapeyron equation of state in this range of temperatures.
Calculate: (a) the entropy variation $S_2 - S_1$ of the gas ; (b) the heat Q_{1-2} that is transferred to the gas during the heating; (c) the increase in the gas entropy $(S_2 - S_1)_e$ due to the flow of heat from the reservoir; (d) the entropy that is generated inside the gas phase $(S_2 - S_1)_i$; (e) the entropy variation of the heat reservoir after the gas expansion; (f) the entropy variation of the whole system gas+reservoir $(S_2 - S_1)_{R+G}$. Comment on the results that were found.
3. A cylinder is divided into two equal-volume chambers A and B. Each chamber contains 1 mole of Helium initially at 20 °C (Fig. 2.16). Chambers A and B are separated by a thermal insulated plate, free to move inside the cylinder, in such a manner as to equalize the pressures in the two chambers. The base plate of the cylinder is a thin thermal contact plate. Consider now that this base plate is put in contact with a thermal reservoir at 200 °C. The Helium in chamber B will expand moving the dividing plate upward.
Since the adiabatic compression of the Helium in chamber A is a very slow process the compression process of He in chamber B can be supposed to be

reversible. Calculate: (a) the entropy variation $(S_2 - S_1)_{A+B}$ of the system A+B, (b) the entropy variation of the system A+B due to the flow of heat from the reservoir, $(S_2 - S_1)_{e,A+B}$; (c) the entropy that is internally produced in system A + B, $(S_2 - S_1)_{i,A+B}$.
Suppose, now, that the compression of the Helium in chamber A is an irreversible process. In this case, it is not possible to know what would be the final equilibrium state without a local non-equilibrium analysis. Nevertheless, it is possible to establish some inequalities.
(d) When He is irreversibly compressed in chamber A the only equilibrium states that are possible to be attained by this gas are the ones with a higher temperature for the same pressure that was reached after the reversible adiabatic compression. So the final pressure in this second process cannot be the same as in the first when He is in the chamber A was supposed to be reversibly compressed, since the total volume did not change. What is it possible to say about the final pressure $P_{2,irr}$ in this second process, it will be greater or smaller than $P_{2,rev}$, i.e., the final pressure in the first process? Try to answer the same question for the final temperature $T^A_{2,irr.}$ of the He in chamber A.
(e) The heat absorbed from the reservoir is used for heating the gas in chamber B and to expand it against chamber A. In the second process, it is not possible to calculate the compression work delivered to chamber A, without a local non-equilibrium analysis. But what about the value of the total heat Q_{1-2} that is absorbed from the heat reservoir? What it is possible to say about Q_{1-2} when the two processes are compared?

4. After the expansion of an ideal gas from the equilibrium state 1 to 2, in thermal contact with a heat reservoir supposed to be at a constant temperature[18]

$$\oint d_i S_R = 0,$$

and all the entropy variation of the reservoir is due to the heat that it exchanges with the gas

$$\oint d_e S_R = \oint \left.\frac{\delta Q}{T}\right)_{bound.},$$

This is only true for idealized heat reservoirs. In the real world, the net transfer of heat after each cycle of the gas system produces non-equilibrium in the temperature field of the liquid bath that is being used with the role of heat reservoir and the internal generation of entropy will be different from zero after each cycle. The analysis of this kind of system requires, nevertheless, a local approach, the entropy variation of the gas is $(S_2 - S_1)$ and the entropy variation of the heat reservoir is $(S_2 - S_1)_R$.

[18] In this exercise the heat reservoir is supposed to be in thermal equilibrium during the entire cycle 1-2-1. Therefore the entropy generated in the reservoir is null.

(i) What is the physical meaning of $(S_2 - S_1) + (S_2 - S_1)_R$? What is necessary for this sum be zero?

(ii) Consider now the isothermal cycle 1-2-1. What alternatives below are the correct ones? Explain your answer.

(a) $\oint dS > 0$, (b) $\oint dS = 0$, (c) $\oint dS_R = 0$, (d) $\oint d\,(S + S_R) \geq 0$.

5. A single mole of an ideal gas is in chamber A with 1ℓ of Fig. 2.42. This chamber is separated from an evacuated chamber B with 9ℓ by a plate p that is fixed on the cylinder wall by two pins. The pins are then removed and the gas has a free expansion occupying all the remaining volume. Both chambers are in perfect thermal contact with an idealized heat reservoir.
Find:

(a) the entropy variation $S_2 - S_1$ of the gas between the initial, 1 and final, 2, equilibrium states. What are the fractions of this entropy variation that are due to heat flow, $(S_2 - S_1)_e$, and to internal generation $(S_2 - S_1)_i$;
(b) the entropy variation of the heat reservoir $(S_2 - S_1)_R$;
Suppose now that the initial pressure P_1 is suddenly restored above the plate p and the gas returns to its initial state after suffering an isobaric compression
Find:
(c) the work W_{2-1} and heat Q_{2-1} exchanged;
(d) the entropy variation of the gas between the equilibrium states 2 and 1 that is due to the heat flow $(S_1 - S_2)_e$ and the one that is due to internal generation;
(e) the entropy variation of the reservoir between the equilibrium states 2 and 1
Finally, consider the entire cycle 1-2-1.
Find:
(f) $\oint \delta Q$ and $\oint \delta W$ for the gas;
(g) $\oint d_e S$ for the gas and reservoir;
(h) $\oint d_i S$ for the gas and reservoir;
(i) $\oint dS$ for the global system gas+reservoir.
(j) What it can be concluded from the above calculations?

2.6 The Meaning of Entropy

Philosophy gets on my nerves. If we analyze the ultimate ground of everything, then everything finally falls into nothingness. But I have decided to resume my lectures again and look the Hydra of doubt straight into the eye. (Ludwig Boltzmann)

The statement of the law of energy conservation in 1847 by Helmholtz meant the acceptance of the corpuscular nature of the matter by the scientific world.

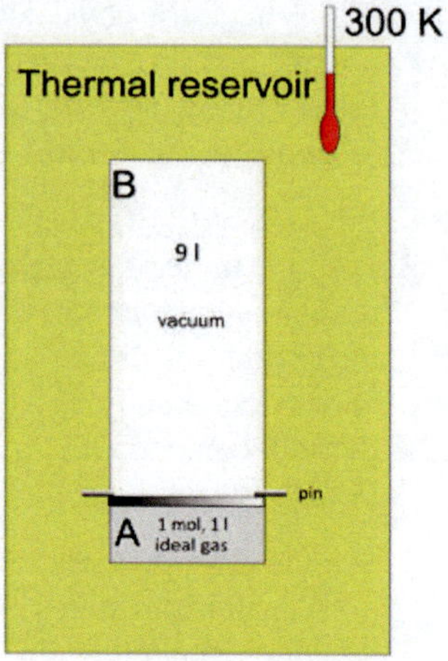

Fig. 2.42 Isothermal expansion against vacuum

References to the concept of atomism and atoms are found in ancient India and ancient Greece. In India the Jain, Ajivika and Carvaka[19] schools of atomism may date back to the fourth century BC.[20] The Nyaya and Vaisheshika schools later developed theories on how atoms combined into more complex objects.[21] In Greece, atomism emerged in the fifth century BC with Leucippus and Democritus.[22]

The particles of chemical matter for which chemists and other natural philosophers of the early nineteenth century found experimental evidence were thought to be indivisible, and therefore were given the name 'atom', long used by the atomist philosophy.

The concept of 'heat' or 'internal energy of a thermodynamic system as the result of the mechanical energy of its individual molecules was long advocated by philosophers and scientists such as Francis Bacon (1561–1626), John Locke (1632–1704), and Thomas Hooke (1635–1703), .

In his 'Hydrodynamica' (Fig. 2.43), Bernoulli (1700–1782) proposed that gases consist of great numbers of molecules moving in all directions, that their impact on a surface causes the gas pressure and that heat transfer occurs from regions with high energetic molecules to those with low energetic molecules.

James Joule equivalence principle and the nineteenth century mechanical theory of heat were highly influenced by the atomistic philosophy.

---

[19] M. Gangopadhyaya (1981). *Indian Atomism: History and Sources*. Atlantic Highlands, New Jersey: Humanities Press. ISBN 0-391-02177-X. OCLC 10916778.

[20] T. McEvilley (2002), *The Shape of Ancient Thought: Comparative Studies in Greek and Indian Philosophies* ISBN 1-58115-203-5, Allwarth Press.

[21] R. King, *Indian philosophy: an introduction to Hindu and Buddhist thought*, Edinburgh University Press, 1999, ISBN 0-7486-0954-7.

[22] C.C.W. Taylor *The atomists, Leucippus and Democritus: fragments, a text, and translation with a commentary*, University of Toronto Press Incorporated 1999, ISBN 0-8020-4390-9.

Fig. 2.43 Front page of Bernoulli's Hydrodynamica

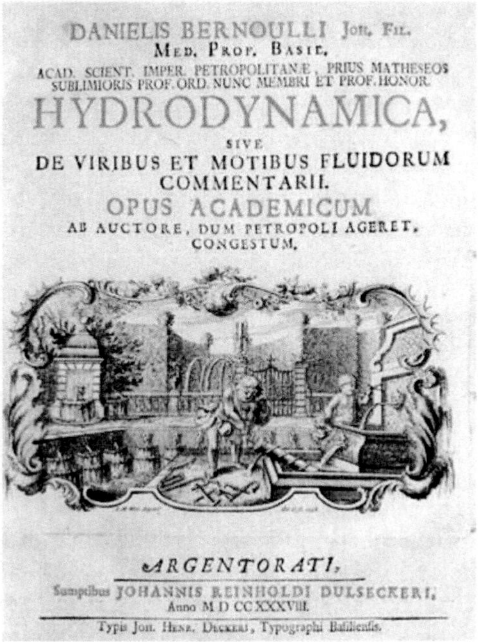
DANIELIS BERNOULLI Joh. Fil.
Med. Prof. Basil.
ACAD. SCIENT. IMPER. PETROPOLITANÆ, PRIUS MATHESEOS SUBLIMIORIS PROF. ORD. NUNC MEMBRI ET PROF. HONOR.
HYDRODYNAMICA,
SIVE
DE VIRIBUS ET MOTIBUS FLUIDORUM COMMENTARII.
OPUS ACADEMICUM
AB AUCTORE, DUM PETROPOLI AGERET, CONGESTUM.
ARGENTORATI,
Sumptibus JOHANNIS REINHOLDI DULSECKERI,
Anno M D CC XXXVIII.
Typis Joh. Henr. Deckeri, Typographi Basiliensis.

But how to conciliate the dissipative second principle of thermodynamics with the conservative laws of mechanics? Indeed, in an isolated universe of particles satisfying Newton's laws, the mechanical energy is preserved and this system is reversible: the previous states can always be attained by reversing the time in the equations of motion.

So, if heat has a mechanical foundation how to explain *irreversibility* based on the mechanical laws of motion?

The answer was only given by Ludwig Boltzmann[23] in the second half of the nineteenth century, by introducing the concept of probability in mechanics.

When the exact position and velocity of each particle of a mechanical system cannot be ascertained without errors, the best it is possible to do is to try to prescribe the *probability* that this particle will have a certain velocity when it is in a position **x** at time t.

As written by Max Planck in the Preface of his book[24]: "*The full content of the second principle can only be understood if we look for its foundation in the known*

[23] L. Boltzmann. *Vorlesungen uber Gas Theory*, (1896) . J.A. Barth, Leipzig [15]. L. Boltzmann. *Vorlesungen uber Gas Theory* (1898) J.A. Barth. Leipzig [16].These books were translated in English by Stephen Brush and edited by University of California Press. Dover editions are presently available .

[24] Planck, M. (1903). *Treatise on Thermodynamics*. Published in German in 1897. Translated to English by Alexander Ogg. London: Longmans, Green & Co. [136].

laws of the theory of probability as they were laid down by Clausius and Maxwell, and further extended by Boltzmann".

Nevertheless, much of the physics establishment did not share Boltzmann's belief in the reality of atoms and molecules, a belief that was however, shared by most chemists since the discoveries of John Dalton in 1808. He had a long-running dispute with the editor of the prominent German physics journal of his day, who refused to let Boltzmann refer to atoms and molecules as anything other than convenient theoretical constructs. Only a couple of years after Boltzmann's death, Perrin's studies of colloidal suspensions (1908–1909), based on Einstein's theoretical studies of 1905, confirmed the values of Avogadro's number and Boltzmann's constant, and convinced the world that the tiny particles really exist.

All the early versions of the mechanical theory of heat assumed that heat was some kind of motion but they were unclear on what kind. In 1857, Clausius made explicit the idea that this motion was a molecular motion. He claimed " *to have held this all along but avoided mentioning these conceptions because I wished to separate the conclusions which are deducible from certain general principles from those which presuppose a particular kind of motion*". On this basis, he developed a dynamic theory of gases. He argued that gases are made up of molecules that move in every direction and at every speed, showing how known properties of the gas, like pressure and temperature, could be related to the mean motion of the molecules. He made calculations about the average speeds and the average distances that molecules would travel before striking another molecule.

Maxwell[25] read these papers and began to play with the model. Maxwell made a fully statistical model of gases, in which the velocities were *distributed according to the same formula as the errors are distributed* in observations (*the normal distribution*). In 1873 Maxwell wrote[26]: "*The modern atomists have adopted a method which is, I believe, new in mathematical physics, though it has long been in use in Statistics. The data of the statistical method as applied to molecular science are the sums of large numbers of molecular quantities. In studying the relation between quantities of this kind, we meet with a new kind of regularity, the regularity of averages (...) If a great many equal spherical particles were in motion in a perfectly elastic vessel, collisions would take place among the particles and their velocities would be altered at every collision, so that after a certain time the vis viva will be divided among the particles according to some regular law, the average number of particles whose velocity lies between certain limits being ascertainable, though the velocity of each particle changes at every collision*".

In 1877, Boltzmann was in Graz, as a professor of experimental physics. He submitted two memoirs to the Academy of Sciences in Vienna. In the first of these, presented on January, 1877, *On Some Problems of the Mechanical Theory of Heat*,[27] Boltzmann made an incisive analysis of the formula $\int \delta Q/T \leq 0$ and argued that its

[25] Maxwell JC (1867) On the dynamical theory of gases. Philos Trans R Soc 157:49–88 [115].

[26] Maxwell, James Clerk (1873). 'Molecules', Nature: 437–441 [116].

[27] *Ludwig Boltzmann (1878). On some problems of the mechanical theory of heat Philosophical Magazine. Series 5. Vol. 6, pp. 236–237. Reprinted from [14].*

validity *was not based on the inherent laws of nature* but rather *on the choice of the initial conditions*. According to him, one had to assume on the basis of the kinetic theory of gases that the perfectly elastic balls representing the molecules always tend to change their actual positions. Consequently, any configuration, however improbable, could conceivably occur as time went on: "*The calculus of probabilities teaches us precisely this: any non-uniform distribution, unlikely as it may be, is not strictly speaking impossible*".

It now remained to give this conclusion a rigorous derivation and quantitative applicability, which Boltzmann did in his memoir submitted to the academy on October 1877, *On the Relation between the Second Law of the Mechanical Theory of Heat and the Probability Calculus with respect to the Propositions about Heat-Equivalence*. He concluded that the entropy of a state is proportional to the probability of the configuration of its component particles. In this way, *entropy became a purely statistical attribute of the overall system*. As the entropy increases, the configuration of the system becomes more probable.

Following Boltzmann's reasoning, let us consider a mechanical system with a great number of particles. Let $f(\mathbf{x}, \xi, t)\Delta\mathbf{x}\Delta\xi$ be the expected number of particles with velocities between ξ and $\xi + \Delta\xi$ that are found at time t in the elementary volume $\Delta\mathbf{x}$ between $\mathbf{x}$ and $\mathbf{x} + \Delta\mathbf{x}$, Boltzmann was able to show that the volumetric integral of the quantity

$$H(\mathbf{x}, t) = \int f \ln f d\xi, \tag{2.16}$$

always decreases with time for isolated systems, in such a manner that the H-function could be identified with the Clausius local entropy (*per particle*) by

$$s = -kH,$$

where k is the Boltzmann constant that can be written as the ratio between the universal gas constant $\mathscr{R}$ and the Avogadro number $\mathscr{N}$.

$$k = \frac{\mathscr{R}}{\mathscr{N}}.$$

Readers are suggested to read Chapter 10 for further details.

The Boltzmann statistical concept of entropy has been extended to discrete systems in Information Theory by Shanon.[28] In this case, the entropy is defined as

$$s = -\sum_i p_i \ln p_i$$

where p_i is the probability of event i.

[28] Shannon, Claude E. (1948). *A Mathematical Theory of Communication*. Bell System Technical Journal 27 (3): 379–423. https://doi.org/10.1002/j.1538-7305.1948.tb01338.x. Ref. [155].

Recent developments in statistical mechanics based on the concept of Tsallis entropy, have intensified the interest in investigating a possible extension of Shannon's entropy to Information Theory [176].

Part II
Thermodynamic Equilibrium

Chapter 3
The Equilibrium State

Abstract This chapter deals with homogeneity, Legendre transforms—leading to the Helmholtz and Gibbs energies and the concept of enthalpy—followed by Maxwell relations and minimum principles and establishing the framework that is needed for the theoretical analysis of thermodynamic systems in equilibrium. A thermodynamic system in equilibrium is homogeneous, meaning that its intensive properties do not vary from point to point inside the system. This property leads to an Euler's relationship between its extensive variables and the Gibbs-Duhem equation. This equation found, simultaneously, by Josiah Willard Gibbs and Pierre Maurice Marie Duhem enables a Lagrangian description of thermodynamic systems, of crucial importance in the analysis of open systems in non-equilibrium states. Minimum principles state that when a restriction that prevents a system from reaching equilibrium is removed, this system will attain an equilibrium state given by the minimization of one of its energies—internal, Helmholtz, Gibbs, or enthalpy—in accordance with the thermodynamic variables that are maintained constant during the equilibration process.

3.1 Introduction

In the first part of this book, the first law of thermodynamics was stated as

$$\delta Q + \delta W = dU, \tag{3.1}$$

meaning that the internal energy of a thermodynamic system—or the mechanical energy of its molecules—is affected both by the addition of heat and by the compression/expansion work that is exchanged between the system and the external environment.

In this second part, we focus on *equilibrium states* and not on the process that leads from one equilibrium state to another. To be clear, consider an isothermal expansion of an ideal gas in contact with a heat reservoir, between any two given pressures P_1 and P_2. This expansion is promoted by successively removing balance weights of mass Δm till pressure P_2 is reached (Fig. 3.1). We know that both the work W_{1-2} and the heat Q_{1-2} are dependent on the mass Δm. Nevertheless, when the process

P. C. Philippi, *Thermodynamics*, https://doi.org/10.1007/978-3-031-49357-7_3

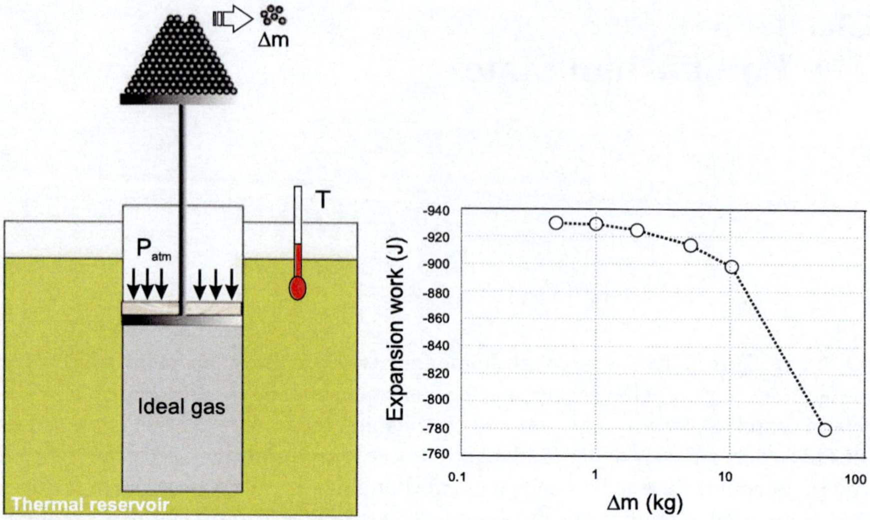

Fig. 3.1 The heat and work that are exchanged are dependent on mass Δm, but the state variables in equilibrium states 1 and 2 are always the same

between states 1 and 2 is reversible, the entropy variation between the initial and final equilibrium state will be always the same and will be given by

$$S_2 - S_1 = \int_1^2 \left.\frac{\delta Q}{T}\right)_{rev},$$

independently of the reversible path we take for going from state 1 to state 2.

Therefore, for calculating the state variables for the different equilibrium states of a thermodynamic system we will only consider reversible processes, for which,

$$\delta Q = TdS,$$
$$\delta W = -PdV.$$

So, the first law of thermodynamics, Eq. (3.1), becomes a differential equation

$$dU = TdS - PdV, \tag{3.2}$$

relating the state variable U with the state variables S and V.

In this representation, we can consider S and V as independent variables and U as the dependent variable. In addition, the absolute temperature T can be written as

$$T = \left.\frac{\partial U}{\partial S}\right)_V, \tag{3.3}$$

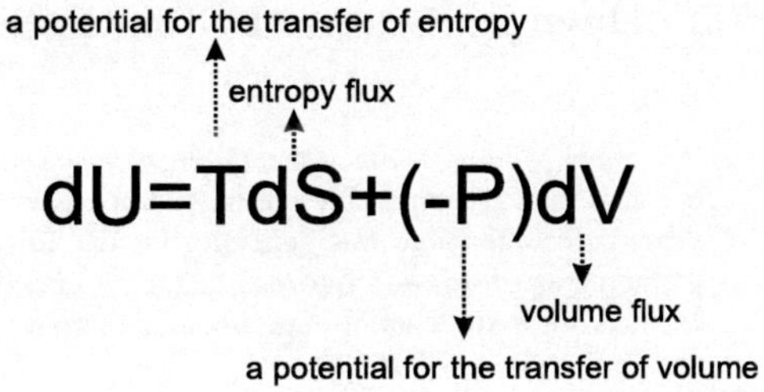

Fig. 3.2 Fluxes and forces in a reversible process

and the pressure P as

$$P = -\left.\frac{\partial U}{\partial V}\right)_S, \tag{3.4}$$

where $\left.\frac{\partial Z}{\partial X}\right)_Y$ means that the partial derivative $\frac{\partial Z}{\partial X}$ is being performed with Y being kept constant. This notation is necessary because, in thermodynamics, contrary to elementary differential calculus, we have a great number of state variables, but only a small number of them are independent.

Before going further let us try to give a *physical* interpretation to the terms that contribute to the variation of the internal energy in Eq. (3.2) in the *reversible world*. When a thermodynamic system changes its internal energy, this change can be attributed to two sources. The first one is the heat $\delta Q = TdS$ it receives, or the flux of entropy through its boundary. In reversible processes there are no internal sources of entropy and δQ is the only source of entropy. On the other hand, the temperature T is the potential (*or force*) for entropy transfer, the entropy flux taking place from higher to smaller temperatures (Fig. 3.2). The second source of internal energy is the compression work. Compression work can be only performed by the transfer of volume and the pressure is the potential for this transfer. Nevertheless, when a system increases its volume because its pressure is higher than the pressure of its neighbor's surroundings, the volume is always transferred from places with lower pressures to places with higher pressure. This explains the negative sign before PdV in Eq. (3.2).

The temperature T and the pressure P can be also *reinterpreted* based on Eqs. (3.3) and (3.4), respectively. Temperature is a measure of the kinetic energy of the molecules, but, based on Eq. (3.3) it can be reinterpreted as the measure of the internal energy variation ΔU of a system when this system increases its entropy by ΔS, in a process where the volume is kept constant. Similarly, the pressure P, which is related to the momentum of the molecules, can be also reinterpreted based on Eq. (3.4).

3.2 Open Systems and Gibbs Potential

It is universally recognized that Gibbs work published by the Connecticut Academy was an event of the first importance in the history of chemistry. Nevertheless, it was a number of years before its value was generally known, this delay was due largely to the fact that its mathematical form and rigorous deductive processes make it difficult to read for anyone, especially for students of experimental chemistry whom it most concerns. J. J. O'Connor and E. F. Robertson, 1997[1]

Lorsqu'il s'était determiné à faire imprimer les résultats de l'une de ses études, Gibbs ne cherchait nullement à leur assurer une diffusion rapide et étendue; les grands recueils scientifiques, grâce auxquels une pensée est bientôt communiquée à tous ceux qui sont capables d'en faire bon usage, n'eurent pas à transmettre les découvertes du professeur du Yale Collège; ces découvertes, il les confia presque toujours à des recueils américains d'une moindre vogue, comme s'il eût éprouvé quelque regret de leur avoir donné libre vol ; ses premiers mémoires de Thermodynamique et, en particulier, son grand travail Sur l′ équilibre des substances hétérogènes parurent dans les Transactions, bien peu répandues, que publiait, depuis peu, à New-Haven, l'Académie des Arts et des Sciences du Connecticut. Il semble parfois qu'en publiant ses travaux, Gibbs eût été possédé du désir de les voir passer inaperçus; s'il en fut ainsi, il fut bien souvent servi à souhait; bien souvent, ses idées demeurèrent ignorées de ceux-là mêmes qui auraient eu le plus grand intérêt à les connaitre. (Pierre Duhem, 1886[2])

Open systems can change their internal energy by mass transfer. Consider Fig. 3.3a. Chamber A is filled with a liquid and is separated from an evacuated chamber B by a metallic plate. When this plate is removed the liquid will evaporate till the vapor pressure P in chamber B reaches the saturation pressure $P_s\,(T)$ (Fig. 3.3b).

Consider now Fig. 3.3c. Each molecule that reaches the liquid surface is able to migrate to the vapor phase. Nevertheless, all the remaining molecules in the liquid phase will try to keep it in this phase with attraction forces that are essentially electrostatic forces.[3] The molecule shown in the sketch has a kinetic energy E_c, which is positive and related to the temperature T. However its potential energy Φ is negative and related to the net attraction force **F** that is pulling it back to the liquid phase, in accordance with

$$\mathbf{F} = -\nabla\Phi.$$

Therefore, this molecule will only jump to the vapor phase when its mechanical energy $E_c + \Phi$ is greater than zero. In other words, when its kinetic energy is high

---

[1] O'Connor, John J.; Robertson, Edmund F. (1997). *Josiah Willard Gibbs*. The MacTutor History of Mathematics archive. University of St Andrews, Scotland. School of Mathematics and Statistics.

[2] Duhem, P., *Josiah Willard Gibbs, à propos de l publication de ses mémoires scientifiques*, A. Hermann, Paris, 1908 [46].

[3] The book of Adamson [2] is suggested for readers interested in better understanding the nature of these intermolecular forces.

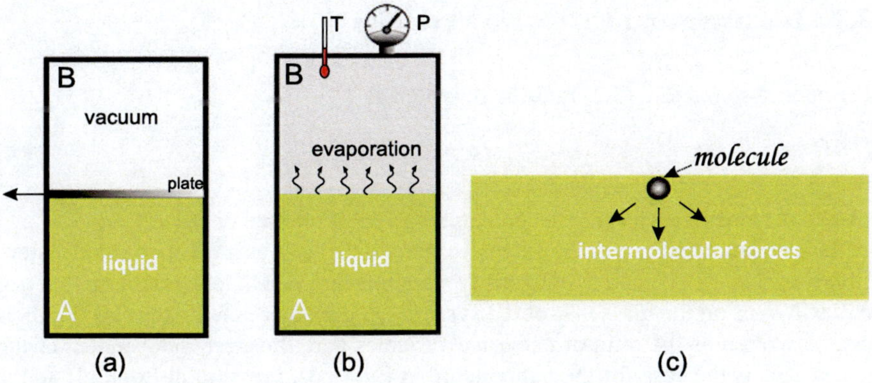

Fig. 3.3 Open systems can change their internal energy by mass transfer

enough to counterbalance the potential energy Φ that is due to its interaction with its neighboring liquid molecules. If this is the case its residual energy will be transferred from the liquid to the vapor phase.

From a macroscopic point of view and considering the reasoning given at the beginning of this section, the variation in internal energy in the vapor phase due to the flux of dN moles from the liquid phase would be written as

$$dU_v = \bar{\mu}_v dN,$$

where μ_v is a sort of *potential* for mass transfer in the vapor phase—the Gibbs or chemical potential.[4] In the same way, the variation of internal energy in the liquid phase due to this transfer will be

$$dU_\ell = -\bar{\mu}_\ell dN.$$

So, the liquid will evaporate till the Gibbs potentials μ_v and μ_ℓ are identical. In this condition, the vapor pressure reaches the saturation value $P_s(T)$.

[4] Josiah Willard Gibbs (1839–1903) was an American scientist who made important theoretical contributions to physics, chemistry, and mathematics. His work on the applications of thermodynamics was fundamental for transforming physical chemistry into a rigorous deductive science. Together with James Clerk Maxwell and Ludwig Boltzmann, he created statistical mechanics (a term that he coined), explaining the laws of thermodynamics as consequences of the statistical properties of ensembles of the possible states of a physical system composed of many particles. He extended the thermodynamic analysis to multi-phase chemical systems and considered a variety of concrete applications. He described that research in a monograph titled '*On the Equilibrium of Heterogeneous Substances*', published by the Connecticut Academy in two parts that appeared respectively in 1875 [59] and in 1877 [60].

3.3 Intensive and Extensive Variables

For open systems, Eq. (3.2) must be rewritten as

$$dU = TdS - PdV + \bar{\mu}dN, \tag{3.5}$$

to take the mass transfer into account as a source of internal energy.

This equation expresses the extensive state variable U as a function of the extensible variables S, V, and N. Extensive variables are dependent on the number of moles N—or on the mass m—of the system. On the other hand, intensive variables can be written as the ratio of extensive variables and, thus, are independent of the mass. This is the case for the temperature in Eq. (3.3), pressure in Eq. (3.4) and is also the case for the Gibbs potential

$$\bar{\mu} = \left.\frac{\partial U}{\partial N}\right)_{S,V}.$$

When using the mass m of the thermodynamic system—instead of the number of moles—Eq. (3.5) would be written as

$$dU = TdS - PdV + \mu dm,$$

where

$$\mu = \left.\frac{\partial U}{\partial m}\right)_{S,V}.$$

3.4 Homogeneity of the Equilibrium State

Thermodynamic systems in equilibrium are homogeneous. So, by partitioning such a system into equal parts, the extensive properties of each part is a corresponding fraction of the total, while the intensive properties remain the same throughout the whole system (Fig. 3.4).

So if λ is the inverse of the number of parts by which the system was partitioned

$$U(\lambda S, \lambda V, \lambda N) = \lambda U(S, V, N). \tag{3.6}$$

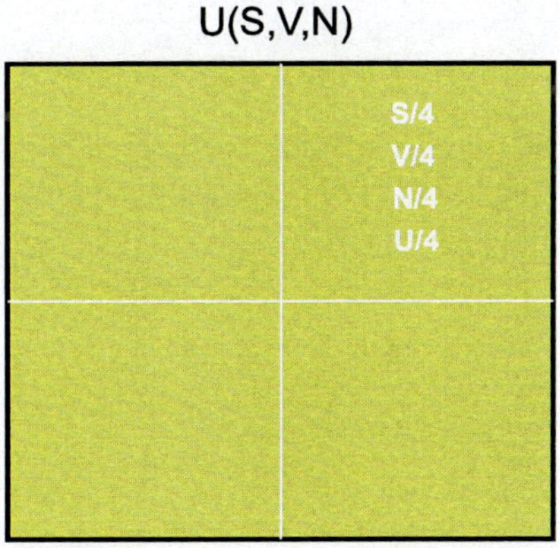

Fig. 3.4 Thermodynamic systems in equilibrium are homogeneous. For a given partition, the extensive properties are divided by the number of parts, whereas the intensive properties remain the same throughout the whole system

Consider now that λ is variable, while the system remains the same for every partitioning. By differentiating Eq. (3.6) with respect to λ

$$S\underbrace{\frac{\partial U\left(\lambda S, \lambda V, \lambda N\right)}{\partial\left(\lambda S\right)}}_{=T}+V\underbrace{\frac{\partial U\left(\lambda S, \lambda V, \lambda N\right)}{\partial\left(\lambda V\right)}}_{=-P}+N\underbrace{\frac{\partial U\left(\lambda S, \lambda V, \lambda N\right)}{\partial\left(\lambda N\right)}}_{=\bar{\mu}}=U.$$

Each derivative in the above equation is an *intensive property* related to the partitioned part where the internal energy is a fraction λU of the total, the entropy is λS, the volume is λV and the number of moles (*or mass*) is λN. Therefore, for a thermodynamic system in equilibrium

$$U = TS - PV + \bar{\mu}N. \tag{3.7}$$

The above equation is a general consequence of homogeneity expressed by the Euler relation, Eq. (3.6). When a thermodynamic system is doubled all the extensive variables (entropy, volume and number of moles) are doubled, including the internal energy, but the intensive variables remain with the same values.

Consider now a process from a given equilibrium state to another one. By differentiating Eq. (3.7) and considering Eq. (3.5), we obtain

$$0 = SdT - VdP + Nd\bar{\mu}.$$

So, if we divide the above equation by the number of moles N, we get the Gibbs-Duhem equation,[5] relating the Gibbs potential with the temperature and pressure

[5] Pierre Maurice Marie Duhem (1861–1916) was a French physicist, mathematician, historian, and philosopher of science. He is best known for his work on chemical thermodynamics, for his philo-

$$d\bar{\mu} = -\bar{s}dT + \bar{v}dP, \tag{3.8}$$

where $\bar{s}$ is the molar entropy and $\bar{v}$ is the molar volume. The Gibbs-Duhem equation enables us to find the chemical potential in terms of the temperature and pressure when the molar entropy $\bar{s}$ and the molar volume $\bar{v}$ are given in terms of T and P.

Consider now the homogeneity relation Eq. (3.7). By dividing this equation by N

$$\bar{u} = T\bar{s} - P\bar{v} + \bar{\mu},$$

and so

$$d\bar{u} = Td\bar{s} - Pd\bar{v} + \underbrace{\bar{s}dT - \bar{v}dP + d\bar{\mu}}_{=0},$$

Therefore

$$d\bar{u} = Td\bar{s} - Pd\bar{v}. \tag{3.9}$$

The above equation is a differential equation that can be solved for expressing the dependence of the intensive variable, the molar internal energy $\bar{u}$ in terms of the molar entropy $\bar{s}$ and the molar volume $\bar{v}$. Although both Eqs. (3.5) and (3.9) were derived for open systems, the former equation means an Eulerian description and the second one, a Lagrangian description. In other words, while Eq. (3.5) applies to *control volumes*, accounting for the influence of mass transfer on the total internal energy of a fixed volume, Eq. (3.9) describes the variation of the internal energy of a fixed mass of fluid—a single mole, in this case—when this single mole is displaced along the flow lines (Fig. 3.5).

3.4.1 Exercises

1. What is the physical meaning of the Euler homogeneity relation Eq. (3.7)? Try to translate this relationship into words. What happens with the intensive and extensive variables of a thermodynamic system in equilibrium when this thermodynamic system is doubled?
2. Write Eq. (3.9) in terms of the specific entropy s, specific internal energy u and mass density ρ. This kind of representation is usual in non-equilibrium thermodynamics when dealing with material derivatives $d\rho/dt$, du/dt, and ds/dt.

sophical writings on the indeterminacy of experimental criteria, and for his historical research into the science of the European Middle Ages. As a scientist, Duhem also contributed to hydrodynamics and to the theory of elasticity. His thermodynamics book was written in 1886: Duhem P (1886) *Le potentiel thermodynamique et ses applications*. A. Hermann, Paris [45].

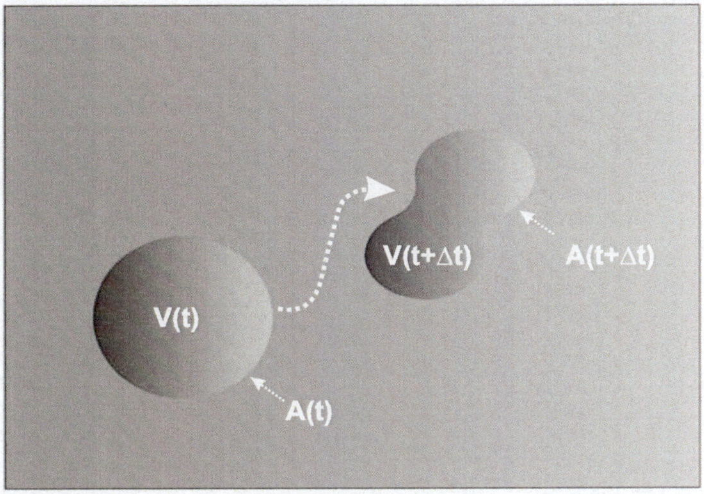

Fig. 3.5 Lagrangian description. Volume V(t) is a *material volume* and although it is deformed along the flow, the mass inside it is constant

3.5 Legendre Transforms of the Internal Energy

The Legendre transforms[6] are commonly used in classical mechanics to derive the Hamiltonian formalism out of Lagrangian formalism and in thermodynamics to derive the thermodynamic potentials.

Consider the function $Y = Y(X)$ shown in Fig. 3.6. We seek for a transformation $Y \to \mathcal{L}$ that preserves all the information given by $Y(X)$, in such a way that Y can be fully retrieved by the inverse transform $\mathcal{L} \to Y$. Figure 3.6 shows that this can be accomplished if, for each point X of the curve $Y(X)$, we take the derivative $Y'(X) = P$ and the ordinate where the tangent to Y at point X crosses the Y-axis, or $\mathcal{L}$. So, from,

$$P = \frac{Y - \mathcal{L}}{X},$$

the Legendre transform $\mathcal{L} = \mathcal{L}(P)$ of $Y = Y(X)$ will be given by

$$\mathcal{L} = Y - PX.$$

[6] Adrien-Marie Legendre (1787). *Mémoire sur l'intégration de quelques équations aux différences partielles*. Mémoires de l' Academie des Sciences. pp. 309–351. [103].

Fig. 3.6 Given a function Y = Y(X), the Legendre transform $\mathcal{L} = \mathcal{L}(P)$ preserves all the information given by the representation Y = Y(X)

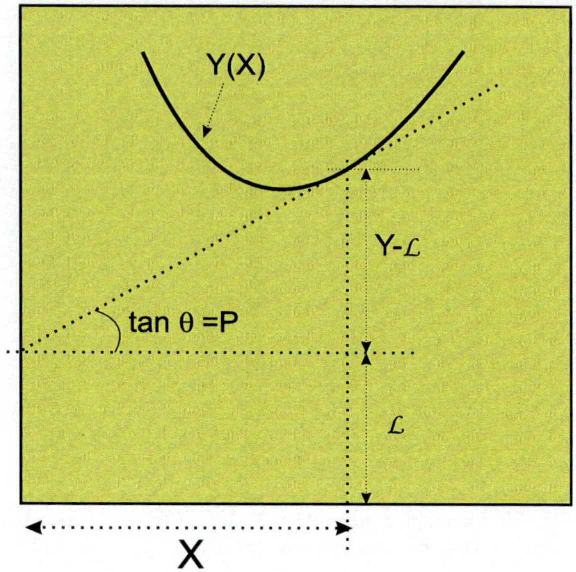

Therefore, in the transformed representation the dependent variable is $\mathcal{L}$ and the independent variable is $P = dY/dX$. This concept can be easily generalized to functions with several independent variables.

3.5.1 *Helmholtz Energy*

The Helmholtz energy was developed by Hermann von Helmholtz and is usually denoted by the letter A (*from the German "Arbeit" or work)*, or the letter F. In physics, the Helmholtz energy is sometimes referred to as the Helmholtz function, Helmholtz free energy, or simply free energy. The Helmholtz energy F is defined as the Legendre transform of the internal energy that replaces the entropy as an independent variable by the temperature $T = (\partial U/\partial S)_{V,N}$,

$$F = U - TS. \tag{3.10}$$

From Eq. (3.5),

$$dF = -SdT - PdV + \bar{\mu}dN. \tag{3.11}$$

So, in the Helmholtz energy representation, the independent variables are T, V, and N instead of S, V, and N. From Eqs. (3.10) and (3.7)

$$F = -PV + \bar{\mu}N. \tag{3.12}$$

In terms of intensive properties, the above equation can be written as

$$\bar{f} = -P\bar{v} + \bar{\mu},$$

and so

$$d\bar{f} = -Pd\bar{v} - \bar{v}dP + d\bar{\mu},$$

or, from the Gibbs-Duhem equation, Eq. (3.8),

$$d\bar{f} = -\bar{s}dT - Pd\bar{v}, \tag{3.13}$$

which is the differential equation for the Helmholtz energy in its Lagrangian differential form.

3.5.1.1 The Meaning of the Helmholtz Energy

Consider an isothermal reversible expansion of a gas as pictured in Fig. 2.4. From the first law

$$\delta Q + \delta W = dU = dF + TdS,$$

or, since $\delta Q = TdS$ because the expansion is reversible

$$\delta W = dF,$$

and

$$W_{1-2} = F_2 - F_1.$$

Since this expansion is reversible, the work delivered by the gas in its isothermal expansion is maximum and this maximum work corresponds to the variation of its Helmholtz energy between the initial and final equilibrium states.

Therefore, since the expansion work W_{1-2} is negative, when the isothermal expansion is arbitrary,

$$|W_{1-2}| \leq |F_2 - F_1|,$$

or,

$$\delta W \geq dF, \tag{3.14}$$

the equality sign applying solely for reversible expansions.

While internal energy can be considered as the available work in adiabatic expansions, the Helmholtz energy can thus be seen as the work that is available (or *free*) in isothermal expansions.

The same reasoning can be used for isothermal compression, leading to the same inequality, Eq. (3.14), which is a general rule in isothermal processes.

Considering the entire cycle

$$\oint \delta W \geq 0,$$

which is the same result presented in Chap. 2 of this book: *In isothermal cycles the absolute value of the compression work is always greater than the absolute value of the expansion work.*

Beyond its meaning as an available work, Helmholtz energy is widely used in surface physics and is directly related to the surface tension of liquids. Readers are invited to read Chap. 6 on surface physics for acquiring a better familiarity with this important thermodynamic concept.

3.5.2 Enthalpy

The word enthalpy is based on the ancient Greek verb enthalpein, which means '*to warm in*'. The name enthalpy is often incorrectly attributed to Benoît Paul émile Clapeyron[7] and Rudolf Clausius[8] through the 1850 publication of their Clausius—Clapeyron relation. However, neither the concept, the word, nor the symbol for enthalpy existed until well after Clapeyron's death.

The earliest writings to contain the concept of enthalpy did not appear until 1875, when Josiah Willard Gibbs introduced "*a heat function for constant pressure*". However, Gibbs did not use the word *enthalpy* in his writings [59, 60];

The actual word first appeared in the scientific literature in a 1909 publication by J. P. Dalton.[9] Proceedings of the Royal Netherlands Academy of Arts and Sciences 11: 863–873 [37]. According to that publication, Heike Kamerlingh Onnes (1853–1926)[10] actually coined the word.[11]

---

[7] Clapeyron, M. C. (1834). *Mémoire sur la puissance motrice de la chaleur*. Journal de l'École polytechnique 23: 153–190. [33].

[8] Clausius, R. (1850). *Ueber die bewegende Kraft der Wärme und die Gesetze, welche sich daraus für die Wärmelehre selbst ableiten lassen*. Annalen der Physik 155: 500–524. https://doi.org/10.1002/andp.18501550403 [34].

[9] J. P. Dalton, (1909). *Researches on the Joule—Kelvin* effect, especially at low temperatures. I. Calculations for hydrogen.

[10] H. Kamerlingh Onnes, Dutch physicist who first liquefied helium in 1908, discovered superconductivity in 1911, and won the Nobel Prize for physics in 1913.

[11] In accordance with Hendrick C. Van Ness [182], Kamerlingh Onnes seems never to have used the word *enthalpy* in any publication. A message from the Kamerlingh Onnes Laboratory at Leiden

Over the years, many different symbols were used to denote enthalpy. In accordance with Irmgard Howard [71] it was not until 1922 that Alfred W. Porter[12] proposed the symbol 'H' as the accepted standard, thus finalizing the terminology still in use today.

The enthalpy is the Legendre transform of the internal energy that replaces the volume with the pressure as the independent variable

$$H = U - (-PV)\,. \tag{3.15}$$

So, from Eq. (3.5),

$$dH = TdS + VdP + \bar{\mu}dN. \tag{3.16}$$

In the enthalpy representation, the independent variables are S, P, and N instead of S, V, and N. From Eqs. (3.15) and (3.7)

$$H = TS + \bar{\mu}N.$$

In terms of intensive properties, the above equation can be written as

$$\bar{h} = T\bar{s} + \bar{\mu},$$

and so

$$d\bar{h} = Td\bar{s} + \bar{s}dT + d\bar{\mu},$$

or, from the Gibbs-Duhem equation, Eq. (3.8),

$$d\bar{h} = Td\bar{s} + \bar{v}dP, \tag{3.17}$$

which shows the enthalpy in its Lagrangian differential form.

3.5.2.1 The Meaning of Enthalpy

Consider the isobaric expansion shown in Fig. 2.40. From the first law of thermodynamics,

---

suggests that he likely proposed the word orally at the first meeting of the Institute of Refrigeration in Paris in 1908. However, nothing appears in the Comptes rendus du Congres international du froid, 1908.

[12] Alfred W Porter was President of the Faraday Society in 1920–1922.

$$\delta Q + \delta W = dU.$$

If there is no imbalance in the pressure field of the gas, a single internal pressure is set up. In addition, the internal and external pressure are identical during the entire process. So

$$dU = dH - PdV - \underbrace{VdP}_{=0} \quad \text{and} \quad \delta W = -PdV.$$

Consequently,

$$\delta Q = dH.$$

Therefore, while the heat that is supplied to a constant volume system corresponds to the variation of its internal energy, the variation of the enthalpy measures the supplied heat when the pressure is maintained constant, and so it is '*a heat function for constant pressure*'.

The concept of enthalpy is widely used in engineering, for solving energy balance problems. In his book, 'Heat and Thermodynamics', Zemansky[13] writes at page 275 of its 5th Edition that enthalpy is a measurement of energy in a thermodynamic system that "*includes the internal energy and the amount of energy required to make room for it by displacing its environment and establishing its volume and pressure*". In effect, consider the air compressor of a turbojet, shown in Fig. 3.7. The control volume V comprises the air that is being compressed and the compressor blades enabling a global analysis by considering the power $\dot{W}_{shaft}$ that is transferred to the compressor by the turbine shaft.

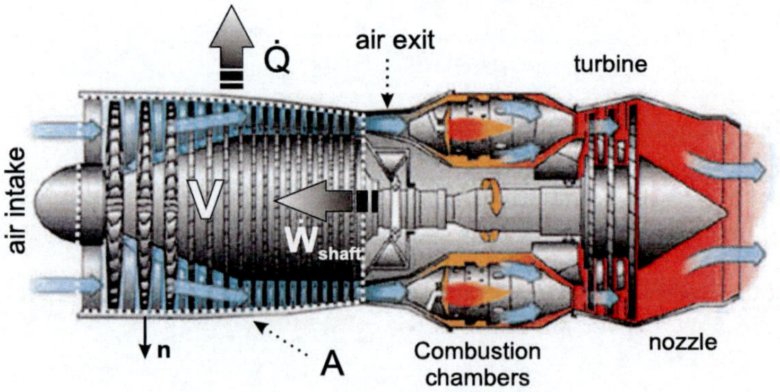

Fig. 3.7 Energy balance for the air compressor of a turbojet

[13] Zemansky, Mark W. (1968). *Heat and Thermodynamics* (5th ed.). New York, NY: McGraw-Hill.

The total energy of this thermodynamic system is given by

$$E = \int_V \rho \left(u + \frac{\mathrm{v}^2}{2} + gZ \right) dV.$$

The time rate variation of the total energy dE/dt is due to the following sources:

(a) the flow of energy across its boundary carried into the system with the fluid

$$-\int_A \rho \left(u + \frac{\mathrm{v}^2}{2} + gZ \right) \mathbf{v}\cdot\mathbf{n}\mathrm{dA}, \tag{3.18}$$

where A refers to the boundary surface, **n** is the unitary vector normal to the boundary surface and oriented toward its exterior;

(b) the pressure work or "*the amount of energy required to make room for the fluid when entering or exiting out of the control volume*" is given by

$$-\int_A P\mathbf{v}\cdot\mathbf{n}\mathrm{dA}; \tag{3.19}$$

(c) the amount of heat that is exchanged between the system and its environment

$$\dot{Q} = -\int_A \mathbf{q}\cdot\mathbf{n}\mathrm{dA}; \tag{3.20}$$

(d) the shaft power that must be taken from the turbine for compressing the air $\dot{W}_{shaft}$.

By assembling Eqs. (3.18–3.20) into a single equation

$$\frac{dE}{dt} = -\int_A \rho \left(h + \frac{\mathrm{v}^2}{2} + gZ \right) \mathbf{v}\cdot\mathbf{n}\mathrm{dA} - \int_A \mathbf{q}\cdot\mathbf{n}\mathrm{dA} + \dot{W}_{shaft}, \tag{3.21}$$

where the internal energy per unit mass u and the pressure work $Pv = P/\rho$ were assembled into a single energy, *the enthalpy*, through

$$\rho h = \rho u + P$$

Equation (3.21) can be further simplified in steady-state conditions, by supposing the potential energy to be negligible and that the compression process is adiabatic. It results

$$\dot{W}_{shaft} = \int_A \rho \left(h + \frac{\mathrm{v}^2}{2} \right) \mathbf{v} \cdot \mathbf{n} dA$$

$$= \dot{m} \left[(h_e - h_i) + \left(\frac{\mathrm{v}_e^2}{2} - \frac{\mathrm{v}_i^2}{2} \right) \right] \tag{3.22}$$

where symbols e and i refer, respectively, to the compressor exit and inlet sections, $h_s \mathrm{v}_s$ for $s = i, e$ are average values of the enthalpy and velocity, respectively. Finally,

$$\dot{m} = -\int_{A_i} \rho \mathbf{v} \cdot \mathbf{n} dA = \int_{A_e} \rho \mathbf{v} \cdot \mathbf{n} dA,$$

is the mass flow rate through the compressor in steady-state operation.

Equation (3.22) means that the power that must be delivered to the air compressor from the turbine is the one that is necessary to increase the enthalpy and kinetic energy of the air that is flowing through the compressor. Since the air can be usually considered as an ideal gas, the enthalpy (*per unit mass*) can be written as

$$h = u + \frac{P}{\rho} = u + RT,$$

in such a way that

$$dh = du + RdT = c_p dT.$$

3.5.3 Gibbs Energy

The Gibbs energy, sometimes called 'available' or 'free' Gibbs energy, was developed in the 1870s by Josiah Willard Gibbs. In 1873, Gibbs described this energy as "*the greatest amount of mechanical work that can be obtained from a given quantity of a certain substance in a given initial state, without increasing its total volume or allowing heat to pass to or from external bodies, except such as at the close of the processes are left in their initial condition*".[14]

The Gibbs energy G is the Legendre transform of the internal energy that replaces the entropy and the volume, as independent variables, by, respectively the temperature and pressure.

$$G = U - (-PV) - TS = H - TS.$$

So, from Eq. (3.5)

$$dG = -SdT + VdP + \bar{\mu}dN.$$

So, when a thermodynamic system is in a state of homogeneous equilibrium,

$$G = \bar{\mu}N,$$

or, in terms of intensive variables

$$g = \bar{\mu},$$

and

$$dg = -\bar{s}dT + \bar{v}dP.$$

3.5.4 Exercises

1. Show that for an ideal gas, the specific entropy can be written as

$$ds = c_p \frac{dT}{T} - R\frac{dP}{P},$$

where $R = \mathscr{R}/M$. With this information, find the entropy s an ideal gas would have with respect to the entropy s_0 of a reference state given by T_0, P_0. Find, also, the change in the chemical potential $\mu\,(T, P) - \mu_0\,(T_0, P_0)$ with respect to this reference state.

[14] J. W. Gibbs (1873), *A Method of Geometrical Representation of the Thermodynamic Properties of Substances by Means of Surfaces* Transactions of the Connecticut Academy of Arts and Sciences 2, pp. 382–404 [61].

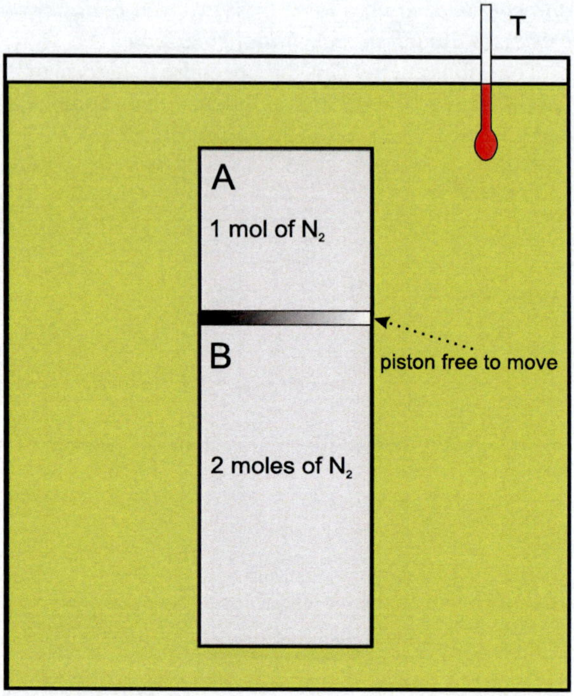

Fig. 3.8 Considering that chambers A and B are in equilibrium with a thermal bath at temperature T, show that the Helmholtz energy $f = F/N$ is the same in the two chambers

2. In Fig. 3.8, chamber B has the double of the volume of chamber A and both chambers contain N_2. Considering that the two chambers are in equilibrium with a thermal bath at a temperature T, show that the Helmholtz energy $f = F/N$ is the same in the two chambers.
   What would be your answer if the N_2 is replaced by O_2 in chamber A?...
3. The thrust of a turbojet is its main design variable and gives the propellant force of the turbojet. This force can be derived from Newton's second law as

$$F = \dot{m}_{nozzle\ exit} \mathsf{v}_{nozzle\ exit} - \dot{m}_{compressor\ inlet} \mathsf{v}_{compressor\ inlet}$$

   (a) Suppose that the compression ratio in the compressor is 8:1 and that the air enters the compressor inlet at the atmospheric pressure and $T = 25^0 C$. What would be the air temperature at the compressor outlet, if the compression is supposed to be isentropic?
   (b) The Boeing 777 has a GE-90 turbojet with a thrust of $4.168 \times 10^5\ N$ in *static conditions.* Considering that the fuel is kerosene with a fuel-air mass ratio of 0.0137, what should be the air mass flow rate when the intake velocity is 40 m/s and the nozzle exit velocity cannot exceed 340 m/s (*the speed of sound*)? What should be, in your calculations, the *minimum* inlet area of

the GE-90 turbojet for enabling such air mass flow rate (*the diameter of the GE-90 turbo-jet is 7.29m*).

(c) The take-off speed of a Boeing 777 is about 100 m/s. Consider that a Boeing 777 with two GE-90 turbojets has a maximum take-off weight of 405,000 Kg. Neglecting friction, what is the minimum take-off length of the Boeing 777?

(d) The compressor exit area of the GE-90 is about 40 inlet area. Estimate the shaft power that the turbojet turbine must deliver to the compressor.

3.6 Maxwell Relations

The Maxwell relations follow directly from the fact that the order of differentiation of an analytic function of two variables is irrelevant (Schwarz theorem). In the case of Maxwell relations the function to be considered is a thermodynamic variable Z and X and Y are two different independent variables for that function,

$$\frac{\partial}{\partial Y}\left(\left(\frac{\partial Z}{\partial X}\right)_Y\right)_X = \frac{\partial}{\partial Y}\left(\left(\frac{\partial Z}{\partial X}\right)_Y\right)_X . \tag{3.23}$$

Now, we rewrite the main differential equations that were derived in this chapter, beginning with Eq. (3.9)

$$du = Tds - Pdv, \tag{3.24}$$

that expresses the specific internal energy u in terms of the specific entropy s and specific volume v, $u = u\,(s, v)$. We have

$$T\,(s, v) = \left(\frac{\partial u}{\partial s}\right)_v , \quad P\,(s, v) = -\left(\frac{\partial u}{\partial v}\right)_s .$$

The specific Helmholtz energy f creplaces the entropy s, as independent variable by the temperature T, resulting in the differential equation

$$df = -sdT - Pdv, \tag{3.25}$$

expressing the Helmholtz energy f in terms of the temperature T and specific volume v, $f = f\,(T, v)$. In this representation the entropy and the pressure are also dependent on T and v,

$$s\,(T, v) = -\left(\frac{\partial f}{\partial T}\right)_v , \quad P\,(T, v) = -\left(\frac{\partial f}{\partial v}\right)_T . \tag{3.26}$$

For the enthalpy $h\,(s, P)$

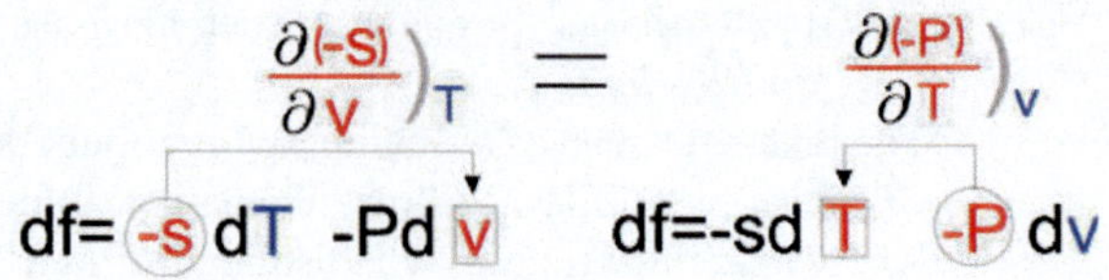

Fig. 3.9 Application of the Schwarz theorem in deriving differential relationships for the thermodynamic properties

$$dh = Tds + vdP, \tag{3.27}$$

with

$$T\,(s, P) = \left(\frac{\partial h}{\partial s}\right)_P, \quad v\,(s, P) = \left(\frac{\partial h}{\partial P}\right)_s.$$

Finally, the differential equation for the Gibbs energy $g\,(T, P)$

$$dg = -sdT + vdP, \tag{3.28}$$

with

$$s\,(T, P) = -\left(\frac{\partial g}{\partial T}\right)_P, \quad v\,(T, P) = \left(\frac{\partial g}{\partial P}\right)_T. \tag{3.29}$$

Now, let us take the derivative of the entropy in Eq. (3.26) with respect to the specific volume

$$\left(\frac{\partial s\,(T, v)}{\partial v}\right)_T = -\left(\frac{\partial}{\partial v}\left(\frac{\partial f}{\partial T}\right)_v\right)_T.$$

By changing the independent variables in accordance with the Schwarz theorem

$$\left(\frac{\partial s\,(T, v)}{\partial v}\right)_T = -\left(\frac{\partial}{\partial T}\left(\frac{\partial f}{\partial v}\right)_T\right)_v.$$

But, in accordance with Eq. (3.26)

$$\left(\frac{\partial f}{\partial v}\right)_T = -P,$$

and so

$$\left(\frac{\partial s}{\partial v}\right)_T = \left(\frac{\partial P}{\partial T}\right)_v. \tag{3.30}$$

Table 3.1 Differential equations for the thermodynamic functions

Thermodynamic function	Differential equation
Internal energy	$du = Tds - Pdv$
Enthalpy	$dh = Tds + vdP$
Helmholtz energy	$df = -sdT - Pdv$
Gibbs energy	$dg = -sdT + vdP$

This is an important relationship since it enables to calculate the entropy $s\,(T, v)$ in terms of the temperature and volume when the equation of state of the thermodynamic system $P\,(T, v)$ is known. Figure 3.9 illustrates the use of Schwarz theorem in deriving differential relationships for the thermodynamic properties. Table 3.1 summarizes the differential equations for the main thermodynamic functions.

3.6.1 *The Entropy as a Function of the Temperature and Volume*

Consider that we want to express the entropy in terms of the temperature and specific volume $s\,(T, v)$.

So,

$$ds = \left(\frac{\partial s}{\partial T}\right)_v dT + \left(\frac{\partial s}{\partial v}\right)_T dv. \tag{3.31}$$

In the first part of this book, we saw that the specific heat capacity at constant volume can be written as

$$c_v = \lim_{\Delta T \to 0} \left(\frac{\delta Q}{m\Delta T}\right)_{N,V}.$$

Nevertheless, when a thermodynamic system is heated at a constant volume and mass, the expansion work is null and

$$\delta Q = dU,$$

and, so, c_v can be written as

$$c_v = \left(\frac{\partial u}{\partial T}\right)_v,$$

or, using Eq. (3.24) and considering that the measurement of the heat capacity must be performed with $\Delta T \to 0$,

$$c_v = T\left(\frac{\partial s}{\partial T}\right)_v.$$

So, replacing the above equation and Eq. (3.30) into Eq. (3.31) it results

$$ds = \frac{c_v}{T}dT + \left(\frac{\partial P}{\partial T}\right)_v dv. \tag{3.32}$$

Therefore if we know the heat capacity c_v in terms of T and v, and the equation of state, it is possible to integrate the above equation and find the entropy s for a thermodynamic state given by its intensive variables, the temperature T and specific volume v.

3.6.2 The Internal Energy as a Function of the Temperature and Volume

From Eqs. (3.24) and (3.32) a differential equation can be derived for the internal energy, considering the temperature T and volume v as the independent variables

$$du = c_v dT + \left(T\left(\frac{\partial P}{\partial T}\right)_v - P\right) dv. \tag{3.33}$$

This equation can be used to write the internal energy u in terms of T and v, when we know the equation of state for the fluid and c_v in terms of T and v.

Let us consider, for instance, that the fluid is an ideal gas. In this case,

$$T\left(\frac{\partial P}{\partial T}\right)_v - P = 0.$$

So, for an ideal gas,

$$du = c_v dT.$$

The above equation does not mean that the internal energy of an ideal gas is only dependent on the temperature. In fact, the heat capacity c_v is in general a function of two state variables, such as T and v or T and P or P and v. In the present case, the independent variables are T and v, and if we want to show that the internal energy of an ideal gas is only dependent on the temperature we must prove that

$$\left(\frac{\partial c_v}{\partial v}\right)_T = 0.$$

See Exercise 2 of the Exercises list.

3.6.3 Exercises

1. Show that when the molar entropy $\bar{s}$ is expressed in terms of the temperature T and pressure P, its differential equation will be written, on a molar basis, as

$$d\bar{s} = \frac{\bar{c}_P}{T} dT - \left(\frac{\partial \bar{v}}{\partial T}\right)_P dP.$$

This equation is frequently used in expansion or compression devices involving an ideal gas when $(\partial \bar{v}/\partial T)_P = \mathscr{R}/P$ and $\bar{c}_p$ is only dependent on the temperature.

2. Show that $\left(\frac{\partial c_v}{\partial v}\right)_T = 0$ for an ideal gas, which equation of state is $Pv = RT$. Show also that the heat capacity c_p does not depend on the pressure for an ideal gas.
3. Using the Maxwell relations, show that

$$dh = c_P \frac{dT}{T} + \left(v - T\left(\frac{\partial v}{\partial T}\right)_P\right) dP.$$

4. **Mathematical skills**. Consider a function $Z = f(X, Y)$. By considering

$$dZ = \left.\frac{\partial Z}{\partial X}\right)_Y dX + \left.\frac{\partial Z}{\partial Y}\right)_X dY,$$

show that when this function has an inverse $Y = f(X, Z)$ then, in each plane where $Z = const.$,

$$\left.\frac{\partial Y}{\partial X}\right)_Z = -\frac{\left.\frac{\partial Z}{\partial X}\right)_Y}{\left.\frac{\partial Z}{\partial Y}\right)_X}.$$

This relationship is of great importance for deriving thermodynamic relations between state variables.
In the same way, the following relations can, almost always, be used for thermodynamic variables

$$\left.\frac{\partial Y}{\partial X}\right)_Z = \frac{1}{\left.\frac{\partial X}{\partial Y}\right)_Z} \quad \text{(inversible functions)},$$

$$\left.\frac{\partial Y}{\partial X}\right)_Z = \left.\frac{\partial Y}{\partial W}\frac{\partial W}{\partial X}\right)_Z \quad \text{(chain rule)}.$$

5. Consider an ideal gas and show why

$$\left.\frac{\partial T}{\partial P}\right)_v, \left.\frac{\partial T}{\partial P}\right)_s, \left.\frac{\partial T}{\partial P}\right)_h,$$

have different meanings. Try to illustrate these meanings using words instead of equations.

6. The speed of sound is given by

$$c_s = \sqrt{\left.\frac{\partial P}{\partial \rho}\right)_s}.$$

Find the speed of sound for an ideal gas.

7. Reducing the pressure can have a refrigerating effect on real gases. What happens with the temperature, when the pressure is reduced following an isenthalpic process? Calculate

$$\left.\frac{\partial T}{\partial P}\right)_h,$$

and base your analysis on the van der Waals equation of state

$$P = \frac{\mathcal{R}T}{\bar{v} - b} - \frac{a}{\bar{v}^2}.$$

3.7 Minimum Principles

As we saw in Chap. 2 of Part I of this book, the entropy measures the tendency of a thermodynamic system to reach the equilibrium state. It attains its maximum value when this equilibrium state is reached; furthermore, we are sure that if this system does not exchange heat with its environment, all the internal processes resulting from internal non-equilibrium in its intensive variables will contribute to increase the entropy.

We can also say that the entropy will not decrease if the system is isolated and, so, does not exchange mass, heat and work with its environment. Therefore, when the internal energy U is constant, all the internal processes will contribute to an increase in the entropy

$$dS \geq 0 \text{ when } U = const.$$

We can illustrate this principle using a very simple example. Consider a thermally insulated system composed of two chambers A and B (Fig. 3.10). These two

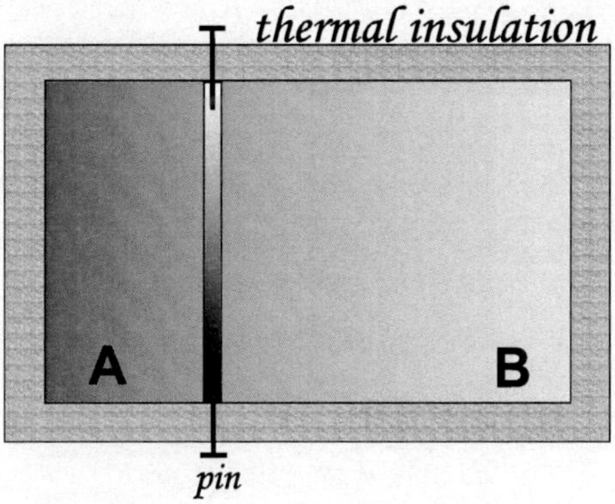

Fig. 3.10 When the internal energy $U = const.$ the internal non-equilibrium will increase the entropy

chambers are separated by a thin plate that does not offer any resistance to the flow of heat between chambers A and B. So, we can consider that, *in equilibrium states of chambers A and B,* $T_A = T_B = T$. Nevertheless, this plate is fixed to the walls by pins, and the pressures P_A and P_B will be different. By letting the plate free, by supposing that we can do that in an infinitesimal process, the volumes V_A and V_B will change by dV_A and $dV_B = -dV_A$, respectively, the two chambers, individually, reaching other equilibrium states. The same comment for the internal energy, because $d\,(U_A + U_B) = 0$.

Since, in equilibrium states, $S_A = S_A\,(V_A, U_A, N_A)$ and $S_B = S_B\,(V_B, U_B, N_B)$, the respective changes in the entropy will be

$$dS_A = \underbrace{\frac{\partial S_A}{\partial V_A}}_{=\frac{P_A}{T}} dV_A + \underbrace{\frac{\partial S_A}{\partial U_A}}_{=\frac{1}{T}} dU_A + \underbrace{\frac{\partial S_A}{\partial N_A} dN_A}_{=0}.$$

In the same way

$$dS_B = \underbrace{\frac{\partial S_B}{\partial V_B}}_{=\frac{P_B}{T}} dV_B + \underbrace{\frac{\partial S_B}{\partial U_B}}_{=\frac{1}{T}} dU_B + \underbrace{\frac{\partial S_B}{\partial N_B} dN_B}_{=0}.$$

Therefore, the change of entropy of the whole system between these two equilibrium states will be

$$dS = dS_A + dS_B = \frac{(P_A - P_B)}{T} dV_A + T \underbrace{(dU_A + dU_B)}_{=0}$$

So, if $P_A > P_B$, dV_A must be positive for assuring $dS > 0$. In addition, $dS = 0$ if and only if $P_A = P_B$, or when the *whole* system A+B is in equilibrium.

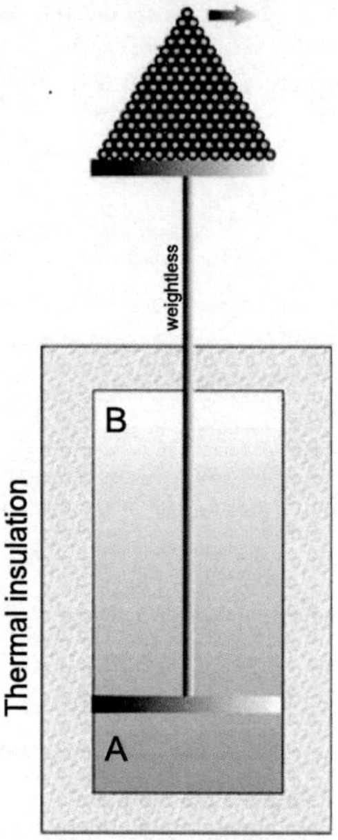

Fig. 3.11 Isentropic expansion for showing the minimum principle for the internal energy

3.7.1 Minimum Principle for the Internal Energy

Consider now the inverse problem pictured in Fig. 3.11. A gas in chamber A is maintained at a high pressure P_A by a certain amount of very small balance weights with mass Δm. The piston is considered to be weightless and the two chambers are separated by a very thin plate that does not offer any resistance to the flow of heat between chambers A and B, in such a way that in equilibrium states *of chambers A and B*, $T_A = T_B = T$.

When a single balance weight is removed from the upper piston plate, the gas in chamber A has an expansion dV_A. Since Δm is very small, we can consider this expansion to be reversible and, so, isentropic since the whole system is separated from the external environment. In equilibrium states $U_A = U_A\,(V_A, S_A, N_A)$ and, so,

$$dU_A = \underbrace{\frac{\partial U_A}{\partial S_A}}_{=T} dS_A + \underbrace{\frac{\partial U_A}{\partial V_A}}_{=-P_A} dV_A + \underbrace{\frac{\partial U_A}{\partial N_A} dN_A}_{=0}.$$

In the same way

$$dU_B = \underbrace{\frac{\partial U_B}{\partial S_B}}_{=T} dS_B + \underbrace{\frac{\partial U_B}{\partial V_B}}_{=-P_B} dV_B + \underbrace{\frac{\partial U_B}{\partial N_B} dN_B}_{=0}.$$

Summing the two above equations

$$dU = T\underbrace{(dS_A + dS_B)}_{=0} - (P_A - P_B)\,dV_A$$

Since $P_A > P_B, dV_A$ will be greater than zero and dU will decrease. Furthermore, $dU = 0$, if and only if $P_A = P_B$, which means the whole system A+B has attained its equilibrium state.

A more theoretical demonstration can perhaps help to convince ourselves that internal non-equilibrium between parts of a thermodynamic system that are themselves in equilibrium, contribute to reducing the internal energy in isentropic processes. In fact, since the internal energy U of the whole system will be dependent on the volume of the chamber A

$$dU = \left(\frac{\partial U}{\partial V_A}\right)_{S,N} dV_A = -\frac{\left(\frac{\partial S}{\partial V_A}\right)_{U,V}}{\left(\frac{\partial S}{\partial U}\right)_{V_A,V}} dV_A \tag{3.34}$$

Since $\left(\frac{\partial S}{\partial U}\right)_{V_A,V}$ is the temperature of the whole system, the sign of $\left(\frac{\partial U}{\partial V_A}\right)_{S,V}$ will be the inverse of $\left(\frac{\partial S}{\partial V_A}\right)_{U,V}$ and we know that in processes where the internal energy U is a constant $\frac{\partial S}{\partial V_A} \geq 0$.

When the system attains the full equilibrium state, in such a way that $P_A = P_B$, we can consider the second derivative of Eq. (3.34)

$$\left(\frac{\partial^2 U}{\partial V_A^2}\right)_{S,V} = -\frac{\left(\frac{\partial^2 S}{\partial V_A^2}\right)_{U,V} T - \left(\frac{1}{\frac{\partial U}{\partial T}}\right)\left(\frac{\partial S}{\partial V_A}\right)_{U,V}}{T^2} > 0$$

because in the equilibrium state, the entropy is a maximum, $\left(\frac{\partial S}{\partial V_A}\right)_{U,V} = 0$ and $\left(\frac{\partial^2 S}{\partial V_A^2}\right)_{U,V} < 0$.

In conclusion:

If a thermodynamic homogeneous system is in equilibrium, its entropy will be a maximum when compared to other states with the same internal energy

Reversely:

The equilibrium value of any unconstrained internal parameter — in this case, the volume V_A— in a system is the one that minimizes the internal energy at constant volume and entropy.

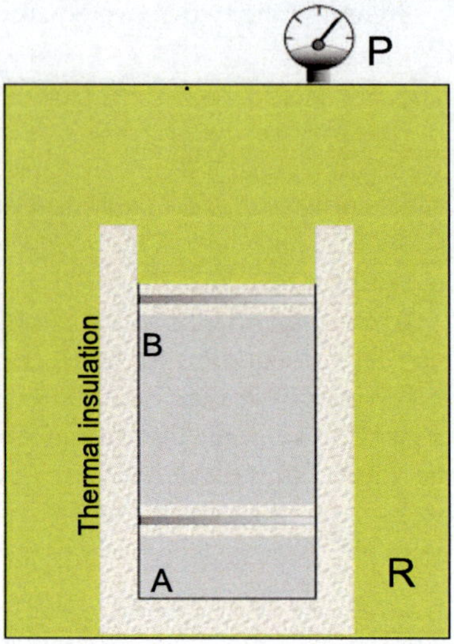

Fig. 3.12 Minimum principle for enthalpy

3.7.2 Minimum Principle for Enthalpy

To illustrate that enthalpy has also a minimum principle consider Fig. 3.12. One mole of a gas is contained in each chamber A and B of a thermodynamic system. Both chambers are insulated as well as the reservoir with which the system can exchange work, but not heat. This "*pressure reservoir*" is sketched by imagining that all pressure fluctuations will be absorbed by the reservoir. In the same way, the upper and lower plates separating the system from the pressure reservoir and chamber A from the chamber B are considered to be free to move along the inner cylinder wall. So, the pressure P is assured to be the same everywhere.

Nevertheless temperature T_B is, initially, greater than T_A and this is the reason the volume of chamber B is greater than the volume of chamber A. Now, suppose that we are capable of removing the thermal insulation separating chamber A from chamber B and, quickly, replace it again. During this process, heat will flow from the chamber B to chamber A, increasing its volume and this will produce a small pressure perturbation, absorbed by the reservoir.

What we can say is that the enthalpy variation between the two equilibrium states of the whole system, at constant pressure, is given by

$$dH = dH_A + dH_B = T_A dS_A + T_B dS_B,$$

Now, the entropy variation of the thermodynamic system will be

$$dS = dS_A + dS_B,$$

and so

$$dH = (T_A - T_B)\, dS_A + T_B dS.$$

Consider that the process is isentropic in such a way that $dS = 0$. We know that

$$dS_A \geq 0$$

because heat will flow from the hot chamber B to the cold chamber A.

When these conditions are satisfied

$$dH \leq 0,$$

and the principle of minimum enthalpy can be enunciated as: "*The equilibrium value of any unconstrained internal parameter — in this case, the entropy S_A— in a system in contact with a pressure reservoir is the one that minimizes the enthalpy at constant pressure and total entropy.*"

3.7.3 Minimum Principle for the Helmholtz Energy

It is difficult to make use of the minimum principle for internal energy or enthalpy in practical applications. In fact, isentropic processes are rather theoretical processes. On the other hand, all we need to perform an isothermal process is a thermal bath.

So, consider the system pictured in Fig. 3.13.

When the pins are removed, enabling chamber A to have a small expansion, we can say that

$$d\,(S + S_R) \geq 0, \qquad d\,(U + U_R) = 0,$$

where S, U are the entropy and internal energy of the thermodynamic system—the two chambers A and B—and R refers to the thermal reservoir.

Since the temperature remains constant, the variation of the Helmholtz energy of the system between the two equilibrium states is

$$dF = dU - TdS,$$

or

Fig. 3.13 Minimum principle for the Helmholtz energy

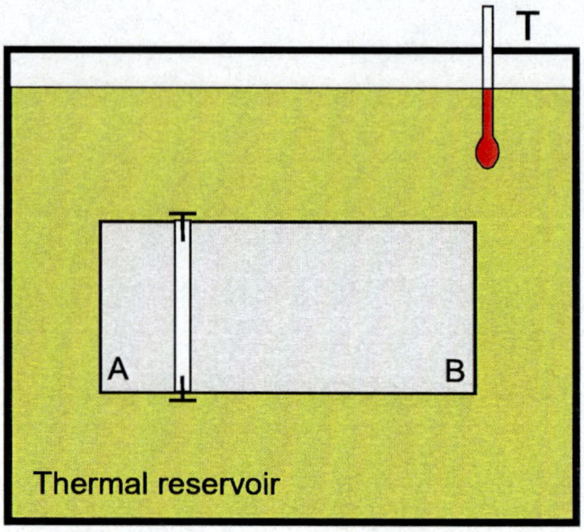

$$dF = -dU_R - TdS.$$

Now

$$dU_R = TdS_R,$$

related to the net heat that is exchanged between the system and the heat reservoir. So,

$$dF = -T\,(dS_R + dS) \leq 0$$

and $dF = 0$ when and only when $dS + dS_R = 0$, or when the whole system has reached its equilibrium state at the reservoir temperature.

The minimum principle for the Helmholtz energy can be also demonstrated by considering that, after the pins are removed and replaced again, the Helmholtz energy of partitions A and B will change by

$$dF_A = -P_A dV_A, \qquad dF_B = -P_B dV_B$$

and since $dV_B = dV - dV_A$,

$$dF = (P_B - P_A)\,dV_A - P_B dV$$

in such way that if $dV = 0$, dF is necessarily non-positive, $dF \leq 0$, because

$$\text{if } P_B \leq P_A \text{ then } dV_A \geq 0$$
$$\text{if } P_B \geq P_A \text{ then } dV_A \leq 0$$

In conclusion, what we have demonstrated is that:

The equilibrium value of any unconstrained internal parameter—in this case, the volume V_A—in a system is one that minimizes the Helmholtz energy at a constant volume and temperature.

Thermodynamic systems are subjected to local non-equilibrium, sketched in this section by using two complimentary partitions A and B that are themselves in equilibrium but at different pressures. When the pins are removed and replaced after a small volume variation—supposing that this is possible—the two partitions attain different equilibrium states at different pressures with respect to the ones they had at the beginning. The use of partitions to demonstrate that in the equilibrium state—when the pressures are the same—the Helmholtz energy has its lowest value was necessary because in equilibrium thermodynamics, we are limited to equilibrium states, i.e., to deal with states where all the intensive variables do not have any spatial variation. This limitation can be avoided with the help of the methods that were developed for describing *non-equilibrium systems* (*Part III of this book*).

Nevertheless, the above assertion can be rewritten—or generalized—for homogeneous system, i.e., without any partitions, in the following form:

If a thermodynamic homogeneous system is in equilibrium, its Helmholtz energy will be a minimum when compared to other (non-equilibrium) states with the same temperature and total volume

3.7.4 Exercises

1. *Minimum principle for the Gibbs energy.* A thermodynamic system is composed of a cylinder with two partitions A and B where two different gases A and B are contained (Fig. 3.14). Plates a and b are free to move inside the cylinder and we can suppose that plate b is permeable to gas A, allowing this gas to mix with gas B in partition B. The pressure is maintained constant by the three balance weights and the atmospheric pressure and the temperature is kept constant with the help of a thermal bath.
   We know that

$$dG = -SdT + VdP + \mu dN.$$

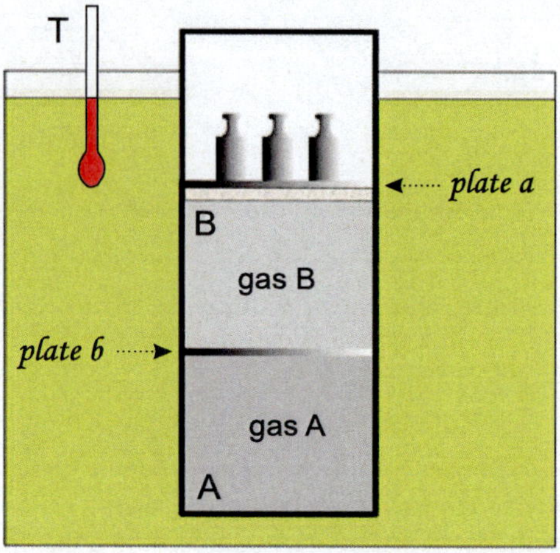

Fig. 3.14 Minimum principle for the Gibbs energy

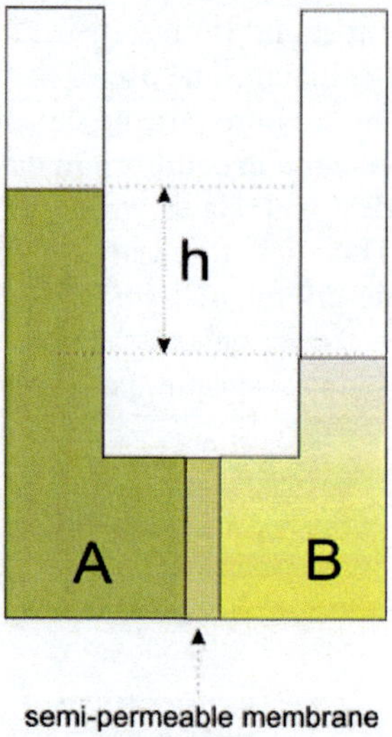

Fig. 3.15 Osmotic pressure

Show that the Gibbs energy has a minimum principle when the temperature T and pressure P are maintained constant.

2. *Osmotic pressure*. A membrane permeable to water separates a solution of water and salt in A from pure water in B (see Fig. 3.15).
In equilibrium, the pressure of the brine P_A will be larger than the water pressure P_B and the difference $P_A - P_B$ is called osmotic pressure. Suppose that the solution of water + salt may be considered an ideal solution with the Gibbs energy given by

$$G(T, P, x_1, \ldots, x_r) = \sum_i N_i g_{i,puro}(T, P) + N\mathscr{R}T \sum_i x_i \ln x_i. \qquad (3.35)$$

(a) Calculate the chemical potential of the water in the solution and in the pure water phase and show that

$$P_A - P_B = \frac{\mathscr{R}T}{v_{water}(T)} \ln x_{water}. \qquad (3.36)$$

(b) In the following show that when x_{salt} is very small

$$P_A - P_B = \frac{\mathscr{R}T}{v_{water}(T)} x_{salt}, \qquad (3.37)$$

which is the van't Hoff's law.[15]

[15] Jacobus Henricus van 't Hoff Jr. (1852–1911) was a Dutch physical chemist. A highly influential theoretical chemist of his time, van 't Hoff was the first winner of the Nobel Prize in Chemistry. His pioneering work contributed to the fundamentals of the modern theory of chemical affinity, chemical equilibrium, chemical kinetics, and chemical thermodynamics.

Chapter 4
Phase Equilibrium

Abstract We begin this chapter by following the efforts of Maxwell to show that the attractive forces between molecules were responsible for vapor condensation, at temperatures below the critical temperature found by Andrews for CO_2. In his 1875 paper '*On the dynamical evidence of the molecular constitution of bodies*', Maxwell recognizes the impact of the 1873 Doctoral Thesis of Johannes Diderik van der Waals for clearing the role of the interaction of real molecules on phase transition and does not omit to mention the important work of Willard Gibbs who, in Maxwell's words, " *has given us a remarkably simple and thoroughly satisfactory method of representing the relations of the different states of matter by means of a model*". Van der Waals equation of state leads to a two-parameter law of corresponding states in the sense that only the critical temperature and pressure are needed for describing the liquid-vapor states of a substance. Most substances require a third parameter for their description. The acentric factor, introduced by Pitzer in 1955, gave rise to a sequence of empirically derived equations of state of substantial practice importance in deriving liquid-vapor phase diagrams. The chapter ends with the presentation of a method for calculating the thermodynamic functions of liquid-vapor systems.

4.1 Andrews, van der Waals and Maxwell

> The molecular theory of the continuity of the liquid and gaseous states form the subject of an exceedingly ingenious thesis by Mr. Johannes Diderik van der Waals, a graduate of Leyden. There are certain points in which I think he has fallen into mathematical errors and his final result is certainly not a complete expression for the interaction of real molecules, but his attack on this difficult question is so able and so brave, that it cannot fail to give a notable impulse to molecular science. It has certainly directed the attention of more than one inquirer to the study of the Low Dutch language in which it is written. (Maxwell 1875)[1]

[1] J.C. Maxwell (1875), *On the dynamical evidence of the molecular constitution of bodies*, Nature 11 357- 359 [117].

P. C. Philippi, *Thermodynamics*, https://doi.org/10.1007/978-3-031-49357-7_4

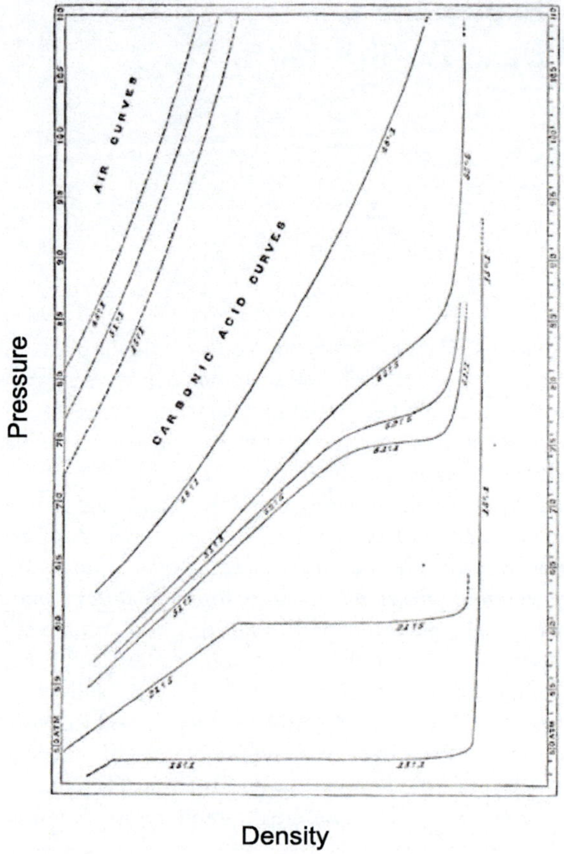

Fig. 4.1 Andrews experimental results for carbon dioxide at page 583 of Ref. [5]

Andrews's experiments published in 1869[2] showed that carbon dioxide has a critical temperature of about 31 °C. Above this temperature, there is one fluid state with a fixed density for each pressure and temperature. Below the critical temperature, there are two densities for each pressure and temperature, the higher being the one of the liquid and the lower corresponding to the vapor in equilibrium with it (Fig. 4.1).

Johannes Diderik van der Waals passed his doctoral examinations in December 1871. Eighteen months later he submitted his thesis '*On the continuity of the gaseous and liquid state*' [181]. It carries the date 14 June 1873, which was the day of his public defense of it. "*Like the early 19th century workers in kinetic theory, he was very much the outsider and brought to the subject a new vision, but unlike them, he was well versed in mathematics and physics and so was able to handle his subject in a way that commanded respect even when it attracted criticism*" [147].

[2] T. Andrews (1869), *On the continuity of the gaseous and liquid states of matter*, Phil. Trans. Roy. Soc. 159 575–90 [5].

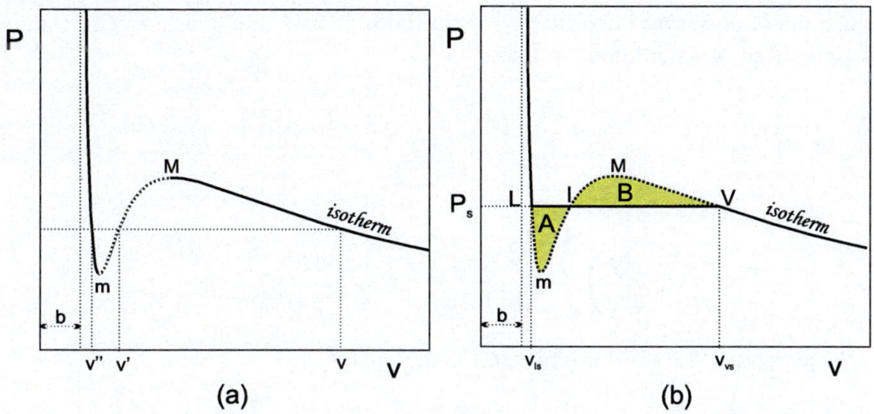

Fig. 4.2 **a** Van der Waals equation of state has 3 roots for each pressure; **b** Maxwell's area rule

He had two problems to solve. First, how to take into account the effect on the pressure of attractive forces of unknown form but, he believes, of essentially short range, that is, of a range comparable with the sizes of the molecules. He was one of the first to emphasize that the cooling of real gases on expansion observed by Joule and Thomson was a direct evidence for the existence of attractive forces in gases. The molecules at the surface of a fluid are pulled inwards and the decay on the pressure, P, due to this attraction effect should be proportional to the probability of two molecules being found together i.e., something proportional to the square of the number density of molecules, n^2.

His second problem was to calculate the amount b by which the observed volume must be reduced by the space taken up by the molecules so as to give an effective volume in which they move. He considered his molecules as hard spheres with a given size.

The equation of state proposed in his thesis had, thus, the form

$$P = \frac{\mathscr{R}T}{\bar{v} - b} - \frac{a}{\bar{v}^2}. \tag{4.1}$$

Van der Waals's equation is a cubic equation in the volume. So, at a fixed pressure and temperature it has three real roots (Fig. 4.2). The lowest v'' and highest v real roots correspond to vapor and liquid states but the third root v' at an intermediate density has no real existence for it is a state for which $(\partial P/\partial v)_T$ is positive, and so is to be considered unstable. Such a state, if formed, would spontaneously break up into a mixture of vapor and liquid.

Indeed, from Eq. (3.13), the differential equation for the Helmholtz energy is

$$df = -Pdv$$

when the temperature is constant. An equilibrium state at a constant temperature requires f to be a minimum, or that

$$\left(\frac{\partial^2 f}{\partial v^2}\right)_T > 0.$$

But

$$\left(\frac{\partial^2 f}{\partial v^2}\right)_T = \frac{\partial}{\partial v}\left(\frac{\partial f}{\partial v}\right)_T = -\left(\frac{\partial P}{\partial v}\right)_T > 0.$$

Therefore, stable equilibrium states must satisfy

$$\left(\frac{\partial P}{\partial v}\right)_T \leq 0,$$

and all the states between m and M in Fig. 4.2 must be considered as '*non-physical*'.

But, how to find the pressure where the vapor phase will change to the liquid state?...

In his 1875 Nature paper [117], Maxwell proposed what is presently known as the '*area rule*': transition at a given isotherm will happen when area B under the isotherm and above a line of constant pressure P_s in Fig. 4.2b is equal to the area A above this same isotherm and below line P_s.

Maxwell's paper is somewhat hard to follow for readers not familiar with kinetic theory. It was entitled "*On the dynamical evidence of the molecular constitution of bodies*". In fact, Maxwell was more interested in the molecular theory than in a practical rule for finding the saturation pressure. He wrote the equation of state as

$$PV = \frac{2}{3}\left(E_c - \sum_{i,j}\frac{1}{2}\mathbf{R}_{ij}\cdot\left(\|\mathbf{r}_j - \mathbf{r}_i\|\right)\right) \tag{4.2}$$

where E_c is the kinetic energy of the molecules, $\mathbf{R}_{ij}$ is the attraction force between any two molecules separated by $\mathbf{r}_j - \mathbf{r}_i$. He writes: "*The quantity* $\frac{1}{2}R_{ij}\cdot\left(\|\mathbf{r}_j - \mathbf{r}_i\|\right)$, *or half the product of the attraction into the distance across which the attraction is exerted, is defined by Clausius as the 'virial of the attraction'... When a gas is in an extremely rarefied condition, the number of particles within a given distance of any one particle will be proportional to the density of the gas. Hence the virial arising from the action of one particle on the rest will vary as the density, and the whole virial in unit of volume will vary as the square of the density*".

So by dividing Eq. (4.2) by the volume Maxwell concluded that

$$P = \frac{1}{3}\rho\bar{\xi}^2 - \frac{2}{3}A\rho^2$$

where $\overline{\xi}$ is the mean molecular speed and A "*is a quantity which is nearly constant for small densities*". The above equation is very similar to the van der Waals equation of state.

At page 359, he finally states his area rule (see Fig. 4.2b): "*Now let us suppose the medium to pass from V to L along the hypothetical curve VMImL in a state always homogeneous, and to return along the straight line LV in the form of a mixture of liquid and vapor. Since the temperature has been constant throughout, no heat can have been transformed into work. Now the heat transformed into work is represented by the excess of the area LmI over VMI. Hence the condition that determines the maximum pressure of the vapor at a given temperature is that the line LV cuts off equal areas from the curve above and below*".

At the end of his paper, he mentions the 1873 paper of Gibbs [61] recognizing that "*The purely thermodynamic relations of the different states of matter do not belong to our subject, as they are independent of particular theories about molecules. I must not, however, omit to mention a most important American contribution to this part of thermodynamics by Prof. Willard Gibbs, of Yale College, U.S., who has given a remarkably simple and thoroughly satisfactory method of representing the relations of the different states of matter by means of a model. By means of this model, problems that had long resisted the efforts of myself and others may be solved at once*".

4.1.1 Exercise

1. We know that

$$d\bar{u} = 0,$$

for an adiabatic free expansion of a gas and that

$$d\bar{u} = \bar{c}_v dT + \left(T \left(\frac{\partial P}{\partial T} \right)_{\bar{v}} - P \right) d\bar{v}$$

Consider 1 mole of carbon dioxide at 320 K occupying a volume of 1 l and that this gas suffers a free adiabatic expansion doubling its initial volume. For CO_2 suppose the van der Waals parameters are known: $a = 0.359$ Pa mol^2/m^6 and $b = 4.27 \times 10^{-5}$ m^3/mol and consider $c_v = 29$ J/ (mol K). Find the CO_2 temperature after the expansion. What would be this temperature if the intermolecular attractive forces between the CO_2 molecules were null?

4.2 Gibbs Potential to Find the Transition Pressure

Table 4.1 shows a sample worksheet for calculating the thermodynamic functions based on the van der Waals equation of state, starting from a reference state $(\overline{v}_0, T)$ and following an isothermal by successive volume increments $\Delta\overline{v}$ of the molar volume.

Pressure P is calculated using Eq. (4.1).

The entropy calculation in the third column is based on Eq. (3.32). In isothermal conditions

$$\overline{s}(T,\overline{v}) - \overline{s}(T,\overline{v}_0) = \int_{\overline{v}_0}^{\overline{v}} \left(\frac{\partial P}{\partial T}\right)_{\overline{v}} d\overline{v}.$$

For a van der Waals equation of state

$$\overline{s}(T,\overline{v}) - \overline{s}(T,\overline{v}_0) = \ln\frac{\overline{v}-b}{\overline{v}_0-b}.$$

The integration of Eq. (3.33) enables to find

$$\overline{u}(T,\overline{v}) - \overline{u}(T,\overline{v}_0) = \int_{\overline{v}_0}^{\overline{v}} \left(T\left(\frac{\partial P}{\partial T}\right)_{\overline{v}} - P\right) d\overline{v}.$$

in the fourth column.

For a van der Waals equation of state

$$T\left(\frac{\partial P}{\partial T}\right)_{\overline{v}} - P = \frac{a}{\overline{v}^2},$$

and

$$\overline{u}(T,\overline{v}) - \overline{u}(T,\overline{v}_0) = a\left(\frac{1}{\overline{v}_0} - \frac{1}{\overline{v}}\right).$$

Table 4.1 Sample worksheet for the calculation of thermodynamic functions

$\overline{v}$	P	$\overline{s}-\overline{s}_0$	$\overline{u}-\overline{u}_0$	$\overline{h}-\overline{h}_0$	$\overline{g}-\overline{g}_0$
$\overline{v}_0$	P_0	0	0	0	0
$\overline{v}_0+\Delta\overline{v}$	...	...	...	...	...
$\overline{v}_0+2\Delta\overline{v}$	...	...	...	...	...
...	...	...	...	...	...

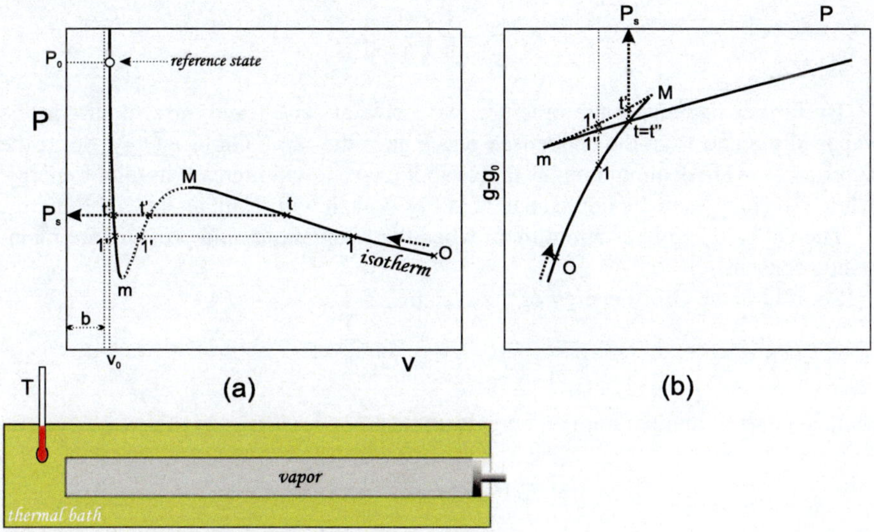

Fig. 4.3 **a** Calculation of the transition pressure using a worksheet. **a** $P - v$ diagram corresponding to a single isotherm. **b** Gibbs energy for different pressures

Calculation of the enthalpy follows

$$\begin{aligned}&\overline{h}(T,\overline{v}) - \overline{h}(T,\overline{v}_0)\\&= \underbrace{[\overline{u}(T,\overline{v}) - \overline{u}(T,\overline{v}_0)]}_{\text{fourth column}} + \underbrace{P}_{\text{second column}} \times \underbrace{\overline{v}}_{\text{first column}} - P_0\overline{v}_0\end{aligned}$$

Finally, the Gibbs energy is given by

$$\begin{aligned}&\overline{g}(T,\overline{v}) - \overline{g}(T,\overline{v}_0)\\&= \underbrace{\left[\overline{h}(T,\overline{v}) - \overline{h}(T,\overline{v}_0)\right]}_{\text{fifth column}} - T\underbrace{(\overline{s} - \overline{s}_0)}_{\text{third column}}\end{aligned}$$

Figure 4.3a shows the $P - \overline{v}$ diagram corresponding to a single isotherm and Fig. 4.3b shows the Gibbs energy $\overline{g}(T,\overline{v}) - \overline{g}(T,\overline{v}_0)$ for different pressures.

Let us consider that we start from the low-pressure state O where the substance is in its vapor state. By increasing the pressure we will find state 1. From O to 1 the Gibbs energy increases due to the compression work that was furnished to the vapor. State 1' is unstable whereas state 1" belongs to the condensed phase, with a very small value of the isothermal compressibility

$$k_T = -\frac{1}{\overline{v}}\left(\frac{\partial \overline{v}}{\partial P}\right)_T.$$

By further increasing the pressure, we arrive at state t and, now, both t in the vapor phase and t'' in the condensed phase have the *same* Gibbs energy. So, if the substance we are dealing remains in its vapor phase it will increase its Gibbs energy, while if it changes to the liquid state t'' in Fig. 4.3a, it will maintain its Gibbs energy.

The Gibbs energy has a minimum when the temperature and pressure are maintained constants.

So, if G is the Gibbs energy of the system

$$G = G_v + G_\ell,$$

then, at constant temperature and pressure

$$dG = \left(\overline{\mu}_v - \overline{\mu}_\ell\right) dN_v = 0,$$

meaning that the Gibbs potentials—or molar Gibbs energies $\bar{g}$—must be identical when phases v and ℓ are in equilibrium. When the pressure is increased from O to M, this state of minimal Gibbs energy is found, as the one represented by point t, when $\bar{\mu}_v$ (or $\bar{g}_v$) and $\bar{\mu}_\ell$ (or $\bar{g}_\ell$) are identical as shown in Fig. 4.3b.

4.2.1 *Maxwell Area Rule*

The Maxwell area rule can be derived from the condition

$$\overline{\mu}_v - \overline{\mu}_\ell = 0,$$

at the transition point. In fact, from the Gibbs-Duhem equation, Eq. (3.8)

$$d\overline{\mu} = \overline{v} dP,$$

at a constant temperature.

Figure 4.4 shows the van der Waals equation of state considering the volume v in terms of the pressure P for a given temperature, By integrating the above equation between points V and L

$$0 = \int_V^L \overline{v} dP,$$

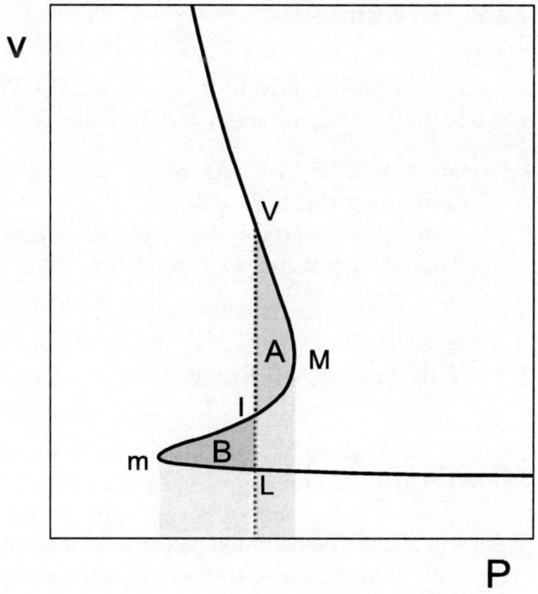

Fig. 4.4 Maxwell area rule derived from the equality of the Gibbs potentials μ_ℓ and μ_v

or

$$\int_V^M \overline{v} dP + \int_M^I \overline{v} dP + \int_I^m \overline{v} dP + \int_m^L \overline{v} dP = 0,$$

or

$$\underbrace{\int_V^M \overline{v} dP - \int_I^M \overline{v} dP}_{\text{area A}} = \underbrace{\int_m^I \overline{v} dP - \int_m^L \overline{v} dP}_{\text{area B}}.$$

The conclusion is that the shaded areas A and B in Fig. 4.4 must be identical at the transition pressure, leading to

$$\int_{\overline{v}_{\ell s}}^{\overline{v}_{vs}} P\left(\overline{v}\right) d\overline{v} = P_s \left(\overline{v}_{vs} - \overline{v}_{\ell s}\right). \tag{4.3}$$

This equation, together with the equation of state is what is needed for finding the transition pressure. Since the resulting system of equations is non-linear, this is usually performed with the help of numerical methods.

4.2.2 *Exercise*

Suppose that carbon dioxide follows a van der Waals equation of state. For CO_2, $a = 0.359\text{Pa moles}^2/\text{m}^6$ and $b = 4.27 \times 10^{-5}\text{m}^3/\text{moles}$.

(a) Using a worksheet such as the one of Table 4.1, find the transition pressure P_s for carbon dioxide at $T = 280$ K.
(b) Use an iteration numerical method for solving the Maxwell area rule, Eq. (4.3) and find the transition pressure P_s for carbon dioxide at this temperature.

4.3 The Transition State

When all the states between V and L in Fig. 4.5 are equilibrium states the pressure will be the transition pressure $P_s(T)$. Therefore if we displace the piston in Fig. 4.5 to the left in a *quasi-static process*, starting from point V, a fraction of the vapor phase will change to the liquid phase, but the pressure will remain constant.

In reality, this piston displacement will produce non-equilibrium states in the liquid-vapor system. In fact, a mass Δm of vapor needs to release heat to be condensed. If the pressure is constant this released heat will correspond to the enthalpy variation between the liquid and vapor states (*the latent heat*),

$$\delta Q = \Delta m \left(h_{vs} - h_{\ell s}\right).$$

Nevertheless, if the time Δt corresponding to the piston displacement is smaller than the time this latent heat needs to be transferred to the heat bath, the vapor pressure close to the piston head will increase above P_s giving rise to a pressure wave that propagates to the left. Pressure P_s will be restored, accompanied by the condensation of the mass fraction Δm of vapor, only after δQ is entirely transferred to the thermal bath. This is the reason why condensation happens adjacent to the wall that separates the vapor from the liquid bath and not as it is drawn in Fig. 4.5, which is only a sketch for showing the amount of liquid and vapor when the piston is in an intermediate position between L and V.

Consider now that the volume of the system liquid + vapor is

$$V = V_v + V_\ell.$$

The volume that is occupied by the vapor is the product of its specific volume by the mass of vapor in the system

$$V_v = m_v v\left(T, P\right).$$

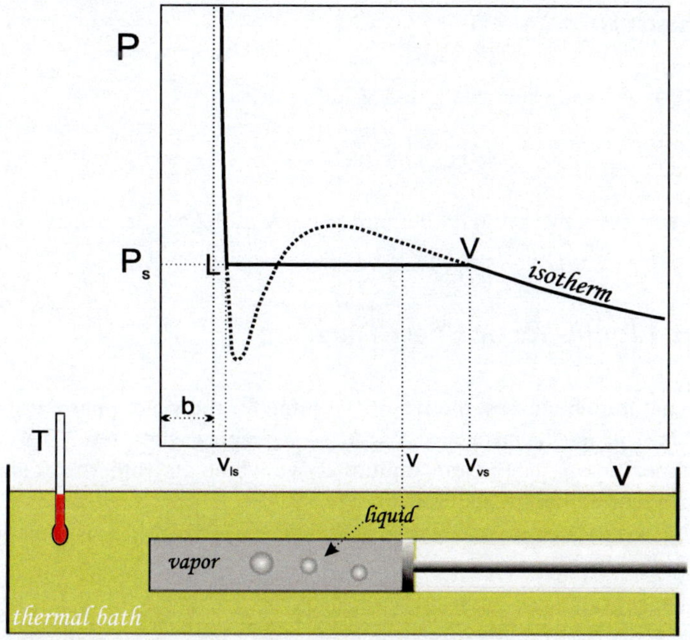

Fig. 4.5 During the phase transition at pressure P_s and temperature T, the molar volume of vapor and liquid do not change and are v_{vs} and $v_{\ell s}$ respectively

For the equilibrium states along the isobaric P_s, the specific volume of vapor is always the same $v_{vs}(T)$ between points V and L because neither T nor P vary along the transition line V-L. The same comment applies to the liquid phase.

So

$$V = m_v v_{vs} + m_\ell v_{\ell s},$$

and if we divide the above equation by the total mass $m = m_v + m_\ell$ we get

$$v = x v_{vs} + (1 - x)\, v_{\ell s},$$

where the mass fraction $x = m_v/m$ is named the '*vapor quality*' in the system.[3]

[3] The name 'vapor quality' is derived from the steam engine. Low-quality steam would contain a high liquid fraction and, therefore, damage the turbine blades more easily.

4.3.1 Exercise

Show that for any state between L and V in Fig. 4.5

$$\mathrm{f} = x\mathrm{f}_{vs}(T) + (1-x)\,\mathrm{f}_{\ell s}(T),$$

where f is any extensive property per unit mass or mole, e.g., s, u, h, g, f.

4.4 The Liquid-Vapor Phase Diagram

As the temperature increases, the specific volume for the vapor phase, v_{vs}, becomes closer to $v_{\ell s}$ and the latent heat $h_{\ell v} = h_{vs} - h_{\ell s}$ approaches zero. We define the critical temperature T_c as the temperature above which an isothermal compression does not produce a phase transition.

The critical point corresponds to the point where $P(\bar{v}, T_c)$ has an inflection point, or, when (Fig. 4.6)

$$\frac{\partial P(\bar{v}, T_c)}{\partial \bar{v}} = \frac{\partial^2 P(\bar{v}, T_c)}{\partial \bar{v}^2} = 0. \tag{4.4}$$

The existence of a critical point was first discovered by Charles Cagniard de la Tour in 1822[4] and named by Dmitri Mendeleev in 1860[5] and Thomas Andrews in 1869 [5].

4.5 Clausius-Clapeyron Equation

The Clausius-Clapeyron equation was initially derived in 1834 by Clapeyron [33] from the analysis of the Carnot cycle for the condensation of steam in thermal equilibrium with the liquid. Clausius refined the equation in 1850 [34] and extended it to other phase transitions (Fig. 4.7).

Let us consider a thermodynamic system where the liquid and vapor phases are in equilibrium at temperature T. In this condition the Gibbs potential will be the same in both phases

$$\mu_v = \mu_\ell.$$

---

[4] Charles Cagniard de la Tour (1822) "*Exposé de quelques résultats obtenu par l'action combinée de la chaleur et de la compression sur certains liquides, tels que l'eau, l'alcool, l'éther sulfurique et l'essence de pétrole rectifiée*", Annales de Chimie et de Physique, 21: 127–132.

[5] In accordance with Landau, L.D. and Lifshitz, E.M. (1980) *Course on Theoretical Physics Vol V, Statistical Physics*, p. 83. Third Edition of Elsevier.

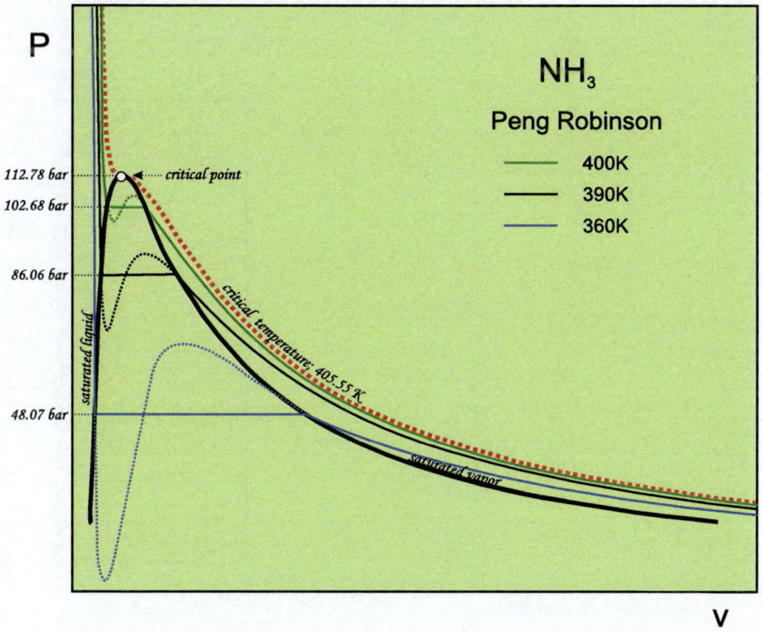

Fig. 4.6 Phase diagram for NH3 showing the isotherms of 360 K (blue), 390 K (black) and 400 K (green) modeled with the Peng-Robinson equation of state. The acentric factor was taken as $\omega = 0.25$. Pressure transition was found using the Maxwell area rule for each one of these isotherms. It is seen that the molar volume variation $v_{\ell v} = v_{vs} - v_{\ell s}$ between the saturated liquid and saturated vapor decreases with the temperature and reaches the zero value at the critical point. Above the critical temperature, there is no phase transition. For NH3 the critical temperature is $T_c = 405.55$ K and the critical pressure is $P_c = 112.78$ bar

Now, suppose that the temperature has an infinitesimal variation dT. Consequently, both potentials will vary, and, since they stay in equilibrium,

$$d\mu_v = d\mu_\ell.$$

Nevertheless, from the Gibbs-Duhem equation, Eq. (3.8)

$$-s_{vs}dT + v_{vs}dP_s(T) = -s_{\ell s}dT + v_{\ell s}dP_s(T),$$

or, in other words, the variation dT will produce a variation dP_s in the saturation pressure because P_s is dependent on the temperature.

From the above equation

$$\frac{dP_s}{dT} = \frac{s_{vs}(T) - s_{\ell s}(T)}{v_{vs}(T) - v_{\ell s}(T)}.$$

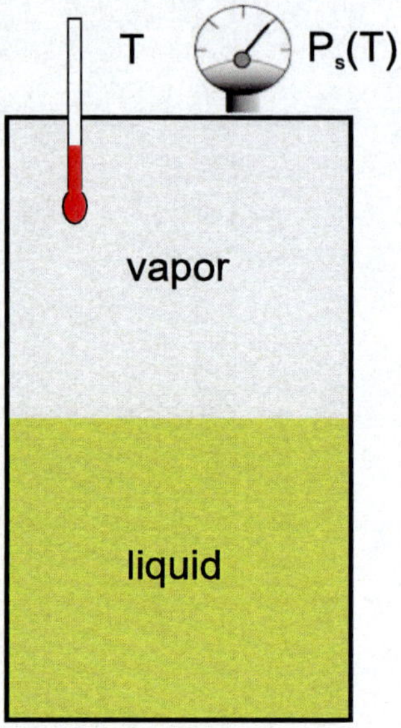

Fig. 4.7 The Clausius-Clapeyron equation

Now, remark that from Eq. (3.17)

$$dh = Tds + vdP,$$

and, since when the vapor state is attained from the liquid state, the pressure and the temperature remain constant

$$dh = Tds,$$

or

$$h_{\ell v} = h_{vs} - h_{\ell s} = T\left(s_{vs} - s_{\ell s}\right),$$

leading to

$$\frac{dP_s}{dT} = \frac{h_{\ell v}(T)}{T\left(v_{vs} - v_{\ell s}\right)}, \tag{4.5}$$

expressing the slope dP_s/dT in terms of the latent heat $h_{\ell v}(T)$.

4.5.1 Exercise

Using Eq. (4.4) show that for the van der Waals equation of state

$$b = \frac{\mathscr{R}T_c}{8P_c}, \quad \bar{v}_c = 3b, \quad a = \frac{27\mathscr{R}^2T_c^2}{64P_c}.$$

4.6 The Law of Corresponding States

The principle of Corresponding States was stated by van der Waals and reads: "*Substances behave alike at the same reduced states. Substances at the same reduced states are at corresponding states.*"

Consider a gas that follows the van der Waals equation of state

$$P = \frac{\mathscr{R}T}{\bar{v} - b} - \frac{a}{\bar{v}^2},$$

and let $P^* = \frac{P}{P_c}$, $T^* = \frac{T}{T_c}$, $v^* = \frac{\bar{v}}{\bar{v}_c}$. With these new variables, the van der Waals equation becomes

$$P^*P_c = \frac{\mathscr{R}T^*T_c}{v^*\bar{v}_c - b} - \frac{a}{(v^*\bar{v}_c)^2}.$$

From Eq. (4.4) it can be found that van der Waals parameters satisfy

$$b = \frac{\mathscr{R}T_c}{8P_c}, \quad \bar{v} = 3b, \quad a = \frac{27\mathscr{R}^2T_c^2}{64P_c},$$

and so

$$P^* = \frac{3T^*}{3v^* - 1}\frac{\mathscr{R}T_c}{P_c\bar{v}_c} - \frac{27}{64}\frac{1}{v^{*2}}\left(\frac{\mathscr{R}T_c}{P_c\bar{v}_c}\right)^2,$$

or, since $\frac{\mathscr{R}T_c}{P_c\bar{v}_c} = \frac{8}{3}$

$$P^* = \frac{8T^*}{3v^* - 1} - \frac{3}{v^{*2}}. \tag{4.6}$$

Therefore, the above reduced equation is a *universal* relation $P^* = P^*(T^*, v^*)$, independent of the substance we are dealing with. In other words, "*the states of all real gases are corresponding for a given reduced temperature and reduced volume*"

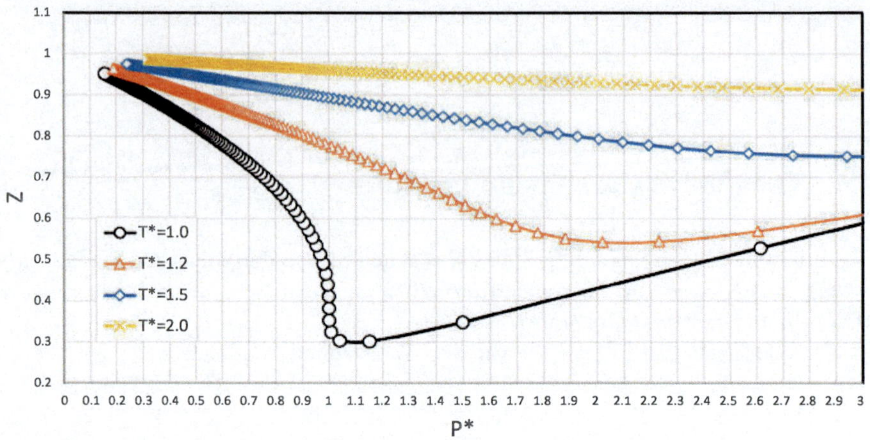

Fig. 4.8 Compressibility factor in terms of the reduced pressure and temperature for the van der Waals equation of state

Now, let us define the compressibility factor,

$$Z = \frac{P\bar{v}}{\mathscr{R}T} = \frac{3}{8}\frac{P^*v^*}{T^*}.$$

Since, for an ideal gas

$$\bar{v}_{ideal} = \frac{\mathscr{R}T}{P},$$

it follows that

$$Z = \frac{\bar{v}}{\bar{v}_{ideal}} = \frac{3}{8}\frac{P^*v^*}{T^*},$$

or that

$$v^* = \frac{8ZT^*}{3P^*}.$$

Replacing the above equation into Eq. (4.6)

$$8T^{*^3}Z^3 - T^{*^2}\left(P^* + 8T^*\right)Z^2 + \frac{27}{8}\left(P^*T^*\right)Z - \frac{27}{64}P^{*^2} = 0. \tag{4.7}$$

The above equation can be solved for finding $Z = Z\left(P^*\right)$ for a given isotherm T^*. This relationship is pictured in Fig. 4.8.

4.7 Acentric Factor

The principle of corresponding states (PCS) that we have just discussed was originally outlined by van der Waals. In reality, it is the simplest version of the principle of corresponding states, and it is referred to as the two-parameter PCS. This is because it relies on two parameters (critical pressure and temperature) for defining a "corresponding state."

With the passing of time, more accurate principles of corresponding state formulations have made use of more than two parameters. In any case, the general statement of PCS remains untouched: "Substances at corresponding states behave alike". What makes the difference is the definition of "*what a corresponding state is*."

The acentric factor "ω" is a concept that was introduced by Pitzer in 1955 [134], [135], and has proven to be very useful in the characterization of substances. It has become a standard for the proper characterization of any single pure component, along with other common properties, such as molecular weight, critical temperature, critical pressure, and critical volume.

Pitzer came up with this factor by analyzing the vapor pressure curves of various pure substances. From thermodynamic considerations, the vapor pressure curve for pure components can be mathematically described by the Clausius-Clapeyron equation, here written on a molar basis,

$$\frac{dP_s}{dT} = \frac{\bar{h}_{\ell v}(T)}{T\bar{v}_{\ell v}(T)},$$

where $\bar{v}_{\ell v} = \bar{v}_{vs} - \bar{v}_{\ell s}$. Now, the compressibility factor variation between the liquid and vapor state, at a temperature T is given by

$$\Delta Z = Z_v - Z_\ell = \frac{P_s(T)\,\bar{v}_{\ell v}(T)}{\mathscr{R}T}.$$

Therefore

$$\frac{dP_s}{P_s} = \frac{\bar{h}_{\ell v}(T)}{\mathscr{R}\Delta Z(T)}\frac{dT}{T^2},$$

or

$$\frac{dP_s^*}{P_s^*} = a(T)\frac{dT^*}{T^{*2}}, \tag{4.8}$$

where

$$a(T) = \frac{h_{\ell v}(T)}{RT_c\Delta Z(T)}.$$

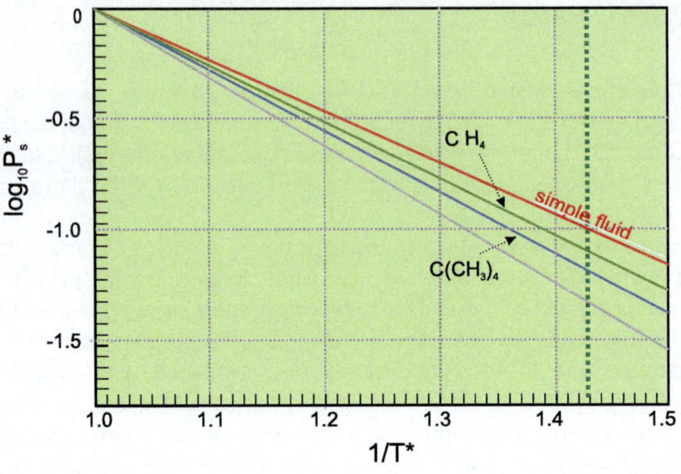

Fig. 4.9 Estimation of the acentric factor based on experimental data for the saturation pressure

Now, suppose that both $h_{\ell v}\,(T)$ and $\Delta Z\,(T)$ decrease at the same rate when the temperature is increased. In this case, parameter a is a constant and Eq. (4.8) can be integrated to give

$$\ln P_s^* = -\frac{a}{T^*} + b.$$

Pitzer wrote the above equation as

$$\log_{10} P_s^* = -\frac{A}{T^*} + B,$$

where $A = a \log_{10} e$ and $B = b \log_{10} e$. If the two-parameter corresponding state principle was to hold true for all substances, the parameters A and B should be the same for all substances. In other words, if the plot is of the form "$\log_{10} P_s^*$" versus "$1/T^*$", all lines should show the same slope A and intercept B. Noble gases (such as Ar, Kr, and Xe) happen to follow the two-parameter corresponding states theory very closely. Hence, they yield themselves amenable to acting as a reference to evaluate "compliance" with the two-parameter equation of state (Fig. 4.9).

Pitzer decided to use noble gases as the base for comparison. Analyzing vapor pressure data for noble gases, Pitzer showed that a value of $\log_{10} P_s^* = -1$ was achieved at approximately $T^* = 0.7$. So he defined the acentric factor as

$$\omega = -\log_{10}\left(P_s^*\right)_{T^*=0.7} - 1$$

Noble gases, being the reference themselves, have an acentric factor value of zero ($\omega = 0$). Substances with an acentric factor of zero are called "*simple*" substances.

Table 4.2 Acentric factors found by Pitzer

Substance	ω
Argaom	−0.002
Krypton	−0.002
Simple fluid	0
Xenon	0.002
Methane	0.13
Nitrogen	0.040
Ethane	0.0944
Propane	0.152
n-Butane	0.201
Benzene	0.215
Carbon dioxide	0.225
n-Pentane	0.252
Ammonia	0.25
Water	0.348

Table 4.3 Saturation pressure for Ethane

$T\ (K)$	100	150	200	250	300	305.33
P_s (MPa)	0.0000111	0.0096780	0.2174	1.3008	4.3559	4.8714

The acentric factor is said to be a measure of the non-sphericity of the molecules. Therefore, the three-parameter corresponding state theory of Pitzer reads: "*Fluids that have the same value of ω will behave alike at the same conditions of reduced pressure and temperature.*"

Table 4.2 shows the acentric factors of some substances, found by Pitzer [135].

4.7.1 Exercises

1. Figure 4.2 shows the Pitzer [134] experimental results for some hydrocarbons, where the base 10 logarithm of the reduced saturation pressure is plotted against the inverse of the reduced temperature. Using the figure as a reference estimate the acentric factor of Methane and Neopentane[6]
2. Table 4.3 shows the experimental values of the saturation pressure for ethane. Estimate the acentric factor ω for ethane based on this data and compare it with

[6] Neopentane, also called 2,2-dimethylpropane, is a double-branched-chain alkane with five carbon atoms. Neopentane is an extremely flammable gas at room temperature and pressure that can condense into a highly volatile liquid on a cold day, in an ice bath, or when compressed to a higher pressure.

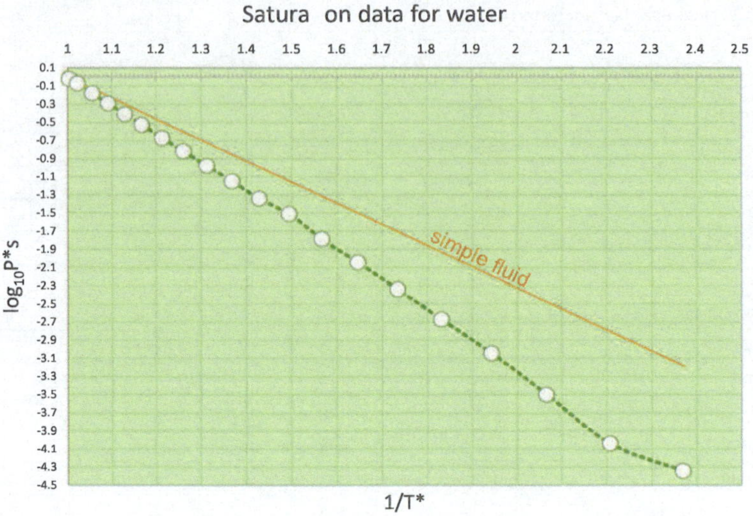

Fig. 4.10 $log_{10} P_s^*$ in terms of the inverse of the reduced temperature T^* for water

the value that was obtained by Pitzer. The critical temperature for ethane is 305.33 K.

3. Figure 4.10 shows the saturation data for water where $\log_{10} P_s^*$ is plotted against $1/T^*$. Find the acentric factor for water and compare with the value found by Pitzer.

4.8 Some Popular Cubic Equations of State

4.8.1 Redlich-Kwong

The van der Waals equation of state does not take into account the temperature dependence of the force parameter a, considered as a constant. When the temperature increases, the cohesion forces become less effective with respect to the dispersion forces and a correction on the van der Waals equation was proposed by Redlich-Kwong in 1949 to take this effect into account [145]

$$P = \frac{\mathscr{R}T}{\bar{v} - b} - \frac{a/\sqrt{T}}{\bar{v}\,(\bar{v} + b)},$$

with

$$a = 0.42747 \frac{\mathscr{R}^2 T_c^{\frac{5}{2}}}{P_c}, \quad b = 0.08664 \frac{\mathscr{R} T_c}{P_c}.$$

4.8.2 Soave

In 1972 G. Soave [159] replaced the $1/\sqrt{T}$ term of the Redlich-Kwong equation with a function $\alpha(T, \omega)$ involving the temperature and the acentric factor (*the resulting equation is also known as the Soave-Redlich-Kwong equation*). The α function was devised to fit the vapor pressure data of hydrocarbons and the equation does fairly well for these substances.

It is written as

$$P = \frac{\mathscr{R} T}{\bar{v} - b} - \frac{a(T_c)\,\alpha(T, \omega)}{\bar{v}(\bar{v} + b)},$$

with

$$a(T_c) = 0.427 \frac{\mathscr{R}^2 T_c^2}{P_c}, \quad b = 0.08664 \frac{\mathscr{R} T_c}{P_c}$$

and

$$\alpha(T, \omega) = \left(1 + \kappa\left(1 - \sqrt{T^*}\right)\right)^2,$$
$$\kappa = 0.48508 + 1.55171\omega - 0.15613\omega^2.$$

4.8.3 Peng-Robinson

The Peng- Robinson equation was developed in 1976 at The University of Alberta by Ding-Yu Peng and Donald Robinson [128] in order to satisfy the following goals: (i) the parameters should be expressible in terms of the critical properties and the acentric factor; (ii) the model should provide reasonable accuracy near the critical point, particularly for calculations of the compressibility factor and liquid density.

It is written as

$$P = \frac{\mathscr{R} T}{\bar{v} - b} - \frac{a(T_c)\,\alpha(T, \omega)}{\bar{v}(\bar{v} + b) + b(\bar{v} - b)}, \tag{4.9}$$

where

$$a\left(T_c\right) = 0.45724\frac{\mathcal{R}^2 T_c^2}{P_c} \quad b = 0.0778\frac{\mathcal{R} T_c}{P_c},$$
$$\alpha\left(T, \omega\right) = \left(1 + \kappa\left(1 - \sqrt{T^*}\right)\right)^2,$$
$$\kappa = 0.37464 + 1.54226\omega - 0.26992\omega^2.$$

For the most part, the Peng- Robinson equation exhibits performance similar to the Soave equation, although it is generally superior in predicting the liquid densities of many materials, especially non-polar ones.

4.8.4 *Exercises*

Carbon dioxide has a measured saturation pressure $P_s = 35.642$ bar at $T = 274$ K.

1. Suppose that carbon dioxide follows a Redlich-Kwong equation of state. For CO_2, $T_c = 304.13$ K, $P_c = 73.77$ bar.
   (a) Using a worksheet such as the one of Table 4.1, find the transition pressure P_s for carbon dioxide at $T = 280$ K.
   (b) Using the Maxwell area rule, Eq. (4.3) find the transition pressure P_s for carbon dioxide at this temperature.
2. Repeat Exercise 1, now supposing carbon dioxide follows a Soave-Redlich-Kwong equation of state. For CO_2, the acentric factor is $\omega = 0.225$ (Table 4.2).

4.9 The Virial Equation of State

The virial equation of state gives the compressibility factor Z as a power series of the reciprocal of the molar volume

$$Z = \frac{P\bar{v}}{\mathcal{R}T} = 1 + \frac{B}{\bar{v}} + \frac{C}{\bar{v}^2} + \cdots, \tag{4.10}$$

where B is the second virial coefficient and C is the third virial coefficient. All virial coefficients are independent of pressure and density and are only dependent on the temperature for pure components. The virial equation can also be written in terms of the pressure

$$Z = \frac{P\bar{v}}{\mathcal{R}T} = 1 + B_P P + C_P P^2 + \cdots, \tag{4.11}$$

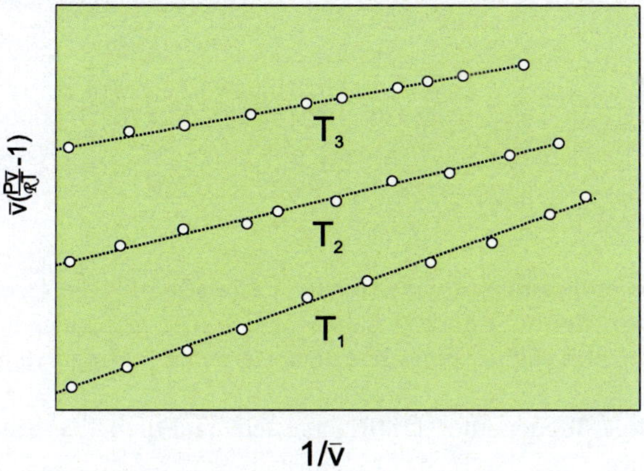

Fig. 4.11 PVT experimental data allows the determination of the virial coefficients B and C for each isotherm. Virial parameter B (T) is found from the interception of the isotherm T with the vertical axis. Parameter C(T) is given by the slope of the isotherm T when $n = 1/\bar{v} \to 0$

where [138],

$$B_P = \frac{B}{\mathscr{R}T}, C_P = \frac{C - B^2}{(\mathscr{R}T)^2}.$$

The virial coefficients can be determined from experimental PVT data, using

$$B = \lim_{n \to 0} \left(\frac{\partial Z}{\partial n} \right)_T, C = \lim_{n \to 0} \frac{1}{2!} \left(\frac{\partial^2 Z}{\partial n^2} \right)_T$$

by rearranging Eq. (4.10) as

$$\bar{v} \left(\frac{P\bar{v}}{\mathscr{R}T} - 1 \right) = B + \frac{C}{\bar{v}}, \tag{4.12}$$

The symbol is used here to denote the inverse of the molar volume $n = 1/\bar{v}$. Figure 4.11 gives a sketch for this determination.

4.9.1 Second Virial Coefficient

Tsonopoulos [177–179] proposed the following correlation for the second virial coefficient

$$B^* = \frac{BP_c}{RT_c} = F_0 \left(T^* \right) + \omega F_1 \left(T^* \right), \tag{4.13}$$

where

$$F_0\left(T^*\right) = 0.1445 - \frac{0.330}{T^*} - \frac{0.1385}{T^{*2}} - \frac{0.021}{T^{*3}} - \frac{0.000607}{T^{*8}},$$
$$F_1\left(T^*\right) = 0.0637 + \frac{0.331}{T^{*2}} - \frac{0.423}{T^{*3}} - \frac{0.008}{T^{*8}}.$$

Tsonopoulos proposition provides a good correlation for the second virial coefficients of *normal* fluids. A normal fluid is one whose molecules are of moderate size, non-polar or else slightly polar, and do not associate strongly (e.g., by hydrogen bonding).

For polar fluids, Tsonopoulos [180] suggested that Eq. (4.13) be amended to read

$$B^* = F_0\left(T^*\right) + \omega F_1\left(T^*\right) + F_2\left(T^*\right), \tag{4.14}$$

where

$$F_2\left(T^*\right) = \frac{a}{T^{*6}} - \frac{b}{T^{*8}}.$$

Parameter a is also dependent on the temperature. Tsonopoulos suggests separate equations, one for each group of different chemical species. For ketones, aldehydes,[7] alkyl nitriles,[8] ethers,[9] and carboxylic-acid esters,[10]

---

[7] Ketones and aldehydes are simple compounds that contain a carbonyl group (a carbon-oxygen double bond). They are considered "simple" because they do not have reactive groups like -OH or -Cl attached directly to the carbon atom in the carbonyl group, as in carboxylic acids containing -COOH. Many ketones are known and many are of great importance in industry and in biology. Examples include many sugars (ketoses) and the industrial solvent acetone, which is the smallest ketone.

[8] A nitrile is any organic compound that has a -C=N functional group. Nitriles are found in many useful compounds, including methyl cyanoacrylate, used in super glue, and nitrile rubber, a nitrile-containing polymer used in latex-free laboratories and medical gloves. Nitrile rubber is also widely used in automotive and other seals since it is resistant to fuels and oils. Organic compounds containing multiple nitrile groups are known as cyanocarbons.

[9] Ethers are a class of organic compounds that contain an ether group—an oxygen atom connected to two alkyl or aryl groups—of general formula R–O–R′. These ethers can again be classified into two varieties: if the alkyl groups are the same on both sides of the oxygen atom, then it is a simple or symmetrical ether, whereas if they are different the ethers are called mixed or unsymmetrical ethers. A typical example of the first group is the solvent and anesthetic diethyl ether, commonly referred to simply as "ether" (CH_3–CH_2–O–CH_2–CH_3). Ethers are common in organic chemistry and pervasive in biochemistry, as they are common linkages in carbohydrates and lignin.

[10] Esters are chemical compounds derived from an acid (organic or inorganic) in which at least one –OH (hydroxyl) group is replaced by an –O–alkyl (alkoxy) group. Usually, esters are derived from carboxylic acid and alcohol. Glycerides, which are fatty acid esters of glycerol, are important esters in biology, being one of the main classes of lipids, and making up the bulk of animal fats and vegetable oils. Esters with low molecular weight are commonly used as fragrances and found in essential oils and pheromones. Phosphoesters form the backbone of DNA molecules. Nitrate esters,

$$a = -2.14 \times 10^{-4}\mu^* - 4.308 \times 10^{-21}\mu^{*^8}.$$

In the above equation

$$\mu^* = 0.9869 \times 10^5 \frac{\mu^2 P_c}{T_c^2}$$

where μ is the molecular dipole moment in Debye, P_c is the critical pressure in bar and T_c is the critical temperature in Kelvin.

For water Tsonopoulos recommends $a = -0.0109$ and $b = 0$.

4.9.2 Third Virial Coefficient

Several corresponding states correlations are available for the third virial coefficients of gases and gas mixtures. Typically these correlations give the reduced third virial coefficient through the generalized correlation,

$$C^* = \frac{C P_c^2}{R^2 T_c^2} = F_0\left(T^*\right) + \omega F_1\left(T^*\right), \tag{4.15}$$

where $F_0\left(T^*\right)$ and $F_1\left(T^*\right)$ are regressed from experimental data.

Orbey and Vera [124] found the following correlations for non-polar gases

$$F_0\left(T^*\right) = 0.01407 + \frac{0.02432}{T^{*2.8}} - \frac{0.00313}{T^{*10.5}},$$
$$F_1\left(T^*\right) = -0.02676 + \frac{0.01770}{T^{*2.8}} + \frac{0.040}{T^{*3}} - \frac{0.003}{T^{*6}} - \frac{0.00228}{T^{*10.5}}.$$

4.9.3 Exercises

1. The Soave-Redlich-Kwong equation

$$P = \frac{RT}{v-b} - \frac{a\left(T\right)}{v\left(v+b\right)},$$

can be rewritten as

$$P = \frac{RT}{v-b} - \frac{\left(b-B\right)RT}{v\left(v+b\right)},$$

---

such as nitroglycerin, are known for their explosive properties, while polyesters are important plastics.

where

$$B = b - \frac{a(T)}{RT}$$

Find the virial coefficient B from the Soave-Redlich-Kwong parameters and compare your results with Tsonopoulos correlation, Eq. (4.13).

4.10 Calculation of State Variables

State variables such as entropy, internal energy, and enthalpy can be calculated for any given state with respect to a reference state given by its temperature T_0 and molar or specific volume v_0.

In Fig. 4.12, consider, for instance, that we want to calculate the entropy $s(T, v)$ with respect to the entropy at the state T_0, v_0, i.e., $s(T, v) - s(T_0, v_0)$. We know that when the entropy is expressed in terms of v and T,

$$ds = c_v dT + \left(\frac{\partial P}{\partial T}\right)_v dv, \tag{4.16}$$

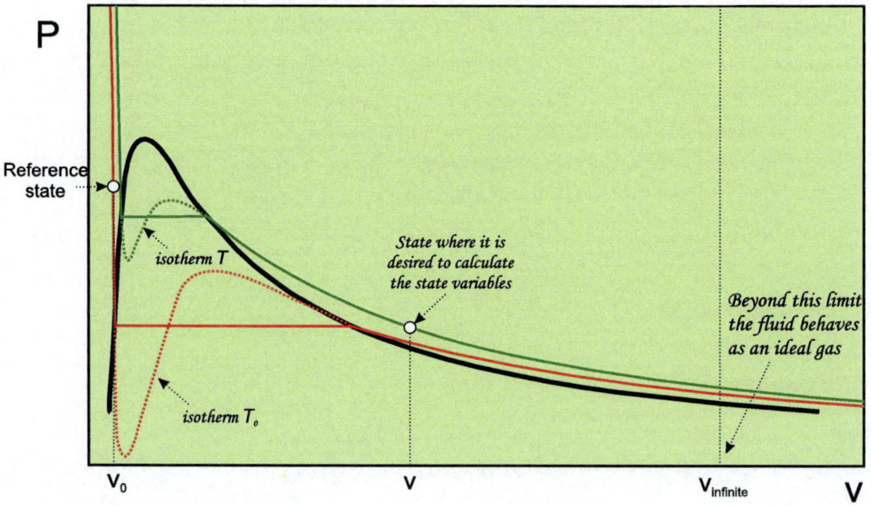

Fig. 4.12 Calculation of state variables

Therefore, all we need for calculating $s(T, v) - s(T_0, v_0)$ is an equation of state and the thermodynamic data for c_v. Nevertheless:

- in general $\left(\frac{\partial P}{\partial T}\right)_v$ is temperature dependent;
- the relationship $c_v(T, v)$ is generally not known in the range of states we desire to calculate the entropy.

Both problems can be solved: i) by calculating the entropy variation separately along the isotherms and following an isochoric line and ii) by extending the integration limit to v_∞ beyond which the vapor behaves as an ideal gas, or, beyond which the compressibility factor Z is very close to 1,

$$|Z - 1| \leq tol,$$

where tol means the allowable error involved in the calculation, below which we consider the results to be satisfactory.

The reason for extending the integration limit to v_∞ is that beyond this limit the molar or specific heat is only dependent on the temperature

$$c_v = c_v^0(T),$$

where the index '0' means that c_v was determined at low pressures $P \to 0$ (*or high volumes* $v \to \infty$). In fact, $c_v^0(T)$ is known for most substances and given in accordance with a polynomial regression in terms of the temperature, e.g.,

$$c_v^0 = A + BT + CT^2 + DT^3.$$

Constants A, B, C and D can be found in many handbooks and textbooks.

Using this reasoning, Eq. (4.16) is first integrated along the isotherm T_0 between v_0 and v_∞ giving

$$s(T_0, v_\infty) - s(T_0, v_0) = \left.\int_{v_0}^{v_\infty} \left(\frac{\partial P}{\partial T}\right)_v dv\right)_{T_0}.$$

Then Eq. (4.16) is integrated along the isovolumetric v_∞ between T_0 and T giving

$$s(T, v_\infty) - s(T_0, v_\infty) = \int_{T_0}^{T} \frac{c_v^0(T)}{T} dT.$$

Finally, Eq. (4.16) is integrated along the isotherm T between v_∞ and v giving

$$s(T, v) - s(T, v_\infty) = \left.\int_{v_\infty}^{v} \left(\frac{\partial P}{\partial T}\right)_v dv\right)_{T}.$$

When the three last equations are added, we get

$$s(T,v) - s(T_0,v_0)$$
$$= \left.\int_{v_0}^{v_\infty}\left(\frac{\partial P}{\partial T}\right)_v dv\right)_{T_0} + \int_{T_0}^{T}\frac{c_v^0(T)}{T}dT + \left.\int_{v\infty}^{v}\left(\frac{\partial P}{\partial T}\right)_v dv\right)_T . \qquad (4.17)$$

4.10.1 Exercises

1. The critical data for ethane is $T_c = 305.33$ K, $P_c = 48.714$ bar, and its acentric factor was measured as $\omega = 0.0944$. The transition pressure for ethane calculated with the Soave-Redlich-Kwong equation gives $P_s = 3.37217$ bar for $T = 210$ K, and $P_s = 28.50510$ bar for $T = 280$ K. The ideal gas molar heat of ethane at constant pressure is given by

$$\bar{c}_p^0 = 6.9 + 17.27 \times 10^{-2}T - 6.406 \times 10^{-5}T^2 + 7.285 \times 10^{-9}T^3.$$

Consider now that you have saturated vapor of ethane at 210 K and that this vapor is compressed reaching a final temperature of 280 K.

(a) Using the Soave-Redlich-Kwong equation of state calculate the molar volume of the saturated vapor of ethane at 210 K.
(b) Consider the compression to be isentropic and, based on Eq. (4.17), find the pressure and molar volume of ethane in its final superheated state.

2. The following experimental data is available for the refrigerant R134a.

$$\bar{c}_p^0 = 19.4 + 0.258531T - 1.29665 \times 10^{-4}T^2$$

$$T_c = 374.23K, \qquad P_c = 40.48bar, \qquad \omega = 0.403$$

Consider that the steady-state expansion of the refrigerant R134a through the expansion valve of a refrigerating cycle is adiabatic an, so, isenthalpic. The refrigerant is in its saturated liquid state at $T = 313$ K at the entrance of the expansion valve and leaves it at a temperature of 253 K.
The transition pressures for R134a that were calculated with the Peng-Robinson equation of state gave $P_s = 1.08323$ bar for $T = 253$ K, and $P_s = 9.44018$ bar for $T = 313$ K

(a) Find the molar volume of R134a for the liquid and vapor phases in the liquid-vapor transition region at 313 K and 253 K

(b) By supposing that the R134a is well described by the Peng-Robinson equation of state and using the data given above, find the vapor content of R134a at the outlet of the expansion valve. Base your calculations of the enthalpy on the same method of Sect. 4.10 (you may consider the saturated liquid at 253K as the reference state).

Chapter 5
Non-ideal Mixtures

Abstract Consider a multicomponent system in the gaseous phase. For low pressures, the gaseous mixture can be considered as ideal with a Clapeyron equation of state $PV = N\mathscr{R}T$. In this state, the mean free path is very high, and the intermolecular electrostatic potentials do not intercept, each component in the mixture behaving as an ideal gas. When the pressure increases or the volume decreases the molecules become closer and intermolecular attraction forces begin to be appreciable, deviating the whole system from the ideal gas behavior. In these conditions, the several components in the mixture may stand as a homogeneous system or segregate into separate phases. In this chapter, we deal with homogeneous mixtures, when the attraction forces between different molecules predominate over the attraction forces between identical molecules and there is no segregation. After exposing the theoretical fundamentals of multicomponent systems, the concepts of fugacity and activity are introduced and used for calculating the bubble and dew points of homogeneous liquid solutions in equilibrium with vapor. The functional group semi-empirical method of Prausnitz and collaborators is also exposed and used in the estimation of the activity coefficients.

5.1 Partial Volumes and the Amagat's Law

A mixture in equilibrium with r-components each one of them labeled as $i = 1, \ldots, r$ can be described by $r + 2$ independent variables, which can be chosen as $T, P, N_1, \ldots, N_r$.

So, the volume can be written as a function of these $r + 2$ variables

$$V = V(T, P, N_1, \ldots, N_r).$$

In an equilibrium state, the homogeneity condition means

$$V(T, P, \lambda N_1, \ldots, \lambda N_r) = \lambda V,$$

P. C. Philippi, *Thermodynamics*, https://doi.org/10.1007/978-3-031-49357-7_5

for any arbitrary partition parameter λ. So, if we consider this partition parameter as the only variable in a homogeneous mixture,

$$\frac{\partial V}{\partial \lambda} = \sum_i \frac{\partial V}{\partial (\lambda N_i)} \frac{\partial (\lambda N_i)}{\partial \lambda} = V.$$

Nevertheless,

$$\frac{\partial (\lambda N_i)}{\partial \lambda} = N_i,$$

and, so, for $\lambda = 1$

$$V = \sum N_i \left(\frac{\partial V}{\partial N_i} \right)_{T,P,N_{j \neq i}}. \tag{5.1}$$

The term $\left(\frac{\partial V}{\partial N_i} \right)_{T,P,N_{j \neq i}}$ is defined as the partial molar volume of component i,

$$\bar{v}_i = \left(\frac{\partial V}{\partial N_i} \right)_{T,P,N_{j \neq i}}, \tag{5.2}$$

and Eq. (5.1) is, here, called Amagat's law, meaning that the total volume of a mixture is the sum of the partial volumes $V_i = N_i \bar{v}_i$ of each component in the mixture.[1]

By partial volume V_i we mean the volume that the N_i moles of component i would occupy in the mixture at the same temperature T and pressure P of the mixture. Amagat's law, as defined by Eq. (5.1) is a general law for homogeneous mixtures and can be used both for ideal and non-ideal systems. In general the partial molar volume $\bar{v}_i$ of each component i is dependent on the amounts $N_1, \ldots, N_r$ of the components that are present in the mixture as well as on the temperature and pressure

$$\bar{v}_i = \bar{v}_i \left(T, P, N_1, \ldots, N_r \right),$$

and this means that a variation in the number of moles of any component j in the mixture will affect the molar volume v_i of each component. This is the case for non-ideal mixtures because v_i is affected by the cross interactions between molecules i and j.

In fact, since ideal gas mixtures are ruled by the Clapeyron equation

$$V = \frac{\left(\sum_i N_i \right) \mathscr{R} T}{P}, \tag{5.3}$$

[1] In accordance with Ref. [138], the Amagat's law was derived based on the partial molar volume of each component in the pure state $\bar{v}_{i,pure}(T, P)$ and not on the molar partial volume $\bar{v}_i$ as defined by Eq. (5.2). In fact, $\bar{v}_i$ is modified by the addition or removal of components and will be only equal to $\bar{v}_{i,pure}$ in a mixture of ideal gases.

the partial molar volume

$$\bar{v}_i = \frac{\mathcal{R}T}{P}$$

is only dependent on the temperature and pressure. So, the partial volume $V_i = N_i \frac{\mathcal{R}T}{P}$ does not depend on the molar amounts of the other components that are present in the mixture and can be interpreted as the volume that component i would occupy if it were *alone* in the mixture at the same temperature and pressure of the mixture.

In the same way, if we consider that each component occupies the whole volume of the mixture, Eq. (5.3) can be rewritten as

$$P = \sum_i P_i, \tag{5.4}$$

where $P_i = N_i \mathcal{R}T/V$ is the partial pressure of component i in the mixture. Equation (5.4) is Dalton's law. In contrast with Amagat's law, Dalton's law only applies to ideal gas mixtures. Figure 5.1 shows a sketch that tries to clarify the meanings of the Amagat and Dalton's laws.

Consider now, as another example, a non-ideal mixture whose equation of state is a virial equation

$$Z = \frac{P\bar{v}}{\mathcal{R}T} = 1 + B\frac{P}{\mathcal{R}T}, \tag{5.5}$$

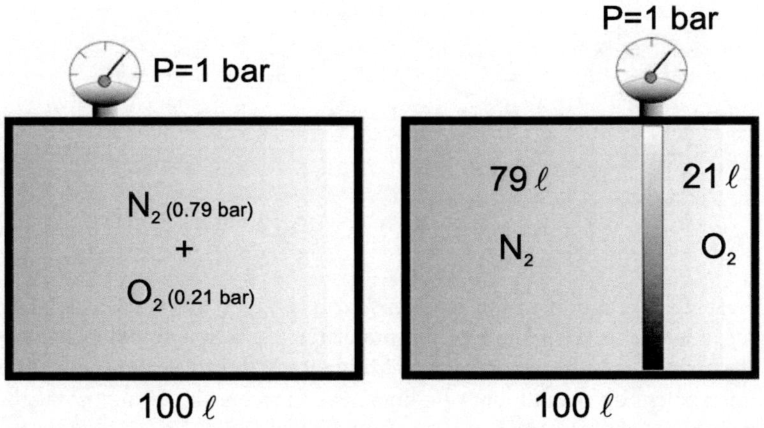

Fig. 5.1 If oxygen in the atmospheric air is separated from nitrogen it will occupy a volume of 21 L for every 100 L of air at atmospheric pressure

where the second virial coefficient is related to the intermolecular attractive forces by

$$B = \sum_{i,j} y_i y_j B_{ij}, \tag{5.6}$$

with

$$B_{ij} = \delta_{ij}\sqrt{B_i B_j}. \tag{5.7}$$

Equation (5.5) can be used for mixtures submitted to low and moderate pressures [138] and parameter $\delta_{ij} = 1$ for $i = j$ but different from zero when $i \neq j$. In fact, parameter δ_{ij} for $i \neq j$, is related to the cross intermolecular interactions between molecules i and j. Experimental values for B_{ij} and models for their estimation may be found in [57, 137, 138].

Now, we rewrite Eq. (5.5) as

$$V = \sum_i N_i \frac{\mathcal{R}T}{P} + \frac{1}{\sum_i N_i} \sum_{j,k} N_j N_k B_{jk}. \tag{5.8}$$

For simplicity, consider a binary mixture with only two components

$$V = (N_1 + N_2)\frac{\mathcal{R}T}{P} + \frac{N_1^2 B_1 + 2N_1 N_2 B_{12} + N_2^2 B_2}{N_1 + N_2}. \tag{5.9}$$

The molar partial volume of component 1 is calculated in accordance with Eq. (5.2) as

$$v_1 = \left(\frac{\partial V}{\partial N_1}\right)_{T,P,N_2} = \frac{\mathcal{R}T}{P} + B_1 + y_2^2\left[2B_{12} - (B_1 + B_2)\right]. \tag{5.10}$$

In the same way

$$v_2 = \left(\frac{\partial V}{\partial N_2}\right)_{T,P,N_1} = \frac{\mathcal{R}T}{P} + B_2 + y_1^2\left[2B_{12} - (B_1 + B_2)\right]. \tag{5.11}$$

Therefore, in addition to the ideal-gas term $\mathcal{R}T/P$, the partial molar volume v_i is dependent on the strength of the molecular attraction forces between the i-molecules which is represented by an additive term B_i in the above equations, but is also dependent on the amount of component j present in the mixture and on the magnitude of the cross-interactions $2B_{12}$ with respect to the self-interactions between identical molecules $B_1 + B_2$.

For diluted mixtures, when $y_2 \to 0$ and $y_1 \to 1$, Eqs. (5.10)–(5.11) become

$$\bar{v}_1 \to \frac{\mathcal{R}T}{P} + B_1,$$

$$\bar{v}_2 \to \frac{\mathcal{R}T}{P} + 2B_{12} - B_1.$$

So, while v_1 is unaffected by the presence of molecules 2 in the mixture, the non-ideal part of the molar partial volume v_2 is dependent on the cross interaction term B_{12} and on the intermolecular forces term B_1 between molecules 1, but not on the self-interaction term B_2. This was to be expected since molecules 2 are diluted and do not interact between themselves.

Another important limiting case of Eqs. (5.10)–(5.11) is represented by systems where the intermolecular forces are similar $B_1 \sim B_2 \sim B_{12}$. In this case

$$\bar{v}_1 \approx \frac{\mathcal{R}T}{P} + B_1,$$

$$\bar{v}_2 \approx \frac{\mathcal{R}T}{P} + B_2,$$

meaning that the molar volume of each component is independent of the presence of the remaining components in the mixture. Mixtures with such properties are called *ideal solutions*, which cannot be confused with mixtures of ideal gases whose molecules do not interact with attractive forces.

5.1.1 Exercise

The equations of state for mixtures are commonly written in a form analogous to the equations of state for pure components. Consider, for instance, the van der Waals equation of state,

$$P = \frac{\mathcal{R}T}{\bar{v} - b} - \frac{a}{\bar{v}^2}$$

For a mixture, parameter b is the mean molecular volume,

$$b = \sum_i y_i b_i$$

and the force parameter a is given by

$$a = \sum_{i,j} y_i y_j a_{ij}$$

where, similarly to Eq. (5.7) parameter a_{ij} takes account of the cross interactions by an interference factor $\delta_{ij} \neq 0$ for $i \neq j$.

$$a_{ij} = \delta_{ij}\sqrt{a_i a_j}$$

Find the molar partial volumes for a binary mixture with a van der Waals equation of state and consider the possible limiting cases, trying to find a physical meaning for the resulting expressions.

5.2 Gibbs–Duhem Equation for Mixtures

For an open multicomponent system, the internal energy can be modified by the addition or removal of each one of the components in the mixture and Eq. (3.5) can be written as

$$dU = TdS - PdV + \sum_i \bar{\mu}_i dN_i.$$

In Sect. 3.4 we found that for a thermodynamic system in equilibrium

$$U = TS - PV + \bar{\mu}N.$$

For a multicomponent system in equilibrium the above homogeneity relationship is written as

$$U = TS - PV + \sum_i \bar{\mu}_i N_i,$$

meaning that for every partition parameter λ, the intensive variables T, P, $\bar{\mu}_i$ remain the same in every partitioned part of the thermodynamic system. Therefore, by differentiating the above equation or by supposing that the multicomponent system changes from an equilibrium state to another

$$dU = SdT - VdP + \sum_i N_i d\bar{\mu}_i + \underbrace{TdS - PdV + \sum_i \mu_i dN_i}_{=dU}.$$

So, we get

$$\sum_i N_i d\bar{\mu}_i = -SdT + VdP \tag{5.12}$$

Nevertheless, as we have just seen in Sect. 5.3, volume V may be written in terms of the molar partial volumes

$$V = \sum_i N_i \bar{v}_i. \tag{5.13}$$

In the same way, considering the entropy S as a function of T, P, $N_1, \ldots, N_r$ and using the homogeneity condition for equilibrium states

$$S = \sum_i N_i \bar{s}_i, \tag{5.14}$$

where

$$\bar{s}_i = \left(\frac{\partial S}{\partial N_i}\right)_{T,P,N_{j \neq i}}, \tag{5.15}$$

is the molar partial entropy of component i.

By inserting Eqs. (5.13) and (5.14) into Eq. (5.12), we find

$$d\bar{\mu}_i = -\bar{s}_i dT + \bar{v}_i dP. \tag{5.16}$$

Equation (5.16) is the Gibbs–Duhem equation for multicomponent systems in equilibrium. On a mass basis, it can also be written as

$$d\mu_i = -s_i dT + v_i dP.$$

5.3 Ideal Gases

In this section, we investigate the thermodynamic properties of mixtures whose components behave as ideal gases. We start by considering a theoretical process where a mixture of ideal gases is formed by adding components to component i without changing the temperature and total pressure (Fig. 5.2).

The effect of this addition is to reduce the pressure of the component i from P to P_i and so, in accordance with Eq. (5.16) its chemical potential will change by

$$d\bar{\mu}_i = \bar{v}_i dP. \tag{5.17}$$

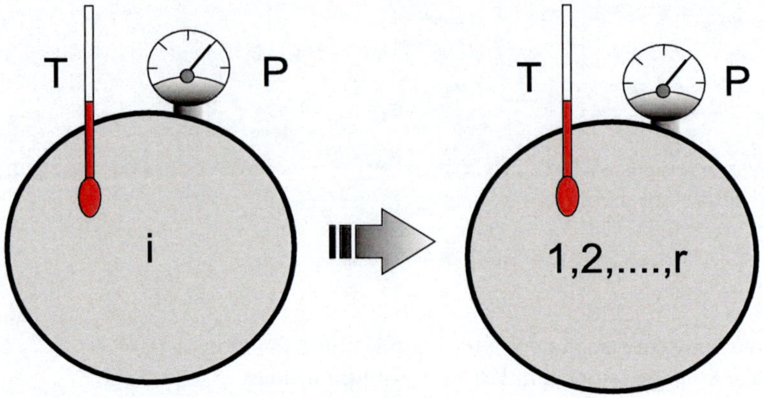

Fig. 5.2 A theoretical process where a mixture of ideal gases is formed by adding $r-1$ components to component i without changing the temperature and total pressure

The molar volume $\bar{v}_i$ is given by

$$\bar{v}_i = \frac{\mathscr{R}T}{P},$$

and so by integrating Eq. (5.17)

$$\bar{\mu}_i\left(T, P, N_1, \ldots, N_r\right) = \bar{\mu}_{i,pure}\left(T, P\right) + \mathscr{R}T \ln y_i, \tag{5.18}$$

where $y_i = P_i/P$ is the molar fraction of component i in the mixture.

Since $y_i \leq 1$, the above equation means that the chemical potential of component i in the mixture will be reduced with respect to its value when in a pure state at the *same* temperature and pressure. Exercise 1 of Sect. 5.3.1 is recommended for people who want to get a deeper understanding of the meaning of the chemical potential.

Now, from the Gibbs–Duhem equation for mixtures, Eq. (5.16), the molar partial entropy of component i can be found by deriving Eq. (5.18) with respect to the temperature,

$$\bar{s}_i = -\left(\frac{\partial \bar{\mu}_i}{\partial T}\right)_{P,N_1,\ldots,N_r} = -\frac{\partial}{\partial T}\left[\bar{\mu}_{i,pure}\left(T, P\right)\right]_P - \mathscr{R} \ln y_i.$$

Nevertheless,

$$-\frac{\partial}{\partial T}\left[\bar{\mu}_{i,pure}\left(T, P\right)\right]_P = \bar{s}_{i,pure}\left(T, P\right),$$

and so

$$\bar{s}_i = \left(\frac{\partial \bar{\mu}_i}{\partial T}\right)_{P,N_1,\ldots,N_r} = \bar{s}_{i,pure}(T, P) - \mathscr{R} \ln y_i .$$

Therefore, in accordance with Eq. (5.14),

$$S = \sum_i S_{i,pure}(T, P) - \mathscr{R}N \sum_i y_i \ln y_i .$$

Therefore, when r components are mixed at a constant temperature and pressure, the entropy of the resulting system is not an additive property and differs from the sum of the entropy of the pure components by a positive quantity, which is called *entropy of mixing*

$$S_{mixing} = -\mathscr{R}N \sum_i y_i \ln y_i .$$

The molar partial volume can be also retrieved from Eq. (5.16) as

$$\bar{v}_i = \left(\frac{\partial \bar{\mu}_i}{\partial P}\right)_{T,N_1,\ldots,N_r} = \frac{\partial}{\partial P}\left[\bar{\mu}_{i,pure}(T, P)\right]_T = \bar{v}_{i,pure}(T, P) .$$

Since $\bar{v}_{i,pure}(T, P) = \mathscr{R}T/P$ in an ideal gas mixture, the total volume of the multicomponent system will be

$$V = \sum_i N_i \frac{\mathscr{R}T}{P} = N\frac{\mathscr{R}T}{P},$$

as expected.

Consider now the internal energy

$$U = \sum_i N_i \bar{u}_i ,$$

or

$$\bar{u} = \sum_i y_i \bar{u}_i ,$$

where

$$\bar{u}_i = \left(\frac{\partial U}{\partial N_i}\right)_{T,P,N_{j\neq i}} ,$$

is the molar partial energy of component i.

From the principle of energy conservation, internal energy is, necessarily, an additive property, meaning that the internal energy of a multicomponent system must be equal to the sum of the internal energies of the individual components, when these components are mixed at constant temperature and pressure

$$\overline{u} = \sum_i y_i \overline{u}_{i,pure},$$

and since $\overline{u}_{i,pure} = \overline{u}_{i,pure}(T)$ for ideal gases,

$$\overline{u}(T, y_1, \ldots, y_r) = \sum_i y_i \overline{u}_{i,pure}(T).$$

Another thermodynamic property of interest is the Helmholtz energy

$$\overline{f} = \overline{u} - T\overline{s}.$$

For a multicomponent system

$$\overline{f} = \sum_i y_i \left(\overline{u}_{i,pure} - T\overline{s}_{i,pure}\right) + \mathscr{R}T \sum y_i \ln y_i,$$

or

$$\overline{f}(T, P, y_1, \ldots, y_r) = \sum_i y_i \overline{f}_{i,pure}(T, P) + \mathscr{R}T \sum y_i \ln y_i.$$

Finally, the Gibbs energy

$$\overline{g}(T, P, N_1, \ldots, N_r) = \sum_i y_i \overline{g}_i.$$

Since $\overline{g}_i = \overline{\mu}_i$, we can use Eq. (5.18) to find

$$\overline{g}(T, P, y_1, \ldots, y_r) = \sum_i y_i \overline{\mu}_{i,pure}(T, P) + \mathscr{R}T \sum_i y_i \ln y_i.$$

5.3.1 Exercises

1. Two spherical chambers each one with a volume of 1 L are separated by a semipermeable membrane (Fig. 5.3). Chamber A contains, initially, 21% of Oxygen and 79% of Nitrogen, and chamber B is filled with Oxygen. The mem-

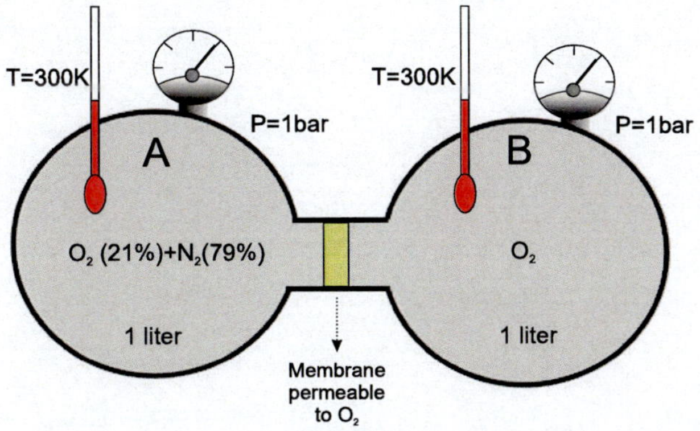

Fig. 5.3 Oxygen will flow from chamber B to chamber A because its chemical potential in chamber B is greater than its chemical potential in chamber A

brane is permeable to O_2. The temperature is the same for the two chambers and equal to 300 K and at time $t = 0$, the pressure is 1 bar in both chambers. Since the chemical potential of the Oxygen in chamber B is greater when compared to the Oxygen in chamber A, there will be a transfer of Oxygen from chamber B to chamber A. Find the pressures P_A and P_B and the molar fraction of Oxygen in chamber A after equilibrium is reached.
Hint:

Use the Gibbs–Duhem equation for pure components

$$d\bar{\mu} = -\bar{s}dT + \bar{v}dP,$$

to find

$$\bar{\mu}_{pure}(T, P_A) - \bar{\mu}_{pure}(T, P_B),$$

after the equilibrium state is reached.

2. In Exercise 1 calculate the entropy variation between the initial and the equilibrium states, (a) in chamber A; (b) in chamber B; (c) in the whole system A + B.
3. The enthalpy of a multicomponent system is given by

$$H = \sum_i N_i \bar{h}_i,$$

where

$$\overline{h}_i = \left(\frac{\partial H}{\partial N_i}\right)_{T,P,N_{j\neq i}} = \left(\frac{\partial U}{\partial N_i}\right)_{T,P,N_{j\neq i}} + P\left(\frac{\partial V}{\partial N_i}\right)_{T,P,N_{j\neq i}}$$

and so

$$\overline{h}_i = \overline{u}_i + P\overline{v}_i$$

Find an expression for the enthalpy of a mixture of ideal gases and show that it is also an additive property.

5.4 The Lewis Concept of Fugacity

The concept of fugacity was introduced by Gilbert Newton Lewis in 1901 [105] as a '*escaping tendency*' of a given molecular species when it migrates from one phase to another till equilibrium is reached as regards the distribution of species.[2]

For introducing the concept of fugacity, consider that a mixture of r-components is subjected to a change of pressure in accordance with the sketch given in Fig. 5.4.

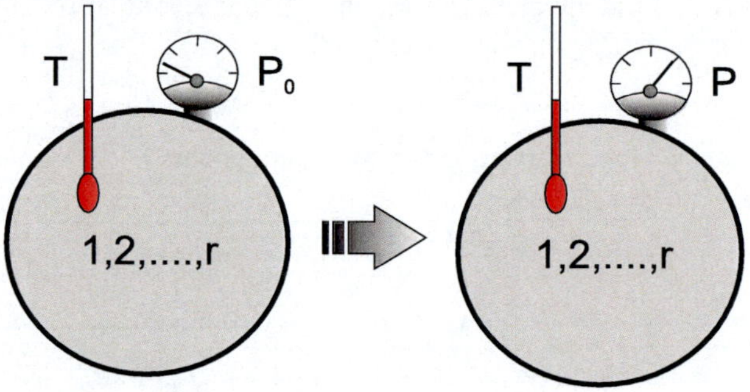

Fig. 5.4 The concept of fugacity

[2] In accordance with Lewis [105] "*If any phase containing a given molecular species is brought into contact with any other phase not containing that species, a certain quantity will pass from the first phase to the second. Every molecular species may be considered, therefore, to have a tendency to escape from the phase in which it is. In order to express this tendency quantitatively for any particular state, an infinite number of quantities could be used, such, for example, as the thermodynamic potential of the species, its vapor pressure, its solubility in water, etc. The quantity which we shall choose is one which seems at first sight more abstruse than any of these but is in fact simpler, more general, and easier to manipulate. It will be called the fugacity, represented by the symbol* ψ *and defined by the following conditions: (1) The fugacity of a molecular species is the same in two phases when these phases are in equilibrium as regards the distribution of that species. (2) The fugacity of gas approaches the gas pressure as a limiting value if the gas is indefinitely rarefied. In other words, the escaping tendency of a perfect gas is equal to its gas pressure*".

Since the temperature remains the same, the change in the chemical potential of the i-component will be given by

$$\overline{\mu}_i\,(T, P, y_1, \ldots, y_r) - \overline{\mu}_i\,(T, P_0, y_1, \ldots, y_r) = \int_{P_0}^{P} \overline{v}_i dP.$$

If the system is a mixture of ideal gases

$$\overline{v}_i = \frac{\mathscr{R}T}{P},$$

and

$$\overline{\mu}_i\,(T, P, y_1, \ldots, y_r) - \overline{\mu}_i\,(T, P_0, y_1, \ldots, y_r) = \mathscr{R}T \ln \frac{P}{P_0}.$$

The fugacity concept of Lewis is now introduced by replacing P and P_0 by ψ_i and $\psi_{i,0}$, respectively, in such a way that, for a non-ideal mixture

$$\overline{\mu}_i\,(T, P, y_1, \ldots, y_r) - \overline{\mu}_i\,(T, P_0, y_1, \ldots, y_r) = \mathscr{R}T \ln \frac{\psi_i}{\psi_{i,0}}.$$

As pointed out by Lewis, the fugacity of a gas *must approach the gas pressure as a limiting value, if the gas is indefinitely rarefied*. In the case of mixtures,

$$\psi_i \rightarrow y_i P \text{ when } P \rightarrow 0.$$

Now, when the temperature is constant

$$d\overline{\mu}_i = \overline{v}_i dP.$$

So by integrating the above equation between P_0 and P

$$\ln \frac{\psi_i}{\psi_{i,0}} = \frac{1}{\mathscr{R}T} \int_{P_0}^{P} \overline{v}_i dP. \tag{5.19}$$

Therefore, fugacity replaces the concept of chemical potential for measuring the *escaping tendency* of a given molecular species, in accordance with

$$\ln \frac{\psi_i}{\psi_{i,0}} = \overline{\mu}_i\,(T, P, y_1, \ldots, y_r) - \overline{\mu}_i\,(T, P_0, y_1, \ldots, y_r)\,.$$

Now, consider rewriting $\frac{\psi_i}{\psi_{i,0}}$ in Eq. (5.19) as

$$\frac{\psi_i}{\psi_{i,0}} = \frac{\psi_i}{y_i P} \frac{P}{P_0} \frac{y_i P_0}{\psi_{i,0}},$$

in such a manner that Eq. (5.19) becomes

$$\ln \frac{\psi_i}{y_i P} + \ln \frac{y_i P_0}{\psi_{i,0}} = \int_{P_0}^{P} \left(\frac{\overline{v}_i}{\mathscr{R}T} - \frac{1}{P} \right) dP.$$

When $P_0 \to 0$

$$\psi_{i,0} \to y_i P_0,$$

and so

$$\ln \phi_i = \int_0^P \left(\frac{\overline{v}_i}{\mathscr{R}T} - \frac{1}{P} \right) dP, \tag{5.20}$$

where

$$\phi_i = \frac{\psi_i}{y_i P}, \tag{5.21}$$

is called the fugacity coefficient and is a measure of the non-ideality of component i in the multicomponent system (for mixtures of ideal gases the fugacity coefficient $\phi_i = 1$).

5.4.1 *Calculation of the Fugacity Coefficient for a Virial Equation of State*

Equation (5.5) can be used for estimating the compressibility Z of a multicomponent system under low and moderate pressures,

$$Z = \frac{P\overline{v}}{\mathscr{R}T} = 1 + B \frac{P}{\mathscr{R}T}, \tag{5.22}$$

where

$$B = \sum_{i,j} y_i y_j B_{ij}.$$

In this case, the molar partial volume of component i is given by

$$\bar{v}_i = \frac{\mathscr{R}T}{P} + 2\sum_j y_j B_{ij} - B,$$

and, from Eq. (5.20)

$$\ln \phi_i = \frac{1}{\mathscr{R}T} \int_0^P \left(2\sum_j y_j B_{ij} - B \right) dP = \frac{P}{\mathscr{R}T} \left(2\sum_j y_j B_{ij} - B \right).$$

As an example, consider a binary mixture. In this case

$$\ln \phi_1 = \frac{P}{\mathscr{R}T} \left(B_1 + y_2^2 \left[2B_{12} - (B_1 + B_2) \right] \right), \tag{5.23}$$

and

$$\ln \phi_2 = \frac{P}{\mathscr{R}T} \left(B_2 + y_1^2 \left[2B_{12} - (B_1 + B_2) \right] \right).$$

But, in accordance with Eqs. (5.10)–(5.11) the terms between the parenthesis in the above equations are related to the molar partial volumes of components 1 and 2 by

$$\bar{v}_1 - \frac{\mathscr{R}T}{P} = B_1 + y_2^2 \left[2B_{12} - (B_1 + B_2) \right],$$

$$\bar{v}_2 - \frac{\mathscr{R}T}{P} = B_2 + y_1^2 \left[2B_{12} - (B_1 + B_2) \right].$$

Therefore, for $i = 1, 2$,

$$\ln \phi_i = \frac{P}{\mathscr{R}T} \left(\bar{v}_i - \frac{\mathscr{R}T}{P} \right) = \frac{P\bar{v}_i}{\mathscr{R}T} - 1,$$

and this is exactly what is wanted with the concept of the fugacity coefficient: *a measure of the deviation from the ideal gas behavior for the components in a non-ideal multicomponent system.* When the molar volume $\bar{v}_i$ is the one an ideal gas

$$\phi_i = 1.$$

Another interesting particular case is one of the ideal solutions when $2B_{12} \sim (B_1 + B_2)$. In this case

$$\ln \phi_i = \frac{PB_i}{\mathcal{R}T}.$$

5.4.2 Calculation of the Fugacity Coefficient for a Cubic Equation of State

For most equations of state, the pressure P is the dependent variable and its dependence with respect to the independent variables is given as

$$P = P\left(T, V, N_1, \ldots, N_r\right),$$

and the calculation of the molar partial volume of each species i requires to invert the derivative in accordance with

$$\bar{v}_i = \left(\frac{\partial V}{\partial N_i}\right)_{T,P,N_{j\neq i}} = -\frac{\left(\frac{\partial P}{\partial N_i}\right)_{T,V,N_{j\neq i}}}{\left(\frac{\partial P}{\partial V}\right)_{T,N_1,\ldots,N_r}}.$$

Therefore, Eq. (5.20) becomes

$$\ln \phi_i = \int_{\infty}^{V} \left[-\frac{1}{\mathcal{R}T} \frac{\left(\frac{\partial P}{\partial N_i}\right)_{T,V,N_{j\neq i}}}{\left(\frac{\partial P}{\partial V}\right)_{T,N_1,\ldots,N_r}} - \frac{1}{P} \right] \left(\frac{\partial P}{\partial V}\right)_{T,N_1,\ldots,N_r} dV,$$

or

$$\ln \phi_i = \int_{V}^{\infty} \left[\frac{1}{\mathcal{R}T} \left(\frac{\partial P}{\partial N_i}\right)_{T,V,N_{j\neq i}} + \frac{1}{P} \left(\frac{\partial P}{\partial V}\right)_{T,N_1,\ldots,N_r} \right] dV. \tag{5.24}$$

Introducing the compressibility factor

$$P = \frac{ZN\mathcal{R}T}{V},$$

we will have

$$\frac{1}{P}\left(\frac{\partial P}{\partial V}\right)_{T,N_1,\ldots,N_r} = -\frac{1}{V} + \frac{1}{Z}\left(\frac{\partial Z}{\partial V}\right)_{T,N_1,\ldots,N_r}$$

So, Eq. (5.24) becomes

$$\ln \phi_i = \int_V^\infty \left[\frac{1}{\mathscr{R}T} \left(\frac{\partial P}{\partial N_i} \right)_{T,V,N_{j \neq i}} - \frac{1}{V} \right] dV + \int_V^\infty \left(\frac{\partial \ln Z}{\partial V} \right)_{T,N_1,\ldots,N_r} dV,$$

or

$$\ln \phi_i = \int_V^\infty \left[\frac{1}{\mathscr{R}T} \left(\frac{\partial P}{\partial N_i} \right)_{T,V,N_{j \neq i}} - \frac{1}{V} \right] dV - \ln Z, \tag{5.25}$$

because when $V \to \infty$, $Z \to 1$.

As an example consider the van der Waals equation with the form,

$$P = \frac{\mathscr{R}T \sum_i N_i}{V - \sum_i N_i b_i} - \frac{\sum_{i,j} N_i N_j}{V^2}. \tag{5.26}$$

The fugacity coefficient in Eq. (5.25) can be calculated for this equation of state as (see Exercise 1 of Sect. 5.4.3),

$$\ln \phi_i = \ln \frac{\bar{v}}{\bar{v} - b} + \frac{b_i}{\bar{v} - b} - \frac{\sum_j y_j a_{ij}}{\bar{v}\mathscr{R}T} - \ln Z. \tag{5.27}$$

Considering the definition of the fugacity coefficient

$$\phi_i = \frac{\psi_i}{y_i P} = \frac{\psi_i \bar{v}}{y_i Z \mathscr{R}T},$$

we have

$$\ln \phi_i = \ln \psi_i - \ln \frac{y_i \mathscr{R}T}{\bar{v}} - \ln Z,$$

and so, Eq. (5.27) may be rewritten in terms of the fugacity ψ_i,

$$\ln \psi_i = \ln \frac{y_i \mathscr{R}T}{\bar{v} - b} + \frac{b_i}{\bar{v} - b} - \frac{\sum_j y_j a_{ij}}{\bar{v}\mathscr{R}T}. \tag{5.28}$$

5.4.3 Exercises

1. Starting from the van der Waals equation of state, Eq. (5.26), derive Eq. (5.27) for the fugacity coefficient ϕ_i.

2. Show that the van der Waals equation of state can be written for moderate pressures as

$$Z = \frac{P\bar{v}}{\mathscr{R}T} = 1 + \left(b - \frac{a}{\mathscr{R}T}\right)\frac{P}{\mathscr{R}T} + O\left(P^2\right)$$

and relate the fugacity coefficient calculated with the above virial equation with the one calculated with Eq. (5.27) for a multicomponent system.

5.5 Ideal Solutions

For a mixture of ideal gases the chemical potential of each component is given by Eq. (5.18)

$$\bar{\mu}_i\left(T, P, N_1, \ldots, N_r\right) = \bar{\mu}_{i,pure}\left(T, P\right) + \mathscr{R}T \ln y_i, \tag{5.29}$$

meaning that, in addition to being dependent on the temperature and pressure, $\bar{\mu}_i$ is only dependent on the molar fraction of the i-component (*and not on the molar fraction of the remaining components*).

An ideal solution is defined as a solution whose chemical potential for each component i is given by

$$\bar{\mu}_i\left(T, P, N_1, \ldots, N_r\right) = \bar{\mu}_{i,pure}\left(T, P\right) + \mathscr{R}T \ln x_i, \tag{5.30}$$

where x_i is the molar fraction of component i in the solution.

Therefore, when components j are totally or partially added or removed from the solution without affecting the temperature, pressure and the total number of moles, the chemical potential $\bar{\mu}_i$ of component i is not affected. This, absolutely, does not mean that the molecules *do not interact* or that the system is ruled by a *Clapeyron equation of state*. In fact, each i-component in the solution is *non-ideal* and they only form an ideal solution due to the particular nature of the electrostatic potential of their molecules.

Ideal solutions are seldom found in nature but from a theoretical point of view they are important as giving a reference framework, enabling us to estimate the deviations from the ideality of real solutions.

From the definition of fugacity,

$$\bar{\mu}_i\left(T, P, x_i\right) - \bar{\mu}_{i,pure}\left(T, P\right) = \mathscr{R}T \ln \frac{\psi_i\left(T, P, x_i\right)}{\psi_{i,pure}\left(T, P\right)},$$

and so

$$\psi_i\left(T, P, x_i\right) = x_i \psi_{i,pure}\left(T, P\right)$$

A common strategy for calculating the fugacity of pure component i is to write it in terms of the fugacity of this component in its state of saturated liquid, based on Eq. (5.19)

$$\ln \frac{\psi_{i,pure}(T, P)}{\psi_{i,pure}\left(T, P_{s,i}\right)} = \frac{1}{\mathscr{R}T} \int_{P_{s,i}}^{P} \overline{v}_i dP, \tag{5.31}$$

where $P_{s,i}(T)$ is the saturation pressure of component i for the temperature T of the system.

Since the compressibility factor of liquids is very small, the molar volume $\overline{v}_i$ can be approximated as a constant and equal to the molar volume of the i-component saturated liquid, $\overline{v}_{\ell s,i}(T)$, in the range between $P_{s,i}$ and P.

Therefore, Eq. (5.31) can be written as

$$\psi_{i,pure}(T, P) = \psi_{i,pure}\left(T, P_{s,i}\right) e^{\frac{\overline{v}_{\ell s,i}}{RT}\left(P-P_{s,i}\right)}. \tag{5.32}$$

Equation (5.32) is often called Poynting correction.

5.5.1 Raoult's Law

J'ai entrepris de rechercher comment varie la tension de vapeur des liquides volatils, quand on y dissout des substances fixes, de différentes natures. Les quantités qu'il faut déterminer dans ce genre de recherches sont la tension de vapeur d'un dissolvant volatil pur et la tension de 1a vapeur du même dissolvant, tenant en dissolution un poids connu de substance fixe, la température restant la même. Je les ai mesurées par la méthode statique ou par la méthode dynamique suivant les cas. F. M. Raoult (1889)[3]

Consider the equilibrium distribution of a component in a multicomponent system between a liquid and a vapor phase (Fig. 5.5a). We seek a simple relation describing the distribution of the components between the phases, i.e., an equation relating x, the mole fraction in the liquid phase, to y, the mole fraction in the vapor phase.

For this purpose, the liquid phase will be considered as an ideal solution and the vapor phase will be considered a mixture of ideal gases.

For each i-component, the equilibrium condition is the equality of the fugacities in the liquid L and vapor V phases,

$$\psi_i^L = \psi_i^V. \tag{5.33}$$

[3] F. M. Raoult. *Recherches expérimentales sur les tensions de vapeur des dissolutions*. J. Phys. Theor. Appl., 1889, 8 (1), pp. 5–20 [142].

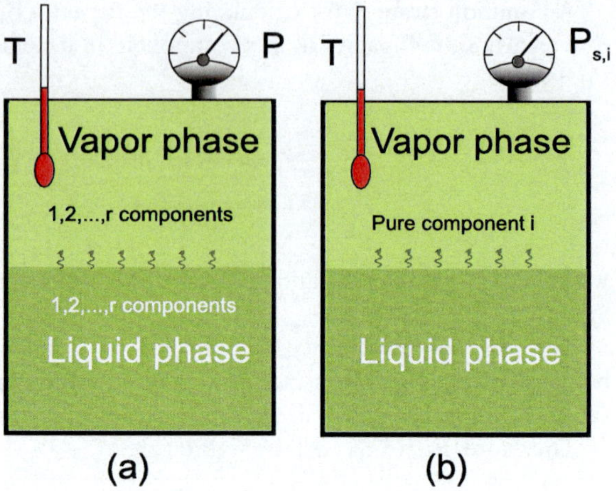

Fig. 5.5 Raoult's law: **a** the liquid phase is considered as an ideal solution and, the gas phase, a mixture of ideal gases; **b** in its saturated condition the fugacity of a liquid is its saturation pressure when the vapor is considered as an ideal gas

Since the vapor phase is considered as a mixture of ideal gases

$$\psi_i^V = y_i P,$$

and if we neglect the Poynting correction in Eq. (5.32), supposing $P \sim P_{s,i}$ for all the r components in the liquid solution,

$$\psi_i^L = x_i \psi_{i,pure}^L \left(T, P_{s,i}\right). \tag{5.34}$$

For finding $\psi_{i,pure}^L \left(T, P_{s,i}\right)$ consider Fig. 5.5b, showing pure component i in its saturated condition. In equilibrium, the fugacities of the liquid and vapor phase are identical

$$\psi_{i,pure}^L = \psi_{i,pure}^V,$$

and since the vapor of component i was considered as an ideal gas, $\psi_{i,pure}^V = P_{s,i}$ and

$$\psi_{i,pure}^L = P_{s,i}. \tag{5.35}$$

Therefore, by inserting Eq. (5.35) into Eq. (5.34), the equilibrium condition for component i, Eq. (5.33) becomes

$$y_i P = x_i P_{s,i}. \tag{5.36}$$

Equation (5.36) was firstly empirically discovered by François-Marie Raoult (1830–1901) in 1887.[4] It states that the total pressure over an ideal solution is the sum of the saturation pressures of its components weighted by their molar fractions [142].

Let us apply Raoult's law for a binary mixture A + B, supposing that component A is more volatile than component B.

The total vapor pressure over the solution will be given by

$$P = x_A P_{s,A}(T) + (1 - x_A)\, P_{s,B}(T) = P_{s,B} + x_A \left(P_{s,A} - P_{s,B}\right), \tag{5.37}$$

meaning that P has a linear variation with the molar fraction x_A.

On the other hand,

$$y_A = \frac{P_A}{P} = \frac{x_A P_{s,A}}{P},$$

and since

$$x_A = \frac{P - P_{s,B}}{P_{s,A} - P_{s,B}},$$

the total vapor pressure can be also expressed in terms of y_A

$$P = \frac{P_{s,B}}{\left(1 - y_A\left(1 - \frac{P_{s,B}}{P_{s,A}}\right)\right)}. \tag{5.38}$$

So for a given pressure P the amount of component A in the liquid solution will be given by its molar fraction, x_A, solution of Eq. (5.37) and the amount of this component in the vapor phase by y_A, solution of Eq. (5.38). This is shown in Fig. 5.6.

5.5.2 *Exercise*

1. Consider that you have a binary mixture of vapors A and B at a constant temperature T and that the amount of A in this mixture is given by $y_{A,1}$ (Fig. 5.7). Component A is more volatile than component B at temperature T. When the pressure is increased, it will reach the value P_1 and the first drops of liquid will appear (this is the reason the continuous black line is called *dew point line*). The amount of A in these liquid drops is given by $x_{A,1}$. By further increasing the pressure both the liquid and the vapor phases become richer in component A, till the pressure attains the value P_2 when the mixture finally condensates and

[4] F. M. Raoult. Comptes rendus des séances de l'Académie des Sciences (6 décembre 1886 et 4 avril 1888.

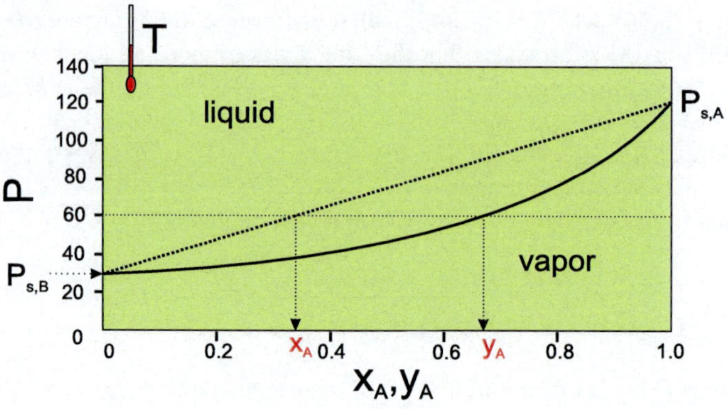

Fig. 5.6 Raoult's law: for a given pressure P the amount of component A in the liquid solution will be given by its molar fraction, x_A, solution of Eq. (5.37) and the amount of this component in the vapor phase by y_A, solution of Eq. (5.38)

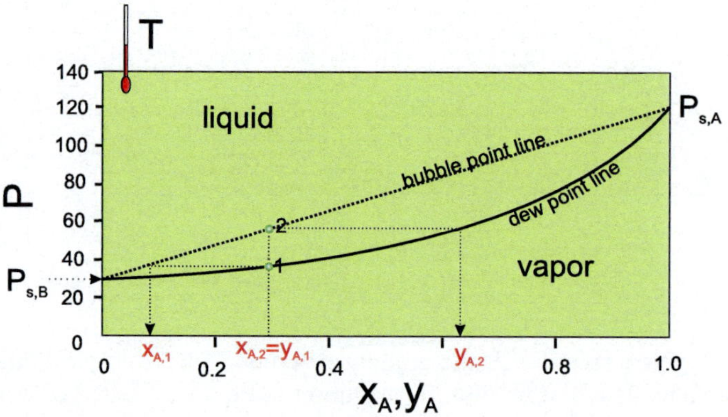

Fig. 5.7 Raoult's law

the molar fraction of A is $y_{A,2}$ in the remaining vapor bubbles. Try to explain the following questions: (a) why the pressure must increase from P_1 to P_2 for the complete condensation of the vapor mixture? … (b) why does component A condensate at pressures smaller than its saturation pressure $P_{s,A}$? … (c) why component B remains in the vapor state at pressures greater than its saturation pressure $P_{s,B}$? … (d) why the first liquid drops that appear in the system when $P = P_1$ are poorer in component A and richer in component B? … (e) why we say that for pressures greater than P_2 all the vapor mixture is in the liquid state? …

5.6 Non-ideal Solutions

For an ideal solution, we saw in Sect. 5.5 that the fugacity may be written as

$$\psi_i\left(T,P,x_i\right)=x_i\psi_{i,pure}\left(T,P\right).$$

For non-ideal solutions, we define the *activity coefficient* γ_i by

$$\gamma_i\left(T,P,x_1,\ldots,x_r\right)=\frac{\psi_i\left(T,P,x_1,\ldots,x_r\right)}{x_i\psi_{i,pure}\left(T,P\right)}. \tag{5.39}$$

In this way, γ_i is a measure of the non-ideality of the solution: when $\gamma_i \sim 1$ for all i, the solution is close to an ideal solution and when γ_i is far from 1, appreciable deviations from ideality must be expected.

As a very simple example consider a non-ideal binary solution with a virial equation of state as in Sect. 5.4.1.

The fugacity coefficient of the first component is given by Eq. (5.23)

$$\ln\psi_1=\ln\left(x_1P\right)+\frac{P}{\mathscr{R}T}\left(B_1+x_2^2\left[2B_{12}-\left(B_1+B_2\right)\right]\right).$$

So,

$$\ln\psi_{1,pure}=\ln P+\frac{PB_1}{\mathscr{R}T},$$

and

$$\ln\gamma_1=\frac{P}{\mathscr{R}T}\left(x_2^2\left[2B_{12}-\left(B_1+B_2\right)\right]\right).$$

Therefore, the activity coefficient will be 1 in the following cases: (a) when $2B_{12}\approx(B_1+B_2)$ meaning that the solution is ideal and (b) when component 1 is in its pure state ($x_2=0$). So, the deviations of the activity coefficient γ_i from 1 may be regarded as due to the *cross interactions* between molecules i and the remaining molecules in the solution.

5.6.1 Gibbs Energy in Excess

The Gibbs energy of a non-ideal solution is given by

$$G\left(T,P,N_1,\ldots,N_r\right)=\sum_i N_i\overline{\mu}_i\left(T,P,N_1,\ldots,N_r\right).$$

We define the Gibbs energy in excess by

$$G^E(T, P, N_1, \ldots, N_r) = G(T, P, N_1, \ldots, N_r) - G^{ideal}(T, P, N_1, \ldots, N_r).$$

Therefore,

$$G^E(T, P, N_1, \ldots, N_r) = \sum_i N_i \overline{\mu}_i^E, \tag{5.40}$$

where

$$\overline{\mu}_i^E = \overline{\mu}_i - \overline{\mu}_i^{ideal}.$$

Now, consider writing $\overline{\mu}_i^E$ in terms of the fugacity. From the definition of fugacity

$$\overline{\mu}_i - \overline{\mu}_i^{ideal} = \mathscr{R}T \ln \frac{\psi_i}{\psi_i^{ideal}} = \mathscr{R}T \ln \left(\frac{\psi_i}{x_i \psi_{i,pure}} \right).$$

But

$$\frac{\psi_i}{x_i \psi_{i,pure}} = \gamma_i,$$

and so the chemical potential in excess is related to the activity coefficient by

$$\overline{\mu}_i^E = \mathscr{R}T \ln \gamma_i.$$

Equation (5.40) can thus be written as

$$G^E(T, P, N_1, \ldots, N_r) = \sum_i N_i \mathscr{R}T \ln \gamma_i. \tag{5.41}$$

Equation (5.41) provides a method for calculating the activity coefficients based on heuristic models for G^E

$$\mathscr{R}T \ln \gamma_i = \left(\frac{\partial G^E(T, P, N_1, \ldots, N_r)}{\partial N_i} \right)_{T,P,N_{j \neq i}} \tag{5.42}$$

5.6.2 Margules Model

The simplest heuristic model for the Gibbs energy in excess was introduced by Max Margules in 1895.[5] For a binary mixture with molecules 1 and 2 of the same molecular size but different polarity,

$$\frac{G^E}{\mathscr{R}T} = A\frac{N_1 N_2}{N_1 + N_2} = NAx_1x_2, \tag{5.43}$$

where $A = A\,(T, P)$ can be considered as an experimental parameter independent of the composition. The model satisfies the condition

$$G^E = 0,$$

when N_1 or N_2 is null. In fact, in this case, the mixture is reduced to a pure component without cross-attraction forces between molecules belonging to different species, which are the sources of non-ideality.

The Margules model can be extended to binary mixtures of components with different molecular sizes in accordance with

$$\frac{G^E}{\mathscr{R}T} = Nx_1x_2\,(A_{21}x_1 + A_{12}x_2)\,.$$

From the above equation

$$\frac{G^E}{\mathscr{R}T} = \frac{N_1 N_2}{N_1 + N_2}\left(A_{21}\frac{N_1}{N_1 + N_2} + A_{12}\frac{N_2}{N_1 + N_2}\right).$$

So, from Eq. ()

$$\ln\gamma_1 = [A_{12} + 2\,(A_{21} - A_{12})\,x_1]\,x_2^2, \tag{5.44}$$

and

$$\ln\gamma_2 = [A_{21} + 2\,(A_{12} - A_{21})\,x_2]\,x_1^2. \tag{5.45}$$

Equation (5.44) enables to find parameter A_{12} from the limiting value of the activity coefficient γ_1 in accordance with[6]

[5] Margules, Max (1895). *Über die Zusammensetzung der gesättigten Dämpfe von Misschungen.* Sitzungsberichte der Kaiserliche Akadamie der Wissenschaften Wien Mathematisch-Naturwissenschaftliche Klasse II. 104: 1243–1278. https://archive.org/details/sitzungsbericht10wiengoog.

[6] Remember that when component 1 is pure $\gamma_1 = 1$. In the same way, $\gamma_2 = 1$ for pure component 2.

$$\lim_{x_1 \to 0} \ln \gamma_1 = A_{12}.$$

In the same way

$$\lim_{x_2 \to 0} \ln \gamma_2 = A_{21}.$$

For similar molecules $A_{12} = A_{21} = A$ and the activity coefficients become

$$\ln \gamma_1 = A x_2^2 \quad \text{and} \quad \ln \gamma_2 = A x_1^2.$$

5.6.3 Exercises

1. Table 5.1 shows the values of the activity coefficients for a water–methanol solution at $T = 50$ °C obtained by using the UNIFAC method of Prausnitz and co-workers [57, 137, 138]. For the calculations, the vapor phase was considered a mixture of ideal gases. Using the data presented in Table 5.1 calculate Margules parameters A_{12} and A_{21}. In the following, calculate the activity coefficients γ for the several molar fractions of methanol presented in Table 5.1 and compare these obtained values with the Margules model with the ones in Table 5.1.
2. Henry's law is a gas law that states that the amount of a dissolved gas labeled i is proportional to its partial pressure in the gas phase,[7]

$$x_i = H P_i.$$

The proportionality factor H is called Henry's law constant. Henry's law was formulated by the English chemist William Henry, who studied the topic in the early nineteenth century. In his publication about the quantity of gases absorbed by water, he described the results of his experiments: "*water takes up, of gas condensed by one, two, or more additional atmospheres, a quantity which, ordinarily compressed, would be equal to twice, thrice, etc. the volume absorbed under the common pressure of the atmosphere.*"
Considering the mixture of vapors in equilibrium with a liquid solution as a mixture of ideal gases and starting from the Margules model for similar molecules given by Eq. (5.43) show that when the dissolved gas is diluted in the liquid phase ($x_i \to 0$), Henry's constant is given by

$$H = \exp\left(\frac{A}{\mathscr{R}T}\right) \psi_{i,pure}(T, P).$$

[7] Henry, William (1 January 1803). *Experiments on the Quantity of Gases Absorbed by Water, at Different Temperatures, and under Different Pressures*. Philosophical Transactions of the Royal Society. London. 93: 29–274. https://doi.org/10.1098/rstl.1803.0004.

Table 5.1 Activity coefficients γ for a water–methanol solution at $T = 50\,^\circ\mathrm{C}$

Molar fraction of methanol	Water	Methanol
0.005	1.0000	2.2415
0.010	1.0001	2.2101
0.050	1.0033	1.9912
0.100	1.0124	1.7809
0.200	1.0436	1.4957
0.500	1.2078	1.1232
0.800	1.4476	1.01479
0.900	1.5423	1.0034
0.990	1.6637	1.0000

5.7 Activity Coefficients from Molecular Groups

It is often convenient to consider a molecule as an aggregate of functional groups; as a result, some thermodynamic properties of pure fluids can be calculated by summing group contributions. Extension of this concept to mixtures was suggested long ago by Langmuir [97], and several attempts have been made to establish group-contribution methods for activity coefficients. Here we will be restricted to the UNIFAC method for activity coefficients γ, which is particularly useful for making reasonable estimates for those non-ideal mixtures for which data are sparse or totally absent.

The UNIFAC method (**UNI**quac **F**unctional-group **A**ctivity **C**oefficients) is a semi-empirical method for the prediction of non-electrolyte activity in non-ideal mixtures that uses the functional groups present on the molecules that make up the liquid mixture to calculate activity coefficients. By using interactions for each of the functional groups present on the molecules, as well as some binary interaction coefficients, the activity of each of the solutions can be calculated. This information can be used to obtain information on liquid equilibria, which is useful in many thermodynamic calculations.

The UNIFAC model was first published in 1975 by Fredenslund, Jones and Prausnitz [57], a group of chemical engineering researchers from the University of California.

The molecular activity coefficient is separated into two parts: one part γ_i^c provides the contribution due to differences in molecular size and shape, and the other γ_i^R provides the contribution due to molecular interactions.

In a multicomponent system, the activity coefficient of (molecular) component i is considered as

$$\ln \gamma_i = \underbrace{\ln \gamma_i^c}_{\text{combinatorial}} + \underbrace{\ln \gamma_i^R}_{\text{residual}}. \tag{5.46}$$

5.7.1 *Combinatorial Activity Coefficient*

In accordance with the UNIFAC method the combinatorial activity coefficient of each molecule takes into account the geometric parameters of its individual groups through the following expression

$$\ln \gamma_i^c = \ln \frac{\phi_i}{x_i} + \frac{z}{2} q_i \ln \frac{\theta_i}{\phi_i} + \ell_i - \frac{\phi_i}{x_i} \sum_j x_j \ell_j, \tag{5.47}$$

where ϕ_i is the volume contribution of component i in the solution

$$\phi_i = \frac{r_i x_i}{\sum_j r_j x_j},$$

and θ_i is the area contribution of component i in the solution

$$\theta_i = \frac{q_i x_i}{\sum_j q_j x_j}.$$

Parameter ℓ_i is given by

$$\ell_i = \frac{z}{2}(r_i - q_i) - (r_i - 1)$$

with $z = 10$.

The volume parameter r_i considers the volume contribution of all functional groups in the molecule

$$r_i = \sum_k \nu_k^i R_k$$

where ν_k^i is the number of groups k in molecule i and R_k is the volume contribution of group k (Fig. 5.8).

The area parameter q_i considers the area contribution of all functional groups in the molecule

$$q_i = \sum_k \nu_k^i Q_k$$

where Q_k is the area contribution of group k (Fig. 5.8).

5.7.1.1 Examples

The butane, Fig. 5.9 is an organic compound with the formula C_4H_{10} that is an alkane with four carbon atoms. Butane is a gas at room temperature and atmospheric pres-

Fig. 5.8 Volume and area group contributions for the combinatorial activity coefficient. Reproduced from Table 8.23 of Ref. [137]

Group Numbers				
Main	Secondary	Name	R	Q
1	1	CH_3	0.9011	0.8480
	2	CH_2	0.6744	0.5400
	3	CH	0.4469	0.2280
	4	C	0.2195	0.0000
2	5	CH_2=CH	1.3454	1.1760
	6	CH=CH	1.1167	0.8670
	7	CH_2=C	1.1173	0.9880
	8	CH=C	0.8886	0.6760
	9	C=C	0.6605	0.4850
3	10	ACH	0.5313	0.4000
	11	AC	0.3652	0.1200
4	12	$ACCH_3$	1.2663	0.9680
	13	$ACCH_2$	1.0396	0.6600
	14	ACCH	0.8121	0.3480
5	15	OH	1.0000	1.2000
6	16	CH_3OH	1.4311	1.4320
7	17	H_2O	0.9200	1.4000
8	18	ACOH	0.8952	0.6800
9	19	CH_3CO	1.6724	1.4880
	20	CH_2CO	1.4457	1.1800

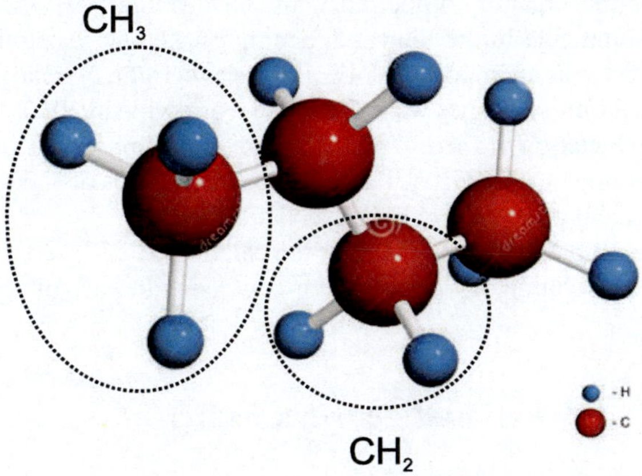

Fig. 5.9 The molecule of n-butane

sure. The term may refer to either of two structural isomers, n-butane or isobutane, or to a mixture of these isomers. In the IUPAC[8] nomenclature, however, "butane" refers only to the n-butane isomer (which is the isomer with the unbranched structure). Butanes are highly flammable, colorless, easily liquefied gases.

We can identify the following molecular groups: 2 groups CH_2 and 2 groups CH_3. The volumetric R and surface Q parameters for the functional groups CH_2 and CH_3 are given in Fig. 5.8.

So

$$r_{\text{butane}} = \sum_k \nu_k^{\text{butane}} R_k = 2 \times 0.6744 + 2 \times 0.9011 = 3.151$$

$$q_{\text{butane}} = \sum_k \nu_k^{\text{butane}} Q_k = 2 \times 0.540 + 2 \times 0.848 = 2.776$$

$$\begin{aligned}\ell_{\text{butane}} &= \frac{z}{2}(r_{\text{butane}} - q_{\text{butane}}) - (r_{\text{butane}} - 1) \\ &= \frac{10}{2}(3.151 - 2.776) - (3.151 - 1) = -0.276\end{aligned}$$

Consider now a solution of n-pentane with acetone (Fig. 5.10). Pentanes are relatively inexpensive and are the most volatile liquid alkanes at room temperature, so they are often used in the laboratory as solvents that can be conveniently and rapidly evaporated. However, because of their non-polarity and lack of functionality, they dissolve only non-polar and alkyl-rich compounds. Pentanes are miscible with most common nonpolar solvents such as chlorocarbons, aromatics, and ethers. Acetone is the organic compound with the formula $(CH_3)_2CO$. It is a colorless, volatile, flammable liquid, and is the simplest ketone. Acetone is miscible with water and serves as an important solvent in its own right, typically for cleaning purposes. About 6.7 million tonnes were produced worldwide in 2010, mainly for use as a solvent and production of methyl methacrylate and bisphenol A. It is a common building block in organic chemistry. Familiar household uses of acetone are as the active ingredient in nail polish remover and as paint thinner.

Tables 5.2 and 5.3 shows the calculation of the combinatorial activity coefficients in a solution of acetone and n-pentane with 25% of acetone using Eq. (5.47).

5.7.2 *Residual Activity Coefficient*

The residual activity coefficient takes the molecular interaction between the molecular groups into account. The molecular interaction parameters are shown in Fig. 5.11, for some groups. Table 8.24 of Poling et al. [137] gives a complete set of values of

[8] International Union of Pure and Applied Chemistry.

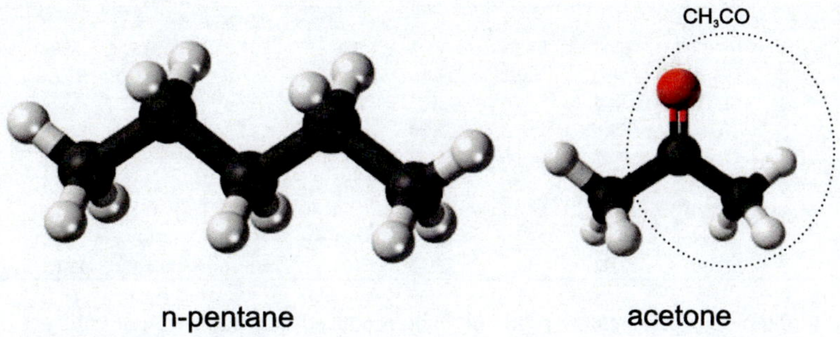

Fig. 5.10 Normal-pentane and acetone molecules

Table 5.2 Calculation of the geometric parameters r, q, and ℓ for the molecules of n-pentane and acetone

	Groups	R	Q	r	q	ℓ
n-pentane	$2CH_3$ + $3CH_2$			3.8254	3.316	– 0.2784
Acetone	$1CH_3$ + $1CH_3CO$			2.5735	2.336	– 0.386
CH_3		0.9011	0.8480			
CH_2		0.6744	0.5400			
CH_3CO		1.6724	1.4880			

Table 5.3 Combinatorial activity coefficient for n-pentane and acetone, based on data given in Table 5.2, when the molar fraction of acetone is $x_{acetone} = 0.25$

	Molar fraction	ϕ	θ	$\ln \gamma_c$	γ_c
n-pentane	0.75	0.8168	0.8098	– 0.0031	0.9969
Acetone	0.25	0.1832	0.1902	– 0.0356	0.9650

the molecular interaction parameters for 50 molecular groups and is suggested for interested readers.

Parameter a_{mn} is considered as a reference when $m = n$, i.e., $a_{mn} = 0$ when $m = n$. Considering, for instance, the molecular groups CH_3 and CH_2, both groups belong to group 1 in accordance with Fig. 5.8 and their interaction parameter is null. In the same way, $a_{19} = 476.4$ K[9] means that the attractive potential between groups 1 and 9 is 476.4 K *larger* than the attractive potential between groups 1 and 1 and $a_{91} = 26.76$ K, means that the attractive potential between groups 9 and 1 is 26.76 K larger than the attractive potential between groups 9 and 9. A negative value for $a_{56} = -137$ K means that the attractive potential between molecules 5 and 6 is 137 K smaller than the attractive potential between molecules 5 and 5.

[9] Parameters are normalized by the Boltzmann constant, so their unit is the Kelvin (K).

Main Group	1	2	3	4	5	6	7	8	9
1	0	86.02	61.13	76.5	986.5	697.2	1318	1333	476.4
2	-35.36	0	38.81	74.15	524.1	787.6	270.6	526.1	182.6
3	-11.12	3.446	0	167	636.1	637.4	903.8	1329	25.77
4	-69.7	-113.6	-146.8	0	803.2	603.3	5695	884.9	-52.1
5	156.4	457	89.6	25.82	0	-137.1	353.5	-259.7	84
6	16.51	-12.52	-50	-44.5	249.1	0	-181	-101.7	23.39
7	300	496.1	362.3	377.6	-229.1	289.6	0	324.5	-195.4
8	275.8	217.5	25.34	244.2	-451.6	-265.2	-601.8	0	-356.1
9	26.76	42.92	140.1	365.8	164.5	108.7	472.5	-133.1	0

Fig. 5.11 Interaction parameters a_{mn} in Kelvin. Reproduced from Table 8.24 of Poling et al. [137]

Table 5.4 Functional groups in the acetone and n-pentane solution

	CH_3	CH_2	CH_3CO
Identifying number	1	2	3
Main group	1	1	9

The effect of temperature on the molecular interaction is taken into account by introducing the interaction energy in a dimensionless form

$$\Psi_{mn} = \exp\left(-\frac{a_{mn}}{T}\right) \tag{5.48}$$

and the residual coefficient of activity is given by

$$\ln \gamma_i^r = \sum_k \nu_k^i \left(\ln \Gamma_k - \ln \Gamma_k^i\right) \tag{5.49}$$

where the symbol k is used for the molecular groups, the gamma factor Γ_k is related to the k-group and is calculated in accordance with

$$\ln \Gamma_k = Q_k \left(1 - \ln \sum_m \theta_m \Psi_{mk} - \sum_m \frac{\theta_m \Psi_{km}}{\sum_n \theta_n \Psi_{nm}}\right)$$

The term Γ_k^i in Eq. (5.49) means that the gamma factor must be calculated in a reference solution of molecules.

For concreteness, consider an acetone-pentane solution. From Fig. 5.8, the functional groups in the solution acetone (labeled as 1) and n-pentane (labeled as 2) are identified in Table 5.4.

The interaction parameters are given below

a_{11}	a_{12}	a_{21}	a_{22}	a_{13}	a_{31}	a_{33}	a_{23}	a_{32}
0	0	0	0	476.40	26.76	0	476.40	26.76

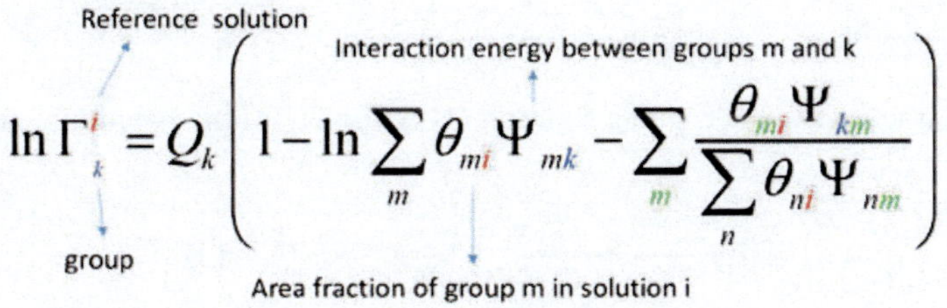

Fig. 5.12 Method for calculating the gamma factors in the reference solutions [137]

5.7.2.1 Calculation of the Gamma Factor in the Reference Solutions

The Γ factors in the reference solutions are calculated in accordance with the method presented in Fig. 5.12 [137].

In our example, there are two molecules present in the solution: the acetone molecule referred to as 1 and the n-pentane molecule referred as 2. So the first reference solution is a solution of acetone, where the two groups CH_3 and CH_3CO are supposed to be the solution components.

Group Fractions

In the acetone solution (labeled as 1) there are two functional groups CH_3 (1) and CH_3CO (3) in equal proportions. So

$$X_{11} = \frac{\nu_1^1}{\nu_1^1 + \nu_3^1} = \frac{1}{2},$$

$$X_{31} = \frac{\nu_3^1}{\nu_1^1 + \nu_3^1} = \frac{1}{2}.$$

In the n-pentane solution (labeled as 2) there are 2 functional groups CH_3 (1) and 3 functional groups CH_2 (2) in proportions

$$X_{12} = \frac{\nu_1^2}{\nu_1^2 + \nu_2^2} = \frac{2}{5},$$

$$X_{22} = \frac{\nu_2^2}{\nu_1^2 + \nu_2^2} = \frac{3}{5}.$$

Area Fractions

The area fractions in the acetone solution (1) are given for the functional group CH_3 (1) by

$$\theta_{11} = \frac{X_{11} \times Q_1}{X_{11} \times Q_1 + X_{31} \times Q_3} = 0.36301,$$

and for the functional group CH_3CO (3), by

$$\theta_{31} = \frac{X_{31} \times Q_3}{X_{11} \times Q_1 + X_{31} \times Q_3} = 0.63699.$$

The area fractions in the n-pentane solution (2) are given for the functional group CH_3 (1) by

$$\theta_{12} = \frac{X_{12} \times Q_1}{X_{12} \times Q_1 + X_{22} \times Q_2} = 0.51146,$$

and for the functional group CH_2 (2), by

$$\theta_{22} = \frac{X_{22} \times Q_2}{X_{12} \times Q_1 + X_{22} \times Q_2} = 0.48854.$$

Let the temperature be $T = 307$ K. The interaction energies Ψ_{mn} between the functional groups at this temperature are given below

Ψ_{11}	Ψ_{12}	Ψ_{21}	Ψ_{22}	Ψ_{13}	Ψ_{31}	Ψ_{33}	Ψ_{23}	Ψ_{32}
1	1	1	1	0.21187	0.91652	1	0.21187	0.91652

Gamma Factors

Following the method presented in Fig. 5.12, in the acetone reference solution, the gamma factors for the group CH_3 (1) and CH_3CO (3) are given by

$$\ln \Gamma_1^1 = Q_1 \left(\begin{array}{c} 1 - \ln(\theta_{11} \times \Psi_{11} + \theta_{31} \times \Psi_{31}) \\ - \left(\frac{\theta_{11} \times \Psi_{11}}{\theta_{11} \times \Psi_{11} + \theta_{31} \times \Psi_{31}} + \frac{\theta_{31} \times \Psi_{13}}{\theta_{11} \times \Psi_{13} + \theta_{31} \times \Psi_{33}} \right) \end{array} \right) = 0.40891,$$

$$\ln \Gamma_3^1 = Q_3 \left(\begin{array}{c} 1 - \ln(\theta_{11} \times \Psi_{13} + \theta_{31} \times \Psi_{33}) \\ - \left(\frac{\theta_{11} \times \Psi_{31}}{\theta_{11} \times \Psi_{11} + \theta_{31} \times \Psi_{31}} + \frac{\theta_{31} \times \Psi_{33}}{\theta_{11} \times \Psi_{13} + \theta_{31} \times \Psi_{33}} \right) \end{array} \right) = 0.13891.$$

In the n-pentane solution, all the gamma factors are unitary because the interaction energies are the reference ones

$$\ln \Gamma_1^2 = Q_1 \left(\begin{array}{c} 1 - \ln(\theta_{12} \times \Psi_{11} + \theta_{22} \times \Psi_{21}) \\ -\left(\frac{\theta_{12}\times\Psi_{11}}{\theta_{12}\times\Psi_{11}+\theta_{22}\times\Psi_{21}} + \frac{\theta_{22}\times\Psi_{12}}{\theta_{12}\times\Psi_{12}+\theta_{22}\times\Psi_{22}} \right) \end{array} \right) = 0,$$

$$\ln \Gamma_2^2 = Q_2 \left(\begin{array}{c} 1 - \ln(\theta_{12} \times \Psi_{12} + \theta_{22} \times \Psi_{22}) \\ -\left(\frac{\theta_{12}\times\Psi_{21}}{\theta_{12}\times\Psi_{11}+\theta_{22}\times\Psi_{21}} + \frac{\theta_{22}\times\Psi_{22}}{\theta_{12}\times\Psi_{12}+\theta_{22}\times\Psi_{22}} \right) \end{array} \right) = 0.$$

Gamma Factors for the Acetone-Pentane Solution

Consider a solution with a very small fraction of acetone, $x_{acetone} = x = 0.047$. Let us calculate the fractions of the 3 molecular groups.

For the CH_3 group present in the two molecules

$$X_1 = \frac{x \times \nu_1^1 + (1-x) \times \nu_1^2}{x \times \nu_1^1 + x \times \nu_3^1 + (1-x) \times \nu_1^2 + (1-x) \times \nu_2^2} = 0.40193.$$

For the CH_3CO present in the acetone (1) molecule

$$X_3 = \frac{x \times \nu_3^1}{x \times \nu_1^1 + x \times \nu_3^1 + (1-x) \times \nu_1^2 + (1-x) \times \nu_2^2} = 9.6728 \times 10^{-3}.$$

For the CH_2 present in the n-pentane (2) molecule

$$X_2 = \frac{(1-x) \times \nu_2^2}{x \times \nu_1^1 + x \times \nu_3^1 + (1-x) \times \nu_1^2 + (1-x) \times \nu_2^2} = 0.58839.$$

So

CH_3 (1)	CH_2 (2)	CH_3CO (3)
$X_1 = 0.40193$	$X_2 = 0.58839$	$X_3 = 9.6728 \times 10^{-3}$

Therefore, the area fractions will be given by,

$$\theta_1 = \frac{X_1 \times Q_1}{X_1 \times Q_1 + X_3 \times Q_3 + X_2 \times Q_2} = 0.50647,$$

$$\theta_3 = \frac{X_3 \times Q_3}{X_1 \times Q_1 + X_3 \times Q_3 + X_2 \times Q_2} = 2.1388 \times 10^{-2},$$

$$\theta_2 = \frac{X_2 \times Q_2}{X_1 \times Q_1 + X_3 \times Q_3 + X_2 \times Q_2} = 0.47214.$$

So

CH_3 (1)	CH_2 (2)	CH_3CO (3)
$\theta_1 = 0.5067$	$\theta_2 = 0.47214$	$\theta_3 = 2.1388 \times 10^{-2}$

The gamma factors are calculated in accordance with

$$\ln \Gamma_k = Q_k \left(1 - \ln \sum_m \theta_m \Psi_{mk} - \sum_m \frac{\theta_m \Psi_{km}}{\sum_n \theta_n \Psi_{nm}} \right).$$

So,

$$\ln \Gamma_1 = Q_1 \left(\begin{array}{c} 1 \\ -\ln(\theta_1 \times \Psi_{11} + \theta_3 \times \Psi_{31} + \theta_2 \times \Psi_{21}) \\ -\left(\begin{array}{c} \frac{\theta_1 \times \Psi_{11}}{\theta_1 \times \Psi_{11} + \theta_3 \times \Psi_{31} + \theta_2 \times \Psi_{21}} \\ + \frac{\theta_2 \times \Psi_{12}}{\theta_1 \times \Psi_{12} + \theta_3 \times \Psi_{32} + \theta_2 \times \Psi_{22}} \\ + \frac{\theta_3 \times \Psi_{13}}{\theta_1 \times \Psi_{13} + \theta_3 \times \Psi_{33} + \theta_2 \times \Psi_{23}} \end{array} \right) \end{array} \right) = 1.37 \times 10^{-3},$$

$$\ln \Gamma_2 = Q_2 \left(\begin{array}{c} 1 \\ -\ln(\theta_1 \times \Psi_{12} + \theta_3 \times \Psi_{32} + \theta_2 \times \Psi_{22}) \\ -\left(\begin{array}{c} \frac{\theta_1 \times \Psi_{21}}{\theta_1 \times \Psi_{11} + \theta_3 \times \Psi_{31} + \theta_2 \times \Psi_{21}} \\ + \frac{\theta_2 \times \Psi_{22}}{\theta_1 \times \Psi_{12} + \theta_3 \times \Psi_{32} + \theta_2 \times \Psi_{22}} \\ + \frac{\theta_3 \times \Psi_{23}}{\theta_1 \times \Psi_{13} + \theta_3 \times \Psi_{33} + \theta_2 \times \Psi_{23}} \end{array} \right) \end{array} \right) = 8.72 \times 10^{-4},$$

$$\ln \Gamma_3 = Q_3 \left(\begin{array}{c} 1 \\ -\ln(\theta_1 \times \Psi_{13} + \theta_3 \times \Psi_{33} + \theta_2 \times \Psi_{23}) \\ -\left(\begin{array}{c} \frac{\theta_1 \times \Psi_{31}}{\theta_1 \times \Psi_{11} + \theta_3 \times \Psi_{31} + \theta_2 \times \Psi_{21}} \\ + \frac{\theta_2 \times \Psi_{32}}{\theta_1 \times \Psi_{12} + \theta_3 \times \Psi_{32} + \theta_2 \times \Psi_{22}} \\ + \frac{\theta_3 \times \Psi_{33}}{\theta_1 \times \Psi_{13} + \theta_3 \times \Psi_{33} + \theta_2 \times \Psi_{23}} \end{array} \right) \end{array} \right) = 2.207.$$

Residual Activity Coefficients

Finally, the residual activity coefficients are calculated for acetone (1) and n-pentane (2), in accordance with

$$\ln \gamma_i^r = \sum_k \nu_k^i \left(\ln \Gamma_k - \ln \Gamma_k^i\right),$$

where the index i refers to the solution component (acetone or n-pentane) and the index k refers to the functional groups that are present in the molecule of the i-component.

For the acetone,

$$\begin{aligned}\ln \gamma_1^r &= \nu_1^1 \left(\ln \Gamma_1 - \ln \Gamma_1^1\right) + \nu_3^1 \left(\ln \Gamma_3 - \ln \Gamma_3^1\right)\\ &= \nu_1^1 \left(1.3694 \times 10^{-3} - 0.40891\right) + \nu_3^1 \left(2.207 - 0.13891\right)\\ &= 1.6605,\end{aligned}$$

and so

$$\gamma_1^r = \exp\left(1.6605\right) = 5.2619.$$

For the n-pentane,

$$\begin{aligned}\ln \gamma_2^r &= \nu_1^2 \left(\ln \Gamma_1 - \ln \Gamma_1^2\right) + \nu_2^2 \left(\ln \Gamma_2 - \ln \Gamma_2^2\right)\\ &= \nu_1^2 \left(1.3694 \times 10^{-3} - 0\right) + \nu_2^2 \left(8.7205 \times 10^{-4} - 0\right)\\ &= 5.3550 \times 10^{-3},\end{aligned}$$

and so

$$\gamma_2^r = \exp\left(5.3550 \times 10^{-3}\right) = 1.0054.$$

Activity Coefficients

The activity coefficients of acetone and n-pentane are calculated in accordance with Eq. (5.46)

$$\ln \gamma_i = \underbrace{\ln \gamma_i^c}_{\text{combinatorial}} + \underbrace{\ln \gamma_i^R}_{\text{residual}}.$$

For acetone, present with a molar fraction $x_{acetone} = 0.047$ the combinatorial activity coefficient was calculated as $\gamma_1^c = 0.9486$ and so

$$\ln \gamma_1 = -0.0527 + 1.6605 = 1.6078.$$

Therefore

$$\gamma_1 = 4.9918.$$

For n-pentane, the combinatorial activity coefficient was calculated as $\gamma_2^c = 0.9999$ for this molar fraction of acetone and so

$$\ln \gamma_2 = -1.01769 \times 10^{-4} + 5.3550 \times 10^{-3} = 5.2532 \times 10^{-3}.$$

Therefore

$$\gamma_2 = 1.0053.$$

5.7.3 *Exercises*

1. Use the functional groups' method presented in Sect. 5.7 for finding the activity coefficients of a solution of acetone and n-pentane at 307 K when the molar fraction of n-pentane is $x_{pentane} = 0.047$.
2. Liquefied petroleum gas (LPG), also referred to as simply propane or butane, are flammable mixtures of hydrocarbon gases used as fuel in heating appliances, cooking equipment, and vehicles. Repeat Exercise 1 for a solution of propane and n-butane at 300 K when the molar fraction of propane is $x_{propane} = 0.1, 0.2, 0.5, 0.8$ and 0.9. What is your conclusion about the deviations from the ideality of such mixtures?

5.8 Phase Equilibrium

Consider a multicomponent system $1, 2, \ldots, r$ at temperature T and pressure P. We seek to find how the N_1 moles of component 1, N_2 moles of component 2, ..., N_r moles of component r, will be redistributed among the liquid ℓ and vapor v phases.

In equilibrium condition the fugacities of each component i must be the same in the liquid and vapor phases

$$\psi_i^\ell = \psi_i^v, \quad \text{for} \quad i = 1, 2, \ldots, r, \tag{5.50}$$

and we can write ψ_i^ℓ in terms of the activity coefficient

$$\psi_i^\ell = \gamma_i^\ell x_i \psi_{i,pure}^\ell (T, P) .$$

On the other hand, using the Poynting correction, the fugacity of pure component i can be found in terms of the

$$\psi_{i,pure}^\ell (T, P) = \psi_{i,pure}^\ell \left(T, P_{s,i}\right) \exp\left(\frac{\overline{v}_{ls,i}}{\mathcal{R}T} \left(P - P_{s,i}\right)\right), \tag{5.51}$$

in accordance with Eq. (5.40). In the above equation, $\psi_{i,pure}^\ell \left(T, P_{s,i}\right)$ is the fugacity of pure component i in its saturated liquid state at T (when the vapor pressure is $P_{s,i}$) and $\overline{v}_{ls,i}$ is the molar volume of component i in its saturated liquid state at temperature T.

The term $\psi_{i,pure}^\ell \left(T, P_{s,i}\right)$ can be written in accordance with Eq. (5.21) as

$$\psi_{i,pure}^\ell \left(T, P_{s,i}\right) = \phi_{i,pure}^\ell \left(T, P_{s,i}\right) P_{s,i},$$

and the fugacity coefficient $\phi_{i,pure}^\ell$ can be found from the equation of state of component i

$$\ln \phi_{i,pure}^\ell \left(T, P_{s,i}\right) = \int_0^{P_{s,i}} \left(\frac{\overline{v}_{i,pure}^\ell}{\mathcal{R}T} - \frac{1}{P}\right) dP.$$

For the vapor phase

$$\psi_i^v = y_i P \phi_i^v (T, P, y_1, \ldots, y_r) , \tag{5.52}$$

where

$$\ln \phi_i^v (T, P, y_1, \ldots, y_r) = \int_0^{P} \left(\frac{\overline{v}_i^v}{\mathcal{R}T} - \frac{1}{P}\right) dP. \tag{5.53}$$

Finally, Eq. (5.50) takes the form

$$\gamma_i^\ell x_i \phi_{i,pure}^\ell \left(T, P_{s,i}\right) P_{s,i} \exp\left(\frac{\overline{v}_{ls,i}}{\mathcal{R}T} \left(P - P_{s,i}\right)\right)$$
$$= y_i P \phi_i^v (T, P, y_1, \ldots, y_r) \tag{5.54}$$

So, if we start considering that we want to find the vapor composition $y_1, \ldots, y_r$ and pressure P from a given composition $x_1, \ldots, x_r$ in the liquid state and temperature T, our task will be to solve the r equations given by Eq. (5.50) for the r unknowns[10]

$$y_1, \ldots, y_{r-1}, P.$$

This is a difficult task and requires solving r non-linear equations, since, in Eq. (5.53), the molar volume of each component in the vapor phase $\overline{v}_i^v$ is dependent on the composition.

To ease this task we rewrite Eq. (5.54) as

$$\gamma_i^\ell x_i P_{s,i} = y_i P K_i, \tag{5.55}$$

where

$$K_i = \frac{\phi_i^v (T, P, y_1, \ldots, y_r)}{\exp\left(\frac{v_{ls,i}}{\mathcal{R}T}(P - P_{s,i})\right) \phi_{i,pure}^\ell (T, P_{s,i})}.$$

When the vapor phase can be considered a mixture of ideal gases,

$$\psi_i^v = y_i P \phi_i^v = y_i P,$$

and so, $\phi_i^v = 1$. On the other hand,

$$\psi_{i,pure}^\ell = P_{s.i} \phi_{i,pure}^\ell = P_{s.i}.$$

Therefore, $\phi_{i,pure}^\ell$ is also 1 and if we neglect the Poynting correction

$$K_i = 1.$$

5.8.1 *Phase Equilibrium at Low Pressures*

A reasonable simplifying assumption for multicomponent systems at low pressures O (1 bar) is to suppose that the vapor phase behaves as a mixture of ideal gases. In this case $\phi_i = 1$ and

$$\psi_i^v = y_i P.$$

---

[10] Remember that

$$\sum_i y_i = 1.$$

Fig. 5.13 Water–methanol activity coefficients at 25 °C in terms of the molar fraction of methanol

Activity coefficients for T=298.15 K		
x	Molecule	
	water	methanol
0.1	1.0126	1.7484
0.2	1.0435	1.4702
0.3	1.0869	1.2997
0.4	1.1398	1.1893
0.5	1.2006	1.1157
0.6	1.2685	1.0664
0.7	1.3431	1.0339
0.8	1.4240	1.0139
0.9	1.5111	1.0032

So, Eq. (5.50) becomes

$$\gamma_i^{\ell} x_i \psi_{i,pure}^{\ell} (T, P) = y_i P, \quad \text{for } i = 1, 2, \ldots, r,$$

and, by neglecting the Poynting correction,

$$\gamma_i^{\ell} x_i P_{s.i} (T) = y_i P, \quad \text{for } i = 1, 2, \ldots, r. \tag{5.56}$$

For concreteness, consider a binary water–methanol system at $T = 25$ °C. By summing Eq. (5.56) over the components

$$\gamma_{meth}^{\ell} x_{meth} P_{s,meth} (T) + \gamma_w^{\ell} (1 - x_{meth}) P_{s,w} (T) = P. \tag{5.57}$$

Therefore, for each molar fraction of methanol in the liquid solution, the activity coefficients of methanol and water can be determined following the method presented in Sect. 5.7 and the pressure can be calculated using the above equation. After the pressure is calculated the molar fraction of methanol in the vapor phase is calculated from Eq. (5.56)

$$y_{meth} = \frac{\gamma_{meth}^{\ell} x_{meth} P_{s.meth} (T)}{P} \tag{5.58}$$

At $T = 25$ °C the saturation pressures of water and methanol are respectively, 24 and 127 mm Hg. Their activity coefficients are presented in Fig. 5.13.

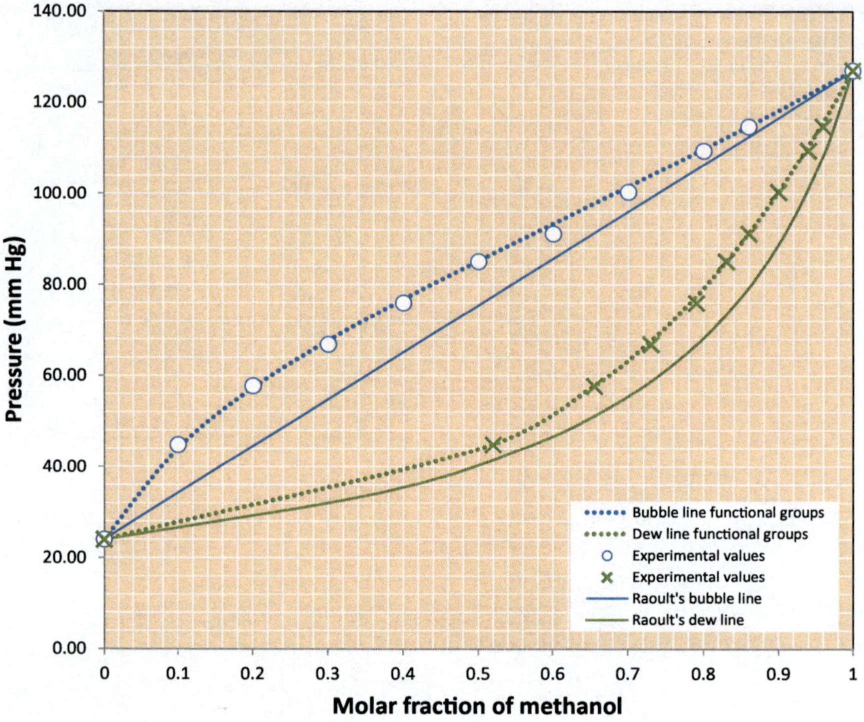

Fig. 5.14 Water–methanol bubble and dew lines at 25 °C calculated from functional groups and compared with experimental values [89] and with Raoult's law

The bubble and dew lines for $T = 25$ °C resulting from the solution of Eqs. (5.57) and (5.58) are presented in Fig. 5.14 and compared with experimental values.[11] In the same figure, we also plot the bubble and dew line when the mixture is supposed to be ideal and follows Raoult's law. Appreciable deviations from ideality can be appreciated in all the ranges of methanol molar fractions.

5.8.1.1 Exercise

Replace the methanol with ethanol in the mixture for a temperature $T = 25$ °C. Calculate the activity coefficients for molar fractions of ethanol that vary between 0 and 1 and, considering the vapor phase as a mixture of ideal gases, draw a phase diagram such as the one shown in Fig. 5.14 based on the functional groups theory. Compare your results with Raoult's law and with available experimental data.

[11] Kontogeorgis GM, V. Yakoumis I, Meijer H, et al. (1999) Multicomponent phase equilibrium calculations for water–methanol–alkane mixtures. Fluid Phase Equilib 158–160:201–209. https://doi.org/10.1016/S0378-3812(99)00060-6 [89].

5.8.2 Phase-Equilibrium Using the Virial Equation of State for the Vapor Phase

When the vapor phase cannot be considered as a mixture of ideal gases, the fugacity coefficient ϕ_i^v given in Eq. (5.52) must be calculated for each component i. In other words, we need to know a suitable equation of state for describing the vapor mixture. The mixture of vapors can be described by the truncated form of the virial equation of state, Eq. (5.22) and the fugacity coefficient can be calculated as

$$\ln \phi_i^v = \frac{P}{\mathscr{R}T}\left(2\sum_j y_j B_{ij} - B\right).$$

For a binary mixture of components labeled as 1 and 2

$$\ln \phi_1^v = \frac{P}{\mathscr{R}T}\left(B_1 + \left(1 - y_1^2\right)\left[2B_{12} - (B_1 + B_2)\right]\right)$$

and

$$\ln \phi_2^v = \frac{P}{\mathscr{R}T}\left(B_2 + y_1^2\left[2B_{12} - (B_1 + B_2)\right]\right).$$

Therefore, the equilibrium condition is given by the following set of non-linear equations

$$\gamma_1^\ell x_1 P_{s,1} = y_1 P K_1, \tag{5.59}$$

$$\gamma_2^\ell (1 - x_1) P_{s,2} = (1 - y_1) P K_2, \tag{5.60}$$

where

$$\ln K_1 = \frac{P - P_{s,1}}{\mathscr{R}T}\left(B_1 - \overline{v}_{ls,1}\right) + \frac{P}{\mathscr{R}T}\left(1 - y_1^2\right)\left[2B_{12} - (B_1 + B_2)\right],$$

$$\ln K_2 = \frac{P - P_{s,2}}{\mathscr{R}T}\left(B_2 - \overline{v}_{ls,2}\right) + \frac{P}{\mathscr{R}T} y_1^2\left[2B_{12} - (B_1 + B_2)\right],$$

for the pressure P and the molar fraction y_1 of component 1 in the vapor phase, when the temperature and the molar fraction x_1 of this component in the liquid phase are known.

The l.h.s. of Eqs. (5.59) and (5.60) are known, when experimental values are disposable or it is possible to estimate the activity coefficients γ_1^ℓ and γ_2^ℓ in terms of the temperature and the molar fraction x_1 of component 1 in the liquid phase. The r.h.s. of these equations are non-linear functions of the pressure P and the molar fraction y_1.

Due to the high non-linearity, the solution procedure is iterative and based on the following steps:

1. Consider $K_1 = K_2 = 1$ and calculate the pressure P and the molar fraction y_1 by solving Eqs. (5.59) and (5.60);
2. With the calculated values of P and y_1, calculate K_1 and K_2;
3. Solve again Eqs. (5.59) and (5.60) with the new values of K_1 and K_2;
4. Compare the new with the old values of P and y_1 till the desired convergence is achieved.

Ideal Solutions

An interesting particular case is represented by ideal solutions. If the solution is ideal, meaning that $2B_{12} - (B_1 + B_2)$ and $\gamma_1^\ell = 1$, then, for $i = 1, 2$,

$$\ln K_i = \frac{P - P_{s,i}}{\mathscr{R}T}\left(B_i - \bar{v}_{ls,i}\right), \tag{5.61}$$

and we obtain

$$y_i P e^{\frac{\left(P-P_{s,i}\right)\left(B_i - \bar{v}_{ls,i}\right)}{\mathscr{R}T}} = x_i P_{s,i},$$

which can be considered an extension of Raoult's law, when the vapor mixture cannot be considered a mixture of ideal gases or, in other words, when the attraction forces between the molecules cannot be neglected but are similar.

5.8.2.1 Exercise

Consider a mixture of ethanol and water at 25 °C. Calculate the activity coefficients γ_i^ℓ using the functional groups theory of Sect. 5.7, and find the virial coefficients B_{water}, $B_{ethanol}$ and $B_{water-ethanol}$, and the vapor pressures for water and ethanol at this temperature that are available in the literature. With these data find the bubble and dew lines of water-ethanol mixtures at 25 °C using the method presented in Sect. 5.8.2. Compare your results with the ones previously obtained when the vapor phase was considered as a mixture of ideal gases.

5.8.3 *Phase Equilibrium Using a Cubic Equation of State for the Vapor Phase*

Although the methodology is the same for all equations of state with the form $P = P(T, v, y_1, \ldots, y_r)$, we restrict our analysis to the Peng–Robinson equation of state and its modification by Stryjek and Vera, in the present section.

The Peng–Robinson equation for mixtures has the same form as in Eq. (4.9)

$$P = \frac{RT}{\bar{v} - b} - \frac{a}{\bar{v}\,(\bar{v} + b) + b(\bar{v} - b)}, \tag{5.62}$$

where the force and volume parameters are dependent on the composition by mixing rules. The simplest ones are[12]

$$a = \sum_{i,j} y_i y_j a_{ij}, \quad a_{ij} = \delta_{ij}\sqrt{a_i a_j}, \quad b = \sum_i y_i b_i,$$

and the pure component force parameters are calculated in accordance with

$$a_i = a_i\left(T_{c,i}\right)\alpha_i\left(T, \omega_i\right),$$

with

$$a_i\left(T_{c,i}\right) = 0.45724\frac{\left(RT_{c,i}\right)^2}{P_{c,i}}, \quad b_i = 0.07780\frac{RT_{c,i}}{P_{c,i}},$$

and

$$\alpha_i = \left[1 + \kappa_i\left(1 - \sqrt{\frac{T}{T_{c,i}}}\right)\right]^2.$$

[12] Other mixing rules based on experimental data have been proposed. See, for instance, Wilczek-Vera G, Vera JH (1987) *A comparative study of mixing rules for cubic equations of state*. Fluid Phase Equilib 37:241–253. https://doi.org/10.1016/0378-3812(87)80054-7, Poling BE, Prausnitz JM, O'Connell JP (2001) *The properties of gases and liquids*. McGraw-Hill, New York and Vera JH, Wilczek-Vera G (2016) *Classical thermodynamics of fluid systems: principles and applications*. CRC-Press, New York.

Parameter κ_i is related to the acentric factor of component i by[13]

$$\kappa_i = 0.37464 + 1.54226\omega_i + 0.26992\omega_i^2.$$

By defining

$$p = \frac{aP}{(RT)^2}, \quad q = \frac{bP}{RT}.$$

Equation (5.62) can be rewritten as

$$Z^3 + AZ^2 + BZ + C = 0, \tag{5.63}$$

where

$$A = (q-1), \quad B = p - q(3q+2), \quad C = q\left(q + q^2 - p\right)$$

This is a very convenient form for a cubic equation of state since it enables us to find the compressibility factor in terms of temperature, pressure and composition, avoiding dependence on the molar volume $\overline{v}$.

In the same way, the fugacity coefficient for the Peng–Robinson equation of state can be found in Eq. (5.27) as

$$\begin{aligned} \ln\phi_i = {} & \frac{q_i}{q}(Z-1) - \ln(Z-q) \\ & + \frac{p}{2\sqrt{2}q}\left(\frac{q_i}{q} - \frac{2}{p}\sum_j y_j p_{ij}\right)\ln\left(\frac{Z + \left(1+\sqrt{2}\right)q}{Z + \left(1-\sqrt{2}\right)q}\right) \end{aligned} \tag{5.64}$$

---

[13] A modification of the Peng–Robinson equation of state proposed by Stryjek and Vera [163–165] reproduces well the vapor pressure data of pure compounds. This cubic equation of state, the PRSV equation has one adjustable parameter per pure compound. When coupled with suitable two-parameter mixing rules, it may be used to reproduce vapor–liquid equilibrium data of mixtures with accuracy comparable to that of some common dual methods using activity coefficients for the liquid phase and equations of state for the vapor phase.

In PRSV EOS, parameter κ is written as

$$\kappa_i = \kappa_{i,0} + \kappa_{i,1}\left(1 + \sqrt{\frac{T}{T_{c,i}}}\right)\left(0.7 - \frac{T}{T_{c,i}}\right),$$

where

$$\kappa_{i,0} = 0.378893 + 1.489715\omega_i - 0.171384\omega_i^2 + 0.0196544\omega_i^3,$$

and $\kappa_{i,1}$ is the single pure compound adjustable parameter.

5.8.3.1 Example

For concreteness, let us consider a methanol–water binary solution and that we want to find the pressure and the molar fraction y_1 of methanol in the vapor phase when this solution is in equilibrium with its vapor at $T = 50$ °C and the molar fraction of methanol in the liquid phase is $x_1 = 0.3$.

The equilibrium condition is given by

$$\gamma_1^\ell x_1 P_{s,1} = y_1 P K_1 \tag{5.65}$$

$$\gamma_2^\ell (1 - x_1) P_{s,2} = (1 - y_1) P K_2 \tag{5.66}$$

In these equations, the only unknowns are the pressure P and the molar fraction y_1. Let us first, calculate the fugacity coefficients for the pure components $\phi_{1,pure}^\ell \left(T, P_{s,1}\right)$ and $\phi_{2,pure}^\ell \left(T, P_{s,2}\right)$ at saturated conditions. For $i = 1, 2$

$$\begin{aligned} \ln \phi_{i,pure}^\ell &= (Z_i - 1) - \ln (Z_i - q_i) \\ &- \frac{p_i}{2\sqrt{2} q_i} \ln \left(\frac{Z_i + \left(1 + \sqrt{2}\right) q_i}{Z_i + \left(1 - \sqrt{2}\right) q_i} \right) \end{aligned} \tag{5.67}$$

where

$$p_i = \frac{a_i P_{s,i}}{(RT)^2}, \quad q_i = \frac{b_i P_{s,i}}{RT},$$

and Z_i is the solution of

$$Z_i^3 + A_i Z_i^2 + B_i Z_i + C_i = 0,$$

where

$$A_i = (q_i - 1), \quad B_i = p_i - q_i (3q_i + 2), \quad C_i = q_i \left(q_i + q_i^2 - p_i\right).$$

For each pure compound (methanol or water), we have three roots for the compressibility factor. Since

$$Z_i = \frac{P_{s,i} \bar{v}_i}{RT},$$

the one that corresponds to the pure component in the liquid phase is the lowest one.

Table 5.5 Peng–Robinson parameters, compressibility factor, and fugacity coefficient for pure components methanol and water in the saturated liquid state at 50 °C

	Methanol	Water
$a_i\ \left(\frac{\text{Jm}^3}{\text{mol}^2}\right)$	1.6621	0.97206
$b_i\ (\text{m}^3/\text{mol})$	4.1016×10^{-5}	1.8919×10^{-5}
$P_{s,i}$ (mm Hg)	417.4	92.5
p_i	1.2646×10^{-2}	1.639×10^{-3}
q_i	8.3845×10^{-4}	8.5706×10^{-5}
A_i	− 0.99916	− 0.99991
B_i	1.0967×10^{-2}	1.4676×10^{-3}
C_i	-9.8995×10^{-6}	-1.3313×10^{-7}
Z_i	9.9228×10^{-4}	9.7141×10^{-5}
$v_{ls,i}\ (\text{m}^3/\text{mol})$	4.8541×10^{-5}	2.1443×10^{-5}
$\phi^{\ell}_{i,pure}(T, P_{s.i})$	0.64058	0.66289

Table 5.5 shows the Peng–Robinson parameters, compressibility factor, and fugacity coefficient for pure components methanol and water in the saturated liquid state at 50 °C.

When we neglect the Poynting correction, parameters $K_1$and K_2 in Eqs. (5.65) and (5.66) are given by

$$K_1 = \frac{\phi^v_1(T, P, y_1)}{\phi^{\ell}_{1,pure}(T, P_{s.1})},$$

$$K_2 = \frac{\phi^v_2(T, P, y_1)}{\phi^{\ell}_{2,pure}(T, P_{s.2})}.$$

In the above equations

$$\ln \phi^v_1 = \frac{q_1}{q}(Z-1) - \ln(Z-q)$$
$$+ \frac{p}{2\sqrt{2}q}\left(\frac{q_1}{q} - 2\frac{y_1 p_1 + (1-y_1)p_{12}}{p}\right)\ln\left(\frac{Z+\left(1+\sqrt{2}\right)q}{Z+\left(1-\sqrt{2}\right)q}\right)$$

$$\ln \phi_2^v = \frac{q_2}{q}(Z-1) - \ln(Z-q)$$
$$+ \frac{p}{2\sqrt{2}q}\left(\frac{q_2}{q} - 2\frac{(1-y_1)\,p_2 + y_1 p_{12}}{p}\right)\ln\left(\frac{Z+\left(1+\sqrt{2}\right)q}{Z+\left(1-\sqrt{2}\right)q}\right)$$

where Z is the root of Eq. (5.63). Parameters q_1 and q_2 are dimensionless molecular volumes

$$q_1 = \frac{b_1 P}{RT}, \quad q_2 = \frac{b_2 P}{RT},$$

parameter q is a weighted average considering the molecular volumes of molecules 1 and 2,

$$q = y_1 q_1 + (1-y_1)\, q_2,$$

parameters $p_{11} = p_1$ and $p_{22} = p_2$ are dimensionless force parameters for the interactions between identical molecules and p_{12} for the interactions between different molecules

$$p_{11} = p_1 = \frac{a_1 P}{(RT)^2}, \quad p_{22} = p_2 = \frac{a_2 P}{(RT)^2} \quad p_{12} = \frac{a_{12} P}{(RT)^2},$$

and parameter p is the weighted average of the above force parameters

$$p = y_1^2 p_{11} + 2y_1(1-y_1)\, p_{12} + (1-y_1)^2\, p_{22}.$$

The solution procedure is the same as the one shown in Sect. 5.8.2 for the virial equation of state. We first consider parameters $K_1 = K_2 = 1$ and calculate P and y_1 by solving Eqs. (5.65) and (5.66). Then parameters K_1 and K_2 are recalculated with these first estimations for P and y_1 and Eqs. (5.65) and (5.66) are solved again. This is repeated till convergence.

5.8.3.2 Exercises

1. Starting from Eq. (5.62), find the form of this equation of state given by Eq. (5.63).
2. Consider a methanol–water mixture in the vapor phase at $T = 50$ °C and $P = 150$ mm Hg when the molar fraction of methanol is $y_{meth} = 0.8$. Supposing that this vapor mixture is well described by the Peng–Robinson equation of state, find the compressibility factor Z and the molar volume of the mixture.
   Answer: The roots are $Z = \{3.8037 \times 10^{-4}, 3.9453 \times 10^{-3}, 0.99535\}$, but only the larger root applies because at this temperature and pressure, the mixture

of methanol–water is in the vapor phase (see Fig. 5.14). The molar volume is $\overline{v} = 0.11291\ m^3/mol$.

3. Solve Eq. (5.62) and find the molar volume for the methanol–water mixture in the same conditions of the previous exercise.
4. Consider a mixture of acetone and n-hexane at $T = 318$ K. Calculate the activity coefficients for molar fractions of acetone varying between 0 and 1 and, considering the vapor phase as a mixture of ideal gases, draw a phase diagram such as the one shown in Fig. 5.14 based on the functional groups theory. In the following, consider that the mixture of vapors follows a Peng–Robinson equation of state, find the fugacity $\phi^{\ell}_{i,pure}\left(T, P_{s,i}\right)$ of these compounds in their saturated states at 318 K and the fugacity of each compound in the vapor phase $\phi^{v}_{i}\ (T, P, y_i)$. Use these results to redraw the phase diagram and compare these last results with the previous ones. Finally, try to compare your results with available experimental results.
5. Consider a mixture of methylcyclohexane and ethylene at 50 °C. Find the activity coefficients based on functional groups and using the PRSV equation of state, find the fugacity $\phi^{\ell}_{i,pure}\left(T, P_{s,i}\right)$ of these compounds in their saturated states at 50 °C and the fugacity of each compound in the vapor phase $\phi^{v}_{i}\ (T, P, y_i)$. Use these results to find K_1 and K_2, the pressure and the molar fraction of ethylene in the vapor phase when the molar fraction of ethylene in the liquid phase is x_{ethy} varies from 0 to 1. Plot these values in a phase diagram such as the one shown in Fig. 5.14 and compare your results with Raoult's law. Functional group parameters are given in Tables 8.23 and 8.24 of [137] and parameters ω and κ_1 are given in [141].

5.9 Summary

In contrast with single-component systems, when the pressure is increased over a mixture of vapors, the vapor–liquid transition pressure does not remain constant during the vapor condensation process at a constant temperature. Therefore, for a given vapor composition, the liquid and vapor will coexist in equilibrium in a range of pressures between the dew and bubble states defined, respectively, as the states where the first drop of liquid appears and the last bubble of vapor condensates.

The calculation of the dew and bubble states is dependent on the knowledge of:

- The activity coefficients γ^{ℓ}_{i} of the several components that form the liquid solution;
- The fugacity coefficient ϕ^{v}_{i} of these components in the vapor state.

The activity coefficient γ^{ℓ}_{i} is a measure of the non-ideality of the liquid solution and the fugacity coefficient ϕ^{v}_{i} is a measure of the non-ideality of the mixture of vapors.

For ideal solutions $\gamma^{\ell}_{i} = 1$ and for mixtures of ideal gases $\phi^{v}_{i} = 1$.

In this simplest situation, the distribution of components among the liquid and vapor phases can be estimated by Raoult's law.

Unfortunately, Raoult's law cannot be used for most liquid solutions which are non-ideal and require experimental values or a method for estimating the activity coefficient.

In this chapter, we considered the molecules as composed of functional groups, and the activity coefficients were estimated by taking into account their geometrical and molecular potential contributions.

On the other hand, the estimation of the vapor fugacity coefficients is dependent on the hypothesis it is possible to make about the thermodynamic state of the vapor phase.

When the pressure is low enough, the vapor phase may be considered a mixture of ideal gases. Otherwise, we should use an equation of state we judge appropriate for the mixture of vapors.

Two alternatives were presented in this chapter for the estimation of the fugacity coefficients based, respectively: (i) on a virial equation of state and (ii) on a cubic equation of state.

Chapter 6
Surface Physics

Abstract Scientific interest in surface physics began very early with the XVIII century efforts trying to explain surface tension and capillarity. Molecules on the surface of a liquid are pulled inwards by the remaining molecules in the liquid phase and this produces a state of tension on the liquid surface. On the other hand, when interacting with a solid surface liquid molecules are subjected to attractive forces from the solid molecules, and, when these forces prevail over the forces between liquid molecules, this will give rise to a contact angle smaller than 90° and to capillary ascension inside small bore tubes. Surface tension forces liquid surfaces to contract to the minimal area and Laplace's law relates the surface tension with the internal pressure that is produced and to the curvature of the liquid drops. Nevertheless, surface tension has a molecular origin and is independent of surface curvature. In this chapter, we try to give a deeper insight into the molecular scale for explaining surface tension under the light of van der Waals theory considering the surface as a region where the fluid density has a strong decay. It is then possible to relate the underlying molecular parameters with the Helmholtz energy in excess with respect to the liquid and vapor bulk phases. In liquid–solid and vapor–solid interaction, interfacial excess energies in the triple line region are also the controlling macroscopic mechanisms for the contact angle when a fluid interacts with a hydrophilic or hydrophobic surface and are important macroscopic parameters in dynamical problems where the contact angle gets away from its equilibrium value. A section is dedicated to exposing Cahn's theory for finding the liquid–solid and vapor–solid interfacial energies. The chapter ends with a brief overview of emulsions.

6.1 Introduction

The cohesive forces among liquid molecules are responsible for the phenomenon of surface tension. In the bulk of the liquid, each molecule is pulled equally in every direction by neighboring liquid molecules, resulting in a null net force. The molecules at the surface do not have the same molecules on all sides of them and therefore are pulled inwards. This creates some internal pressure and, also, forces liquid surfaces to contract to the minimal area. Surface tension is responsible for

P. C. Philippi, *Thermodynamics*, https://doi.org/10.1007/978-3-031-49357-7_6

the shape of liquid droplets. Although easily deformed, droplets of water tend to be pulled into a spherical shape by the imbalance in cohesive forces of the surface layer. In the absence of other forces, including gravity, drops of virtually all liquids would be spherical.

Another way to view surface tension is in terms of energy. A molecule in contact with a neighbor is in a lower state of energy than if it were alone (*not in contact with a neighbor*). The interior molecules have as many neighbors as they can possibly have, but the boundary molecules are missing neighbors (*compared to interior molecules*) and therefore have higher energy. For the liquid to minimize its energy state, the number of higher energy boundary molecules must be minimized. The minimized quantity of boundary molecules results in a minimal surface area. As a result of surface area minimization, a surface will assume the smoothest shape it can. Since any local curvature in the surface shape results in a greater area, a higher energy will also result. Consequently, the surface will push back against any curvature singularity.

Research in surface physics was first undertaken to explain capillarity. In fact, capillarity seemed to contradict the laws of fluid equilibrium. In consequence, many worthless theories have been proposed with a view to explain apparent anomalies.

In accordance with Bashfort [8], "*after long groping in the dark, it was found to be desirable to discover by experiment what were the actual phenomena which required explanation*". Hauksbee[1] found that the height to which a fluid would rise in a capillary tube of a given radius was the same for all thicknesses of the tube wall. From this, it was apparent that the attracting force of the tube was situated at or near the inner surface of the tube. But Hauksbee does not appear to have taken account of the mutual attractions of the fluid particles.

James Jurin,[2] a physician educated at Cambridge and Leiden made an important experiment that added a new fact to those discovered by Hauksbee; the height to which water rose in a tube depended only on the diameter at the position of the meniscus. A tube that was wide at the bottom but narrow at the top could therefore hold in suspension a greater volume of water than one of a uniform bore. This fact undermined Hauksbee's not very coherent explanation that the rise was due to a diminution of the "*gravitating force*" by a horizontal attraction of the whole of the glass wall. Jurin claimed to have found "*the real cause of that phenomenon, which is the attraction of the periphery, or section of the surface of the tube, to which the upper surface of the water is contiguous and coheres*". He expounded six propositions, such as, for example, that water particles attract water but not as strongly as they are attracted to glass, whereas mercury (quicksilver) attracts mercury more strongly than mercury is attracted to glass. He established that *the depression of mercury in a capillary tube, like the rise of water, is the reciprocal of the bore*.

---

[1] Francis Hauksbee (1710) '*An Account of an Experiment Touching the Ascent of Water Between Two Glass Planes, in a Hyperbolic Figure*'. Phil. Trans. 27, pp. 539–540 [64].

[2] James Jurin (1717) '*An Account of Some New Experiments, relating to the Action of Glass Tubes upon Water and Quicksilver*'. James Jurin, M. D. Reg. Soc. and Coll. Med. Lond. Soc. Phil. Trans. 30, pp.1083–1096 [82].

Clairaut[3] was the first to attempt to explain capillary phenomena on right principles, by referring them to the mutual attraction of the particles of the fluid, and to the attraction of the particles of the solid on the particles of the fluid and supposing these attractions to depend upon the same function of the distance. After this precise and promising start his analysis quickly goes astray. He assumes that all the liquid in a tube of, say, one-twentieth of an inch in diameter is within the attractive range of the glass walls, an assumption tentatively made by Hauksbee in spite of the evidence of his own experiments.

Shortly afterwards Segner[4] introduced the supposition that forces of attraction of both the particles of the solid and of the fluid "*were not sensible at sensible distances*". He concluded that these forces gave a constant tension to the free capillary surfaces, and tried to calculate the forms of sessile drops of fluid with a view to compare them with their measured forms. But in his calculations, he took into account only the curvature of the vertical sections made by a plane passing through the axis of the drop. In addition, his measurements of the actual forms appear not to have been very precise.

In 1805 Thomas Young wrote[5]: "*It has already been asserted, by Mr. Monge and others, that the phenomena of capillary tubes are referable to the cohesive attraction of the superficial particles only of the fluids employed*" … "*Segner, who appears to have been the first that maintained a similar opinion, has shown in what manner the principle may be deduced from the doctrine of attraction, but his demonstration is complicated, and not perfectly satisfactory; in applying the law to the forms of drops, he has neglected to consider the very material effects of the double curvature, which is evidently the cause of the want of a perfect coincidence of some of his experiments with his theory*".

Young criticizes Segner's notion of a tension only on surfaces of variable curvature and couples this idea with the assertion that there is a fixed angle of contact between any given pair of liquid and solid, an assertion which he describes as "*one observation, which appears to be new, and which is equally consistent with theory and with experiment*". He uses these two facts to produce the first satisfactory phenomenological treatment of capillary rise. He writes: "*It is well known, and it results immediately from the composition of forces, that where a line is equably distended, the force that it exerts, in a direction perpendicular to its own, is directly as its curvature; and the same is true of a surface of simple curvature; but where the curvature is double, each curvature has its appropriate effect, and the joint force must be the sum of the curvatures in any two perpendicular directions*".

Young's work was and would have been seen by his predecessors as a triumphant success. The next advance came at once; it required Laplace's combination of physical insight and a mathematical grasp which grew from the resurgence of French mathematics at the turn of the XVIII century.

---

[3] M. Clairaut (1743), '*Theorie de la figure de la Terre*', David Fils, Paris.

[4] I. A. Segner (1751), '*De figuris superficierum fluidarum*', Comment. Soc. Reg. Sci. Gottingensis 1, pp. 301–372.

[5] T. Young (1805), 'An essay on the cohesion of fluids', Phil. Trans. Roy. Soc. 95, pp. 65–87.

6.2 Laplace

> In the field of capillarity it is usual to consider together the work of Young and Laplace, and it is true that they both obtained some of the same important results within a year of each other. Their aims and methods were, however, quite different. In reading Young we are reading 18th-century natural philosophy; in reading Laplace we are reading 19th-century theoretical physics. This 'sea-change' in the early years of the new century is as dramatic as that of the scientific revolution of the 17th century, and was due to the efforts of the great French school of mathematical physics of that time. This is not the place to discuss the origin of this second revolution but to concentrate only on how it led to a revival of the subject of cohesion and to a second period of advance. The man responsible was Laplace. Rowlinson [147]

Laplace, in his '*Traité de Mécanique Céleste*' published in 1805[6] is considered to be the first who formally derived the law relating the pressure with the surface tension that is presently known as Young–Laplace's law.

In his 156 pages dedicated to the study of the surface tension phenomena, he gave a very detailed treatment of the forces responsible for capillarity: "*Jurin analyse avec exactitude, toutes les forces qui peuvent concourir á élever l'eau dans un tube de verre. Mais sa théorie, exposée avec l'élégance qui caractérise son bel Ouvrage, laisse á desirer l'explication de la loi de cette ascension qui, d'après l' expérience, est en raison inverse du diamètre du tube*".

He, also, attributed the hydrophilic or hydrophobic behavior of solid surfaces to the short range interaction between the fluid molecules and the solid surface, responsible for the contact angle "*Ainsi l'attraction des tubes capillaires n'a d'influence sur l'élévation ou sur l'abaissement des fluides qu'ils renferment qu' en déterminant l' inclinaison des premiers plans de la surface du fluide intérieur, extrêmement voisins des parois du tube, inclinaison dont dépend la concavité ou la convexité de cette surface, et la grandeur de son rayon*".

Figure 6.1 shows Figure 6 of Laplace's supplement 'Sur l'action capillaire' [99].

For deriving Laplace's law consider a liquid droplet (phase α) in mechanical equilibrium with a phase β (*a gas, another liquid or its own vapor*), in absence of gravity (Fig. 6.2). In these conditions, the droplet surface will be spherical. In fact, as commented above, whereas, in the bulk phases, the intermolecular forces are balanced, on the surface, the liquid molecules are subjected to an unbalanced field of intermolecular forces. These molecules will attract themselves with electrostatic forces, being as closest as possible one from another and the sphere offers the minimal superficial area for this purpose.

Suppose now that, by any means, we try to detach the surface element shown in Fig. 6.2 from the liquid droplet. The remaining liquid molecules will react to this detachment with a force $\gamma 2\pi\ (R \sin\theta)$ directed outward the surface element and acting along the perimeter of this element shown in Fig. 6.2. Thus γ is considered

---

[6] P. S. Laplace (1805), *Sur l' action capillaire*, Supplement au Dixième Livre du Traité de Mécanique Céleste, Courcier, Paris, pp. 398–553 [99].

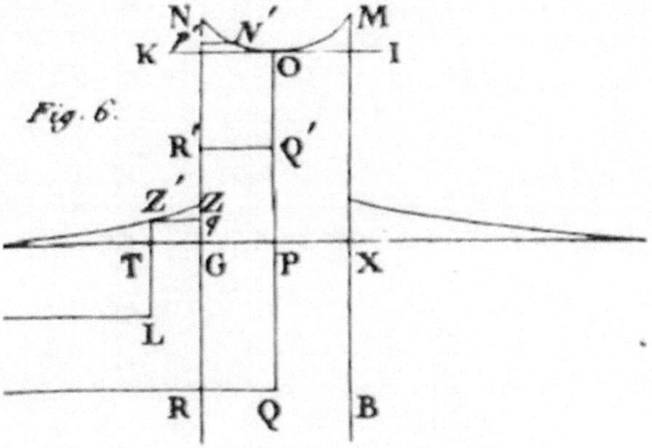

Fig. 6.1 Figure 6 of Laplace's supplement *Sur l'action capillaire*

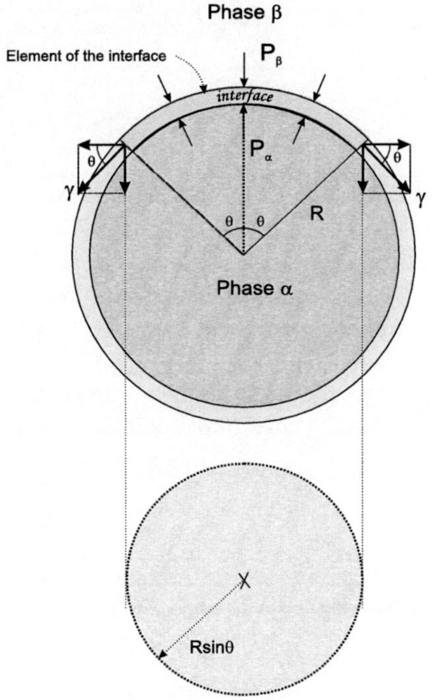

Fig. 6.2 Derivation of Laplace's law

to be related to the—short-range—attraction forces between contiguous molecules along the perimeter of the detached surface element.

While the horizontal component of this force is null, its vertical component $\gamma 2\pi\,(R\sin\theta)\sin\theta$ will be balanced by the pressure difference $P_\alpha - P_\beta$ on the two faces of the curved surface, in such a way that

$$\left(P_\alpha - P_\beta\right)\pi\,(R\sin\theta)^2 = \gamma 2\pi\,(R\sin\theta)\sin\theta,$$

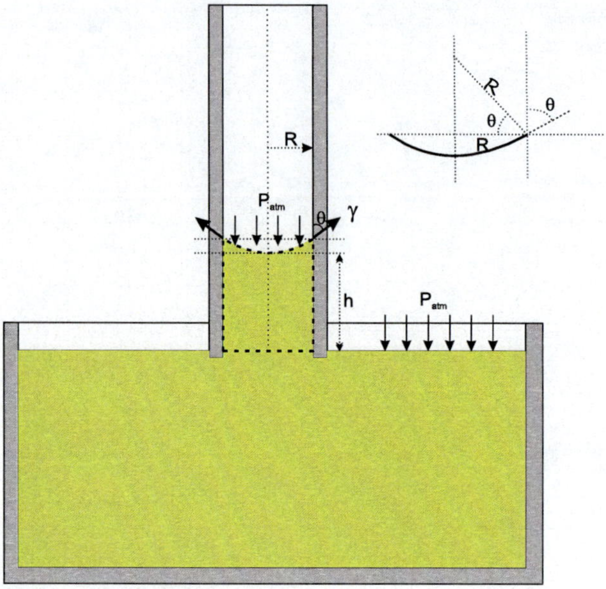

Fig. 6.3 Capillary rise inside a cylindrical tube

or

$$P_\alpha - P_\beta = \gamma \frac{2}{R}. \tag{6.1}$$

The above equation can be generalized for any curved interface by taking into account its curvature radius $\mathcal{R}$, given by R_1 and R_2 in accordance with

$$\frac{1}{\mathcal{R}} = \frac{1}{R_1} + \frac{1}{R_2}.$$

Equation (6.1) is the famous Laplace's equation, proposed in 1805 [99], which relates the pressure imbalance in curved surfaces of liquids with the interface tension.

6.2.1 *Exercises*

1. Consider water rising inside a capillary cylindrical tube with radius R (Fig. 6.3). By making a balance of forces for the volume indicated by the dashed lines show that, in equilibrium, the height h of the liquid column is given by Jurin's law [82],

$$h = \frac{2\gamma \cos\theta}{\rho g R},$$

where θ is the contact angle between the liquid and the solid surface, g is the local acceleration due to gravity and ρ is the density of water. What is the liquid pressure on the top of the liquid column?

2. Starting from the Laplace's equation, Eq. (6.1), and supposing the meniscus surface to be spherical with a radius of curvature $\mathcal{R}$, retrieve the Jurin's law of Exercise 1.
3. Consider now that during the capillary rising the height $h = h(t)$ and that the liquid rises into the tube due to the imbalance of capillary F_{cap}, gravitational F_g and viscous F_{visc} forces in such a way that

$$\frac{d}{dt}\left(\varrho\pi R^2 h \frac{dh}{dt}\right) = F_{cap} - F_g - F_{visc},$$

in accordance with Newton's second law. Find a differential equation for $h(t)$ and try to solve this (nonlinear) equation[7] using, e.g. a Runge–Kutta method, plotting h in terms of the time t. What are your conclusions? By considering a Hagen–Poiseuille flow, the viscous force is given by

$$F_{visc} = 8\pi\eta h \frac{dh}{dt}$$

where η is the dynamic viscosity of the liquid.

6.3 The Form of a Sessile Drop

> Many years have elapsed since this work was commenced, and it is even now only partially completed. My object was to test the received theories of Capillary Action, and through them the assumed laws of molecular attraction, on which they are founded. To this end it was proposed to compare the actual forms of drops of fluid resting on horizontal planes they do not wet, with their theoretical forms. After some trials, a satisfactory micrometrical instrument was constructed for the measurement of the forms of drops of fluid, but my attempts to calculate their forms as surfaces of double curvature failed entirely, and my undertaking must have ended here if I had depended upon my own resources. But at this point Professor J. C. Adams furnished me with a perfectly satisfactory method of calculating by quadratures the exact theoretical forms of drops of fluids from the Differential Equation of Laplace, an account of which he has now had the kindness to prepare for publication.
> Francis Bashforth (1883), "*An attempt to test the theories of capillary action by comparing the theoretical and measured forms of drops of fluids*", Cambridge University Press, p. 158 [8]

As it was commented at the beginning of this chapter, a liquid drop deposed on a horizontal surface will have a spherical form when it not subjected to gravity.

Consider, now, a liquid drop that is deposed on a horizontal surface and under the action of gravity g (Fig. 6.4). Considering radial symmetry the problem is to find the

[7] The Bosanquet equation [18].

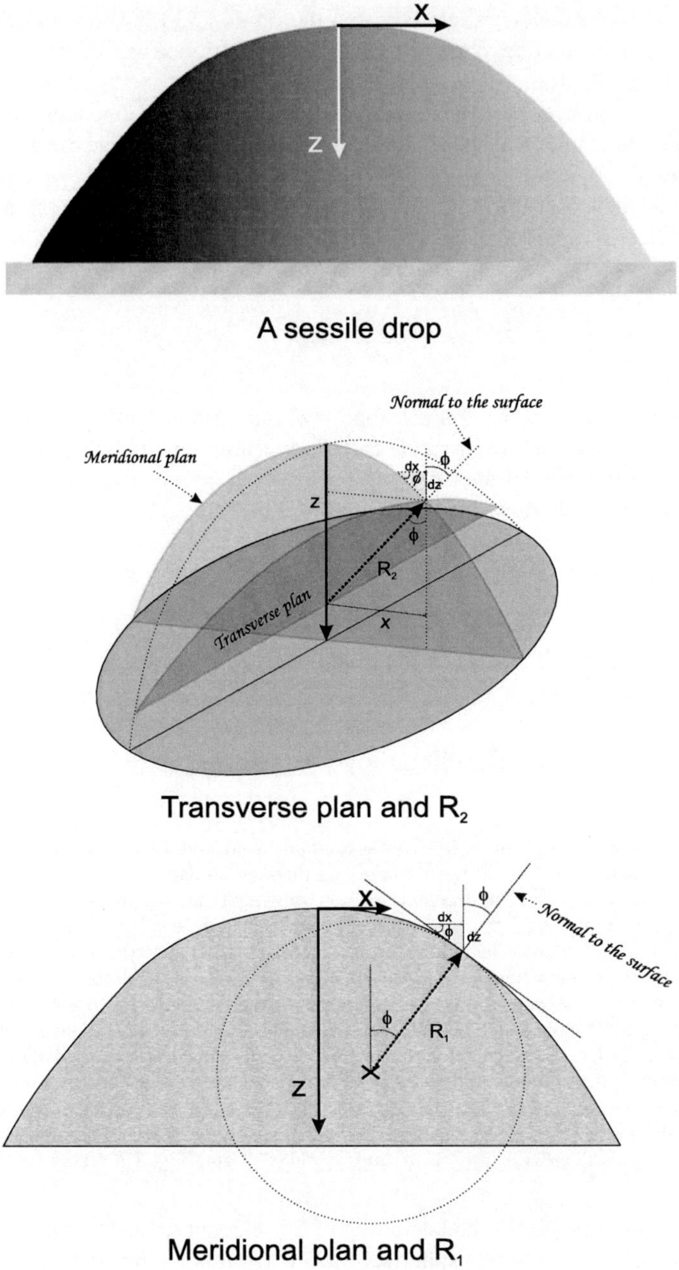

Fig. 6.4 A sessile drop and its meridional and transverse plans

equation of the drop surface, i.e., the relationship between the vertical coordinate z and the horizontal coordinate x under equilibrium conditions in the meridional plan shown in the figure.

For each point in the surface, we have two curvature radii. R_1 is the curvature radius on the meridional plan and R_2 is the curvature radius on the transverse plan. The projection of R_2 on the horizontal direction of the meridional plan is the ordinate x. So

$$x = R_2 \sin \phi,$$

where ϕ is the angle between the normal to the surface and the vertical direction. Nevertheless,

$$\tan \phi = \frac{dz}{dx},$$

and so

$$\sin \phi = \frac{\frac{\sin \phi}{\cos \phi}}{\sec \phi} = \frac{\tan \phi}{\left(1 + \tan^2 \phi\right)^{\frac{1}{2}}}.$$

It results

$$\frac{1}{R_2} = \frac{\frac{dz}{dx}}{x\left(1 + \left(\frac{dz}{dx}\right)^2\right)^{\frac{1}{2}}}.$$

On the other hand, the curvature radius R_1 is given by

$$\frac{1}{R_1} = \frac{d\phi}{ds} = \frac{d\phi}{dx}\frac{dx}{ds}.$$

But

$$\frac{dx}{ds} = \cos \phi = \frac{1}{\sec \phi} = \frac{1}{\left(1 + \left(\frac{dz}{dx}\right)^2\right)^{\frac{1}{2}}},$$

and

$$\frac{d\phi}{dx} = \frac{d}{dx}\left(\arctan \frac{dz}{dx}\right) = \frac{\frac{d^2z}{dx^2}}{1 + \left(\frac{dz}{dx}\right)^2}.$$

It results that

$$\frac{\frac{d^2z}{dx^2}}{\left(1 + \left(\frac{dz}{dx}\right)^2\right)^{\frac{3}{2}}} + \frac{\frac{dz}{dx}}{x\left(1 + \left(\frac{dz}{dx}\right)^2\right)^{\frac{1}{2}}} = \frac{\Delta P}{\gamma}. \tag{6.2}$$

The pressure difference can be written as

$$\Delta P = (\rho_\ell - \rho_g)\, gz + C,$$

where C is a constant that can be determined considering that when $z \to 0$ the drop surface becomes spherical with $R_1 = R_2 = b$. So

$$C = \frac{2\gamma}{b},$$

and Eq. (6.2) becomes

$$\frac{\frac{d^2z}{dx^2}}{\left(1+\left(\frac{dz}{dx}\right)^2\right)^{\frac{3}{2}}} + \frac{\frac{dz}{dx}}{x\left(1+\left(\frac{dz}{dx}\right)^2\right)^{\frac{1}{2}}} = \frac{(\rho_\ell - \rho_g)\, gz}{\gamma} + \frac{2}{b}. \tag{6.3}$$

Numerical solutions of the sessile drop problem are known for more than a century [8, 126] and results of this theory are found in many applications. For example, a technique to measure surface tension experimentally is based on comparing measured and calculated profiles of a drop of suitable size [23].

In addition to numerical solutions, analytical solutions were provided by some authors in particular situations. See, e.g., [32].

The solution of Eq. (6.3) is specially important for measuring the surface tension by the pendant drop method. Song and Springer [160] developed a computer-aided method based on the comparison between the theoretical profile given by Eq. (6.3) and the profile that is extracted from a pendant liquid drop (Fig. 6.5) using image analysis techniques. In this case, the drop apex is chosen at the lower point of the drop and the ordinate z is directed upward.

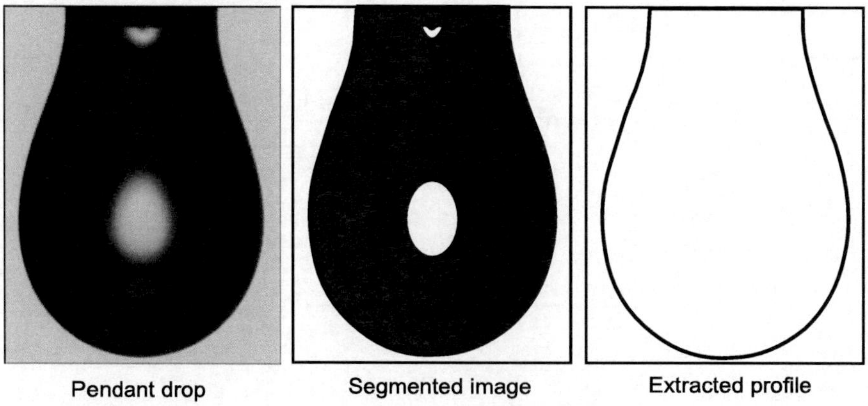

Fig. 6.5 Measurement of the interface tension by the pendant drop method

6.4 Surface Tension and the Helmholtz Energy

A fluid interface separating two phases a and b is a region where some thermodynamic properties have a strong variation. This is the case of the density when phase a is a liquid and phase b its vapor. From a molecular standpoint, the interface is a region where molecules of phase a are simultaneously subjected to intermolecular attraction forces from its own phase a and from the nearby molecules of phase b. When the kinetic energy of a molecule in phase a is large enough to compensate for its potential energy—which is always negative—this molecule will change to phase b. The same comment for molecules in phase b.

In this section, we are not concerned with phase transition, but with the mechanical properties of fluid interfaces. So we will suppose that there is no mass transfer between the two phases or that the two phases are already in equilibrium with respect to mass transfer, i.e., that the Gibbs potentials μ_i^a and μ_i^b are the same in the two phases for $i = a, b$. In this aspect, we can consider that the interface has an energy excess with respect to the bulk phases. In fact, the potential energy related to the intermolecular attraction between molecules in phase a and a single molecule a in the interface can be quantified as

$$\phi_{aa}(\mathbf{x}) = \int_{\|\mathbf{r}-\mathbf{x}\| \geq \sigma_a} \phi_{aa}^{(1-2)}(\|\mathbf{r}-\mathbf{x}\|)\, n_a(\mathbf{r})\, d\mathbf{r} \leq 0, \tag{6.4}$$

where $\phi_{aa}^{(1-2)}(\|\mathbf{r}-\mathbf{x}\|)$ is the potential energy related to the intermolecular attraction between any two molecules a, supposed to be only dependent on the distance $\|\mathbf{r}-\mathbf{x}\|$ that separates a molecule a in $\mathbf{r}$ from the molecule a in the position $\mathbf{x}$ of the interface, σ_a is the molecular diameter of molecule a and n_a is the number of molecules a per unit volume.

In the bulk phase n_a is constant, but, in the interface, n_a decreases and $\phi_{aa} \to 0$. This gradient in the potential energy is responsible for a force on molecule a

$$\mathbf{g}_{aa} = -\frac{\nabla \phi_{aa}}{m},$$

directed toward the bulk phase a. Using the same reasoning we can conclude that phase b will attract the molecule a with a force

$$\mathbf{g}_{ab} = -\frac{\nabla \phi_{ab}}{m},$$

and this force is responsible for the reduction of the interface tension γ_{ab} with respect to the surface tension γ_a, i.e., with respect to the surface force that arises, when a liquid is in equilibrium with its own vapor.

Since the potential energy is always negative, a reduction on its absolute value means an excess in energy. For the molecules in the interface to minimize this energy excess, the number of higher energy boundary molecules must be minimized.

From a macroscopic point of view, the Helmholtz energy of this two-phase system can be written as

$$F = F_a + F_b + F_{ab},$$

where F_{ab} is the Helmholtz energy in excess in the interface. In this case, our thermodynamic system is considered to be a drop of phase a immersed on phase b.

If a small perturbation results on an volume increase dV_a and, consequently, a volume decrease $dV_b = -dV_a$ at constant temperature,

$$dF_a = -P_a dV_a,$$
$$dF_b = P_b dV_a,$$

and

$$dF_{ab} = f_{ab} dA_{ab},$$

where f_{ab} is the Helmholtz energy per unit area of the interface.

Since in equilibrium, F must be a minimum when the total volume and temperature are kept constants

$$(P_b - P_a)\, dV_a + f_{ab} dA_{ab} = 0,$$

and we retrieve Laplace's law Eq. (6.1)

$$P_a - P_b = \frac{\gamma_{ab}}{\mathcal{R}},$$

where $\gamma_{ab} = f_{ab}$ and

$$\frac{1}{\mathcal{R}} = \frac{dA_{ab}}{dV_a}.$$

To convince ourselves that the interface tension is the Helmholtz energy in excess in the interface, let us derive Laplace's law based, this time, on the principle of maximum entropy, instead of the principle of minimum Helmholtz energy. In this case, we must keep the internal energy constant and the entropy variations that follows a small volume variation of phase a will be

$$dS_a = \frac{1}{T_a} dU_a + \frac{P_a}{T_a} dV_a,$$
$$dS_b = \frac{1}{T_b} dU_b + \frac{P_b}{T_b} dV_b,$$
$$dS_{ab} = s_{ab} dA_{ab}.$$

On the other hand

$$dU = dU_a + dU_b + u_{ab} dA_{ab} = 0,$$

as a condition for the entropy to be maximum in equilibrium.

Therefore

$$dS = \left(\frac{1}{T_a} - \frac{1}{T_b}\right) dU_a + \left(\frac{P_a}{T_a} - \frac{P_b}{T_b}\right) dV_a - \left(\frac{u_{ab}}{T_b} - s_{ab}\right) dA_{ab}$$

In equilibrium conditions

$$T_a = T_b = T,$$

and

$$dS = 0.$$

So

$$P_a - P_b = \left(\underbrace{u_{ab} - T s_{ab}}_{= f_{ab}}\right) \frac{dA_{ab}}{dV_a}$$

as it was to be demonstrated.

6.4.1 *Exercise*

Show that Laplace's equation can be also derived using the minimum principle for the internal energy (*at constant entropy*).

6.5 A Deeper Insight into the Physics of Interfaces

To get a deeper insight into the physics of interfaces let us go back to Eq. (6.4) expressing the potential energy of a single molecule a in the interface, supposing, now, that phase a is a liquid and phase b its vapor. So omitting the index a and b let $n(\mathbf{r})$ be developed in a Taylor series around $n(\mathbf{x})$

$$n(\mathbf{r}) = n(\mathbf{x}) + \eta_\alpha \partial_\alpha n + \frac{1}{2} \eta_\alpha \eta_\beta \partial_{\alpha\beta} n + \cdots,$$

where $\eta = \mathbf{r} - \mathbf{x}$. Inserting this expansion into Eq. (6.4), since $\phi^{(1-2)}(\|\eta\|)$ is an even function of η,

$$\phi(\mathbf{x}) = -2an(\mathbf{x}) - \kappa\nabla^2 n + \cdots \tag{6.5}$$

where

$$a = -\frac{1}{2}\int_{\|\eta\|\geq\sigma_a} \phi^{(1-2)}\|\mathbf{x}\|\, d\eta \geq 0,$$
$$\kappa = -\int_{\|\eta\|\geq\sigma_a} \phi^{(1-2)}(\|\eta\|)\frac{1}{6}\eta^2 d\eta \geq 0,$$

are molecular parameters that depend on the interaction potential between the molecules. In fact, parameter a can be shown to be related to the attractive part of the van der Waals equation of state and parameter κ to the surface tension.

The force acting on a single molecule a and directed toward the liquid bulk phase will, thus, be given by

$$\mathbf{g} = \frac{2a}{m}\nabla n + \frac{\kappa}{m}\nabla\nabla^2 n,$$

and the force per unit volume acting on n liquid molecules can be written as

$$\begin{aligned}\rho\mathbf{g} &= 2an\nabla n + \kappa n\nabla\nabla^2 n\\ &= \nabla\left(an^2 + \kappa n\nabla^2 n + \frac{\kappa}{2}(\nabla n)^2\right) - \nabla\cdot(\kappa\nabla n\nabla n).\end{aligned}$$

In general the radius of curvature $\mathcal{R}$ is much larger than the interface thickness and this last expression can be simplified if we consider that the interface is very thin and the density variation happens along the normal direction x, neglecting the curvature $\frac{1}{\mathcal{R}}$ of the interface,

$$\rho g_x = \frac{d}{dx}\left(an^2 + \kappa n\frac{d^2 n}{dx^2} - \frac{\kappa}{2}\left(\frac{dn}{dx}\right)^2\right).$$

When the interface is in equilibrium and at rest this force creates a gradient in the density and, consequently, in the ideal-gas pressure $P_0 = nkT$, where $k = R/\mathcal{N}_{Avogrado}$ is the Boltzmann constant

$$\rho g_x = \frac{d(nkT)}{dx}, \tag{6.6}$$

in such a way that

$$\underbrace{\frac{d}{dx}\left(\kappa n\frac{d^2n}{dx^2}-\frac{\kappa}{2}\left(\frac{dn}{dx}\right)^2\right)}_{\text{surface force}}=\underbrace{\frac{d}{dx}\left(nkT-an^2\right)}_{\text{pressure gradient}}. \tag{6.7}$$

The above equation means that the surface force is locally equilibrated, in the interface, by a pressure gradient.[8] We can identify the term between the parenthesis in the r.h.s. of the above equation as the pressure P when the molecules are considered as material points subjected to intermolecular attractive forces.

By integrating Eq. (6.7)

$$\kappa n\frac{d^2n}{dx^2}-\frac{\kappa}{2}\left(\frac{dn}{dx}\right)^2=P+C,$$

where C is an integration constant that is to be assigned to the saturation pressure in the bulk phases, since there are no gradients in the bulk phases (*the liquid and vapor phase were assumed to be in equilibrium*)

$$C=-P_s(T).$$

Therefore,

$$\kappa n\frac{d^2n}{dx^2}-\frac{\kappa}{2}\left(\frac{dn}{dx}\right)^2=P-P_s(T). \tag{6.8}$$

The Helmholtz energy per unit volume can be written as

$$\psi=nf,$$

and if we identify, in accordance with van der Waals, the Helmholtz energy in excess in the interface as [181]

$$\psi_{exc}=\frac{\kappa}{2}\left(\frac{dn}{dx}\right)^2. \tag{6.9}$$

Equation (6.8) can be written as

$$n\frac{d\psi_{exc}}{dn}-\psi_{exc}=P-P_s(T),$$

or, by dividing the above equation by n^2,

[8] A more rigorous derivation of Eq. (6.7) can be given, considering it as a particular case of the momentum balance equation for two-phase systems. Interested readers are invited to read Sect. 8.2.

$$\frac{1}{n}\frac{d\psi_{exc}}{dn} - \frac{1}{n^2}\psi_{exc} = \frac{1}{n^2}\left(P - P_s\left(T\right)\right),$$

or

$$\frac{d}{dn}\left(\frac{\psi_{exc}}{n}\right) = \frac{1}{n^2}\left(P - P_s\left(T\right)\right).$$

Now, we integrate the above equation between the saturated number density of molecules in the vapor phase $n_{vs}\left(T\right)$ and an arbitrary number density n inside the interface,

$$\psi_{exc}\left(n, T\right) = n\int_{n_{vs}}^{n} \frac{1}{n'^2}\left(P\left(n', T\right) - P_s\left(T\right)\right) dn'. \tag{6.10}$$

Therefore, we get an equation expressing the Helmholtz energy in excess in the interface in terms of the equation of state of the substance with which we are dealing. This result is not surprising since is based on the equilibrium between surface forces and pressure gradient inside the interface, expressed by Eq. (6.7). So the effect of the surface forces is measured by the pressure gradient they produce across the interface. In the bulk liquid and vapor phases this pressure gradient is null and, also, the excess energy.

In fact, when the number density n becomes $n_{\ell s}$ in the liquid bulk phase $\psi_{exc} = 0$ and so

$$\int_{n_{vs}}^{n_{\ell s}} \frac{1}{n'^2}\left(P\left(n', T\right) - P_s\left(T\right)\right) dn' = 0,$$

or, by replacing n by $v = 1/n$ as the integration variable,

$$\int_{v_{\ell s}}^{v_{vs}} P dv' = P_s\left(v_{vs} - v_{\ell s}\right),$$

which is the well-known *Maxwell area rule*, enabling to find the saturation pressure $P_s\left(T\right)$ corresponding to a given equation of state $P\left(v, T\right)$.

Equation (6.10) can be written using dimensionless variables

$$\psi^*_{exc} = n^*\int_{n^*_{vs}}^{n^*} \frac{1}{\left(n'\right)^2}\left(P^*\left(n'\right) - P^*_s\left(T^*\right)\right) dn', \tag{6.11}$$

where

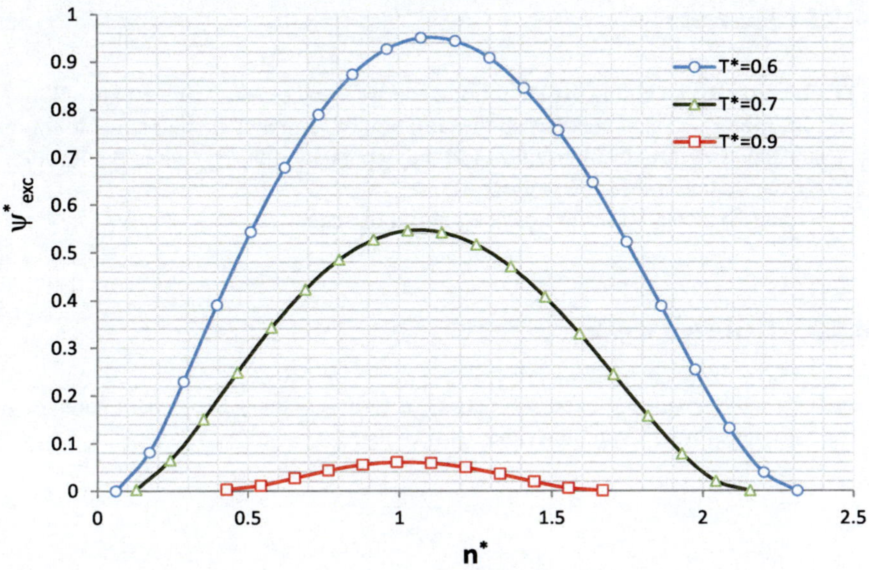

Fig. 6.6 Helmholtz excess energy in the liquid–vapor interface, calculated with Eq. (6.10) for different reduced temperatures. The van der Waals equation of state was used for the calculation

$$\psi_{exc}^* = \frac{\psi_{exc}}{P_c}, \quad n^* = \frac{n}{n_c}, \quad P^* = \frac{P}{P_c}, \quad T^* = \frac{T}{T_c}.$$

Although Eq. (6.7) was derived considering a system of material points with added attractive forces from the surface, more rigorous derivations can show that pressure P can be written in terms of n and T, using any arbitrary equation of state [156].

Figure 6.6 shows the excess function that is obtained with the van der Waals equation of state, which can be written in dimensionless form as

$$P^* = \frac{8n^*T^*}{3 - n^*} - 3n^{*^2}. \tag{6.12}$$

Results show that the Helmholtz energy in excess in the interface, ψ_{exc}^*, is always non-negative and becomes zero in the bulk phases in equilibrium when $n^* = n_{vs}^*$ and $n^* = n_{\ell s}^*$. Furthermore, as would be expected, a temperature increase reduces ψ_{exc}^*, which converges to zero when T approaches the critical temperature T_c.

6.5.1 Exercises

1. Equation (6.11) giving the Helmholtz energy in excess $\psi^*_{exc}(n^*, T^*)$ in a liquid–vapor interface is dependent on the equation of state that we find suitable for the substance we are dealing with. Find the expression for ψ^*_{exc} when the Redlich–Kwong equation of state is used.
2. Repeat Exercise 1 for the Soave equation of state.

6.5.2 Density Profile

Equation (6.9) gives the excess function in terms of the surface parameter κ. In dimensionless form, it can be written as

$$dx^* = \frac{dn^*}{\sqrt{2\psi^*_{exc}}},$$

where $x^* = \frac{x}{n_c\sqrt{\frac{\kappa}{P_c}}}$.

Therefore, by integrating the above equation between a point $\overline{x}$ in the interface where the density $n^* = \frac{n^*_{vs}+n^*_{\ell s}}{2}$ and an arbitrary point in the interface

$$x^* - \overline{x}^* = \int_{\frac{n^*_{vs}+n^*_{\ell s}}{2}}^{n^*} \frac{dn'}{\sqrt{2\psi^*_{exc}(n')}},$$

it is possible to find the density profile $n^*(x^*)$ along the interface thickness.

This profile is shown in Fig. 6.7 for a van der Waals equation of state. The pressure difference $P^* - P^*_s$ is also shown in the same figure. In the vapor phase, the pressure is the saturation pressure P_s. Starting from the vapor phase, as we leave this phase the pressure has a slight increase due to the increase in the density n. Nevertheless, as we approach the liquid phase, the attractive term an^2, in the van der Waals equation of state, becomes dominant with respect to the dispersion term $nkT/(1 - bn)$ and the pressure decays to a minimum. After this point, the denominator $1 - bn$ approaches zero and the dispersion term begins to prevail over the attraction term restoring the saturation pressure when x approaches the liquid phase.

6.5.2.1 Exercise

Find the liquid–vapor density profile for water at a reduced temperature $T^* = 0.6$, using a Soave equation of state.

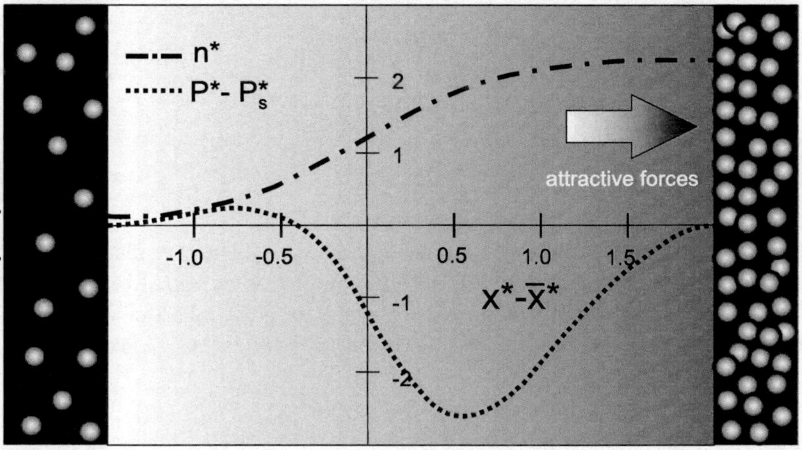

Fig. 6.7 Figure shows the density profile n^* and the pressure $P^* - P_s^*$ along the interface thickness

6.5.3 Interface Thickness

The interface thickness is commonly defined as in Fig. 6.8,

$$\left(\frac{dn}{dx}\right)_{x=\overline{x}} = \frac{n_{\ell s} - n_{vs}}{\lambda},$$

or using Eq. (6.9)

Fig. 6.8 Definition of the interface thickness

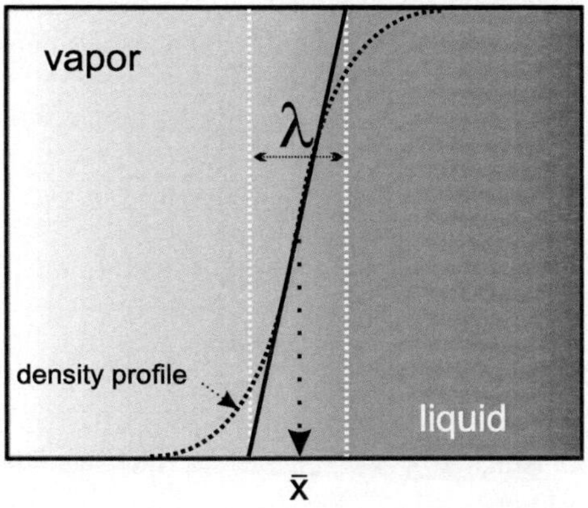

$$\lambda = \frac{n_c}{\sqrt{\frac{2P_c}{\kappa}}} \frac{n_{\ell s}^* - n_{vs}^*}{\sqrt{\psi_{exc}^* \left(\overline{x} \right)}}.$$

6.5.4 Surface Tension

Consider an isothermal one-dimensional liquid–vapor system at rest with a plane interface orthogonal to the x-axis. The Helmholtz free energy in excess Ψ_{exc} in the interface per unit area orthogonal to the x-axis must take the van der Waals contribution into account [181], i.e., the effect of strong density gradients on the local excess function $\psi_{exc}(n, T)$,

$$\Psi_{exc} = \int_{-\infty}^{\infty} \left(\underbrace{\psi_{exc}(n, T) + \frac{\kappa}{2} \left(\frac{dn}{dx} \right)^2}_{f\left(n, \frac{dn}{dx}, T\right)} \right) dx, \tag{6.13}$$

where

$$\psi_{exc}(n, T) = \psi(n, T) - \psi_0 \tag{6.14}$$

In the above equation $\psi_0 = \psi_0(n_\infty, T)$ is the Helmholtz free energy of the homogeneous phases and n_∞ is the density of the homogeneous phase when $x \to -\infty$, i.e., the vapor phase, where $n_\infty = n_{vs}$, or when $x \to \infty$, the liquid phase, where $n_\infty = n_{\ell s}$.

The problem we want to solve here is to find the excess energy $\psi_{exc}(n, T)$ in *equilibrium conditions*, in such way that the integral in Eq. (6.13) is a minimum. This problem can be solved using variational calculus (see Appendix A). Since the integrand f in Eq. (6.13) does not depend explicitly on x, the Euler–Lagrange equations reduces to

$$n' \frac{\partial f}{\partial n'} - f = C,$$

where $n' = \frac{dn}{dx}$ and C is a constant. In the present case,

$$\kappa \left(\frac{dn}{dx} \right)^2 - \psi_{exc}(n, T) - \frac{\kappa}{2} \left(\frac{dn}{dx} \right)^2 = C.$$

Since $\psi_{exc} \to 0$ and $\frac{dn}{dx} \to 0$ when $x \to \pm\infty$, we can conclude that the solution we seek is

$$\psi_{exc}(n, T) = \frac{\kappa}{2}\left(\frac{dn}{dx}\right)^2.$$

In equilibrium conditions, the minimal Helmholtz free energy in excess is the surface tension γ and so, from Eq. (6.13),

$$\gamma = \kappa \int_{-\infty}^{\infty} \left(\frac{dn}{dx}\right)^2 dx = \kappa \int_{n_{vs}}^{n_{\ell s}} \left(\frac{dn}{dx}\right) dn, \tag{6.15}$$

or

$$\gamma = n_c\sqrt{2\kappa P_c} \int_{n_{vs}}^{n_{\ell s}} \sqrt{\psi^*_{exc}} dn^*.$$

This last equation can be rewritten as

$$\kappa = \frac{\gamma^2}{I^2 n_c^2 2P_c}, \tag{6.16}$$

where

$$I = \int_{n^*_{vs}}^{n^*_{\ell s}} \sqrt{\psi^*_{exc}} dn^* \tag{6.17}$$

Since the excess function ψ^*_{exc} can be determined for a given equation of state, Eq. (6.16) relates the surface tension γ with the molecular parameter κ and can be used to find the value κ of from experimental values of γ.

6.5.5 Sample Case: Water

Table 6.1 summarizes the main thermodynamic properties of water obtained with a Peng–Robinson equation of state,

$$P = \frac{nRT}{1 - bn} - \frac{a\alpha(T, \omega)}{1 - (bn)^2 + 2bn},$$

where n is the number of moles per unit volume (*otherwise, we would use the Boltzmann constant k, instead of R, in the case n was assigned to the number of molecules per unit volume*),

Table 6.1 Surface properties of water

T^*	γ (N/m)	n^*_{vs}	$n^*_{\ell s}$	I	$\kappa\left(\frac{\text{N}}{(\text{kg/m}^3)^2}\right)$	λ (Å)
0.4220	0.07560	0.00002	3.75343	8.26090	4.86366×10^{-17}	2.69
0.4622	0.07183	0.00009	3.69318	7.52090	5.29792×10^{-17}	2.97
0.5008	0.06773	0.00034	3.62967	6.82980	5.71060×10^{-17}	3.27
0.5997	0.05568	0.00400	3.43707	5.15170	6.78435×10^{-17}	4.17
0.7001	0.04183	0.02193	3.18226	3.55090	8.06069×10^{-17}	5.49
0.8005	0.02752	0.07806	2.83531	2.07030	1.02591×10^{-16}	7.92
0.9009	0.01250	0.23065	2.32105	0.78810	1.46081×10^{-16}	13.92
0.9395	0.00700	0.35270	2.03483	0.38959	1.87617×10^{-16}	20.44
0.9596	0.00370	0.44937	1.84592	0.21757	1.67950×10^{-16}	30.00
0.9797	0.00201	0.59576	1.60197	0.07998	3.66696×10^{-16}	50.72

$$\alpha = \left(1 + \varkappa\left(1 - \sqrt{T^*}\right)\right)^2, \quad a = 0.45724\frac{R^2T_c^2}{P_c}, \quad b = 0.0778\frac{RT_c}{P_c},$$

with

$$\varkappa = 0.37464 + 1.54226\omega - 0.26992\omega^2,$$

considering an acentric factor $\omega = 0.344$. The critical properties of water are $T_c = 647.3$ K, $P_c = 221.2$ bar.

The table was constructed using Maxwell's area rule, to finding the saturation densities n^*_{vs} and $n^*_{\ell s}$ and the integral I in the fifth column was calculated using Eq. (6.17). The excess energy ψ^*_{exc} was calculated using Eq. (6.11) and is given, in this case, by

$$\begin{aligned}\psi^*_{exc} &= E \times \alpha\left(T^*, \omega\right) \times n^* \times \left[\begin{array}{c}\ln \frac{1-Bn^*}{1+An^*} \\ \\ -\ln \frac{1-Bn^*_\infty}{1+An^*_\infty}\end{array}\right] \\ &+ D \times n^* \times T^* \times \ln\left[\frac{n^*\left(1 - Cn^*_\infty\right)}{n^*_\infty\left(1 - Cn^*\right)}\right] \\ &+ \left(1 - \frac{n^*}{n^*_\infty}\right) \times \left[\begin{array}{c}D \times \frac{n^*_\infty T^*}{(1-Cn^*_\infty)} \\ \\ -F \times \frac{\alpha(T^*,\omega)n^{*2}_\infty}{1-C^2n^{*2}_\infty+2Cn^*_\infty}\end{array}\right]\end{aligned} \tag{6.18}$$

where $A = 0.58443$, $B = 0.10027$, $C = 0.24208$, $D = 3.1116$, $E = 6.4655$, $F = 4.4269$. The density n^*_∞ corresponds to the dimensionless density of the bulk phases, indifferently n^*_{vs} or $n^*_{\ell s}$.

Parameter κ was calculated from Eq. (6.16), based on experimental values of γ. The expression

$$\kappa = -5 \times 10^{-17} \ln \gamma - 9 \times 10^{-17}$$

gives a reasonable approximation for expressing κ in terms of γ and this approximation can be useful in practical calculations.

The interface thickness λ is presented in the last column of Table 6.1, given in Angstroms. Interfaces are very thin for temperatures far from the critical point and their $\mathcal{O}$(Å) thicknesses are a challenge for a continuous approach like the one that was used in this section. Molecular dynamics simulation is suggested to interested readers (see, e.g., Ref. [169]). The interface thickness, λ has experimentally been measured via both ellipsometry [85] and X-ray reflectivity [20]. Braslau et al. [20] found an interface thickness for the water liquid–vapor interface of about 3 Å at room temperature.

Molecular dynamics simulations appear to confirm the adequacy of the continuum approach (Fig. 6.9) for estimating the liquid–vapor interface thickness. Nevertheless, both MD and Peng–Robinson models underestimate the interface thickness, when compared with the experimental ellipsometry results of Kinosita et al. [85]. Nevertheless, experiments are rather difficult in this kind of measurement and this underestimation cannot be considered as conclusive. In fact, this subject is still challenging and full of open questions.

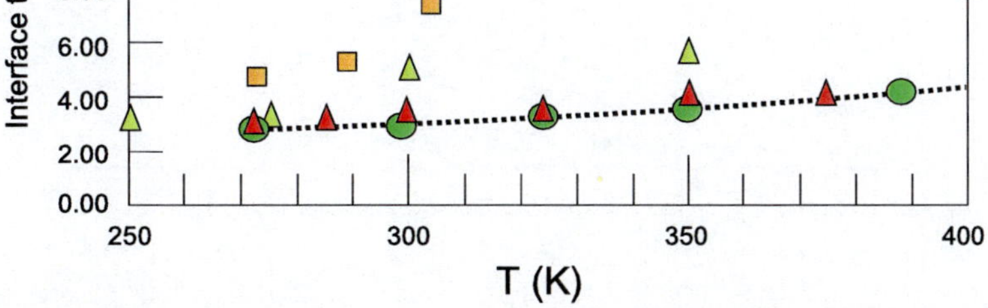

Fig. 6.9 Comparison of the continuum (PR) and molecular dynamics (MD) approach with ellipsometry experimental results for estimating the interface thickness of water. MD simulation results are from Matsumoto and Kataoka [112] and Taylor et al. [169]. Experimental results are from Kinosita and Yokota [85]

6.5.6 *Exercise*

Table 6.1 for water was built based on the Peng–Robinson equation of state and experimental values of the surface tension γ. Consider now carbon dioxide. Surface tension data for CO_2 can be found in, e.g., the Dortmund Data Bank (DDB). Use the Soave equation and find the interface thickness in terms of the reduced temperature for the carbon dioxide.

6.6 Interaction Between Fluids and Solid Surfaces

It is frequently stated that Thomas Young established the relationship between the contact angle and the surface energies in his '*An essay on the cohesion of fluids*' published by the Royal Society in 1805 [189]. Undoubtedly, this paper clearly states the significance of the contact angle, but as commented by Vogle [183] "*Close reading fails to detect clear statements that can be formulated as the familiar. So it is tempting to conclude that the Young equation, as we currently enjoy it, is more of a digestion or a composite of Laplace and Young and possibly Dupré*".

In the present section, the henceforth called Young–Laplace–Dupré equation is derived by using the approach given by Gibbs [59, 60], by considering the interface as a region where the free energy has an excess with respect to the bulk phases. So, consider a system as shown in Fig. 6.10, consisting of three phases ℓ, v and s and their interfaces ℓv, ℓs and vs. The interface is supposed to be spherical when at rest and in the absence of gravity effects.

The Helmholtz free energy of this system is given by

$$F = \underbrace{F_s + F_\ell + F_v}_{\text{bulk phases}} + \underbrace{F_{s\ell} + F_{sv} + F_{\ell v}}_{\text{interfaces}}$$

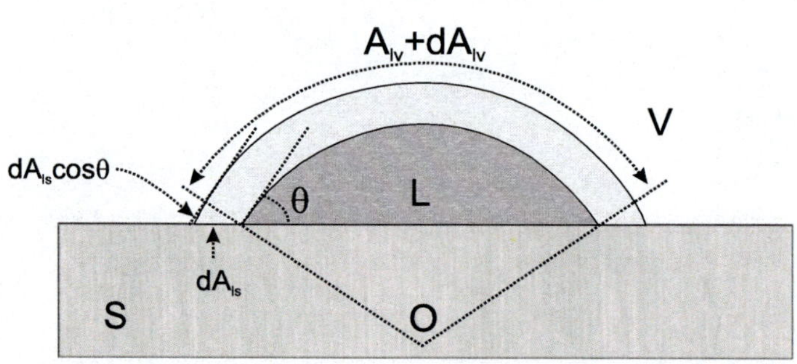

Fig. 6.10 Contact angle between a fluid and a solid surface

So by supposing, as depicted in Fig. 6.10 that the liquid drop suffers a small expansion at a constant temperature, the system will undergo a variation in its free energy, given by

$$F = dF_s + dF_\ell + dF_v + dF_{s\ell} + dF_{sv} + dF_{\ell v}.$$

As it is well known from equilibrium thermodynamics,

$$dF = -SdT - PdV + \mu dm,$$

for the bulk phases and for an interface $\alpha\beta$

$$dF_{\alpha\beta} = \gamma_{\alpha\beta} dA.$$

So, by considering that $dm = 0$, meaning that the involved phases do not exchange mass or that this mass exchange is negligible during the droplet expansion, and taking into account that

$$dV_s = 0, \quad dV_v = -dV_\ell,$$

and that each increase in the liquid–solid contact area corresponds to an equal decrease in the vapor–solid contact area

$$dA_{vs} = -dA_{\ell s},$$

we can conclude that

$$dF = (P_v - P_\ell)\,dV_\ell + \gamma_{\ell v} dA_{\ell v} + (\gamma_{\ell s} - \gamma_{vs})\,dA_{\ell s} + \gamma_{\ell v} dA_{\ell s}\cos\theta.$$

In equilibrium conditions $dF = 0$. In addition, from the Young–Laplace–Dupré equation the first two terms of the second member cancel themselves, since in mechanical equilibrium

$$(P_\ell - P_v) = \gamma_{\ell v}\frac{dA_{\ell v}}{dV_\ell} = \frac{\gamma_{\ell v}}{\mathcal{R}}, \tag{6.19}$$

where $\mathcal{R}$ is the curvature radius. Therefore,

$$(\gamma_{\ell s} - \gamma_{vs})\,dA_{\ell s} + \gamma_{\ell v} dA_{\ell s}\cos\theta = 0.$$

This last equation leads to the Young–Laplace–Dupré equation that expresses the equilibrium contact angle in terms of the surface energies $\gamma_{\ell s}$, γ_{vs}, and $y_{\ell v}$

$$\cos\theta = \frac{\gamma_{vs} - \gamma_{\ell s}}{\gamma_{\ell v}}.$$

The adhesion energy E_{adh} will be here defined as[9]

$$E_{adh} = \gamma_{vs} - \gamma_{\ell s}$$

and the ratio between E_{adh} and the cohesion energy, $E_{coh} = \gamma_{\ell v}$, is related to the wettability of the solid surface with respect to the fluid in contact with it. When $E_{adh} > E_{coh}$ the fluid is considered to completely spread on the solid surface without attaining any equilibrium state at a given equilibrium contact angle. When $1 \geq E_{adh}/E_{coh} \geq 0$ the fluid attains an equilibrium state with a contact angle $0 \leq \theta \leq 90°$. When the adhesion energy is a negative fraction of the cohesion energy, the cohesion prevails and the liquid attains an equilibrium contact angle $90° \leq \theta \leq 180°$.

The theoretical limit $\theta = 90°$ corresponds to an adhesion energy $E_{adh} = 0$, when both γ_{vs} and $\gamma_{\ell s}$ solid–fluid surface energies are zero. In other words, the fluid molecules do not interact with the solid surface in this limit.

6.6.1 Exercise

Figure 6.11 shows a balance of forces for an element *e* of the surface of a liquid droplet supposed to be spherical and deposed on a solid surface. The surface forces act only on the perimeter area of this element $2\pi R \sin\alpha$. In fact, when we try to detach *e* from the liquid droplet, the remaining liquid molecules, on their side, will react trying to prevent this detachment with a force $\gamma_{\ell v} 2\pi R \sin\alpha$ in the perimeter area, directed along the liquid surface and outward from *e*. These forces are null in the lower and upper surfaces of *e*, because the potential energy related to the intermolecular interaction has a null gradient when we move into the liquid phase or into the vapor phase. Using the same reasoning as above, make a force balance in the ℓs and vs interfaces and try to find the Young–Laplace–Dupré law as given in Eq. (6.19).

[9] There is some controversy in the definitions of the *adhesion energy* [183]. The adhesion energy is here defined as the *adhesion tension* or $\gamma_{vs} - \gamma_{\ell s}$ because it represents the net effect of the solid surface on the fluid molecules and because it appears to be more universal than the usual concept of *adhesion work*. When two liquids 1 and 2 interact with a solid surface, liquid 1 will be adherent to the surface when $\gamma_{1s} - \gamma_{2s}$ is positive. Otherwise, liquid 2 will be adherent to the surface.

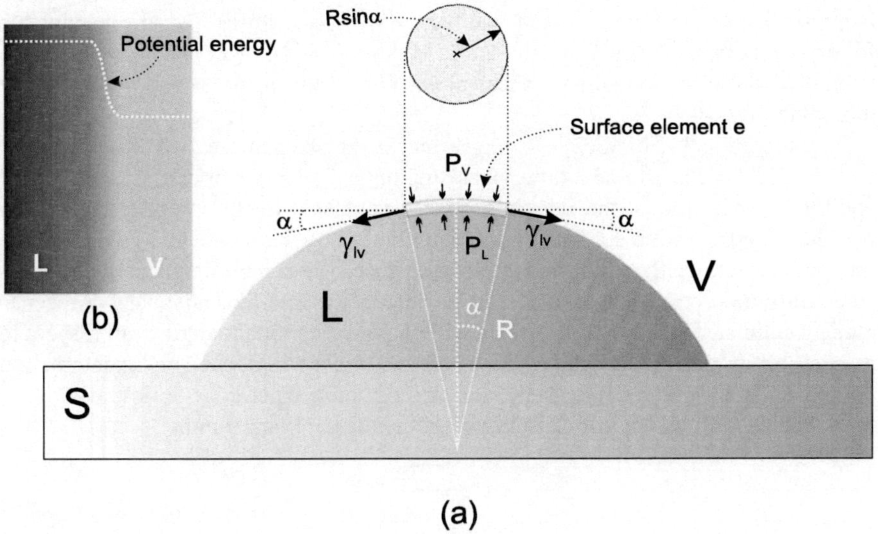

Fig. 6.11 **a** Force balance on a surface element *e* of a liquid droplet. **b** Potential energy related to the intermolecular interaction

6.6.2 Dynamic Contact Angle

Interest in gaining a deeper understanding of the physical processes that control the fluid–fluid interface displacement at the pore level is currently increasing due to (a) the growing possibility of direct simulation of immiscible displacement in complex geometry offered by the significant development of computers and numerical methods that has occurred over the recent years and (b) the increasing accuracy of three-dimensional representations of the porous structure promoted by the development of three-dimensional reconstruction methods [4, 55, 77, 106, 118] and by the increasing spatial resolution that it is possible to achieve with microtomography-based methods [6, 56, 65].

Classically, the immiscible displacement of fluids is approached by considering that the transition layer has a null thickness and by performing a momentum balance around this layer. In this way, in addition to the intrinsic physical properties of the involved fluids, the interface tension is another important macroscopic parameter.

The main physical mechanisms that control the interface displacement are the fluid–fluid and the fluid–solid intermolecular potentials, so, the correct knowledge of this process requires downscaling to the molecular scale. Intermolecular potentials may be of many different types, including polar forces among permanent dipoles, induction forces, and dispersion forces among symmetrical molecules. In the transition region between two immiscible fluids, a molecule is predominantly subjected to attractive fields from its own phase that act as a potential barrier and give rise to the fluid–fluid interface tension. In addition, molecules that are found in this transition layer are subject to collisions that tend to mix the two fluids and are responsible for diffusion. Consequently, the thickness of the transition layer is controlled by the strength and interaction length of intermolecular cohesion forces and by the binary diffusivity, $\mathcal{D}_{12}$.

The theoretical difficulty is strongly increased when these two fluids interact with a solid surface around a three-phase region, called triple or contact line. Under equilibrium conditions, the interfacial energies related to the interaction between two fluids and the solid surface are the main macroscopic parameters governing the interaction between the fluids and the solid surface. These interfacial energies lead to an equilibrium contact angle. Nevertheless, when the interface advances or recedes along a solid surface, dynamic effects will change the contact angle with respect to its equilibrium value. The changes in the contact angle will depend on the interaction potentials, on the surface roughness, and on the chemical heterogeneities, in addition to the imposed flow conditions summarized by the capillary number,

$$Ca = \frac{\mu \mathsf{v}}{\gamma}$$

where μ is the viscosity of the displaced fluid, v is the interface velocity and γ is the interface tension.

Figure 6.12a and b give a sketch of the variation of the dynamic contact angle in accordance with the flow direction. When a wetting fluid displaces a non-wetting fluid, inside a channel, the contact angle increases, but when the displaced fluid is the wetting one, the contact angle is reduced.

It is usual, in studies of capillary-rise dynamics, to consider that the contact angle varies according to the expression:

$$\cos\theta_d = \cos\theta - aCa^b \tag{6.20}$$

where θ_d is the dynamic contact angle and a and b are constants. In Fig. 6.12c it is shown the dependence of $\cos\theta$ on the capillary number obtained from lattice-Boltzmann simulations. The graph shows that the expression given by Eq. (6.20) represents adequately the contact angle dependence on Ca. In fact, Eq. (6.20) fits very well with simulation results when $a = 18.21$ and $b = 0.88$.

Finally, Fig. 6.12d shows the streamlines simulated with the lattice-Boltzmann method, for a capillary rising process. The flow near the interface can be very complex, with the formation of vortices near the triple line as can be seen in the figure. Reference [43] is suggested for further details (Fig. 6.12).

Therefore, due to the dependence of the dynamic contact angle on both the flow conditions that are imposed and on the geometry and properties of the solid surface, the equilibrium, or even the dynamic contact angles cannot be used as a parameter in dynamic simulations.

Fluid invasion into a tube as a result of unbalanced capillary forces is a classical problem in the literature devoted to fluid dynamics. It is of great scientific interest and is important in various applications, such as ink printing, dyeing of textiles,

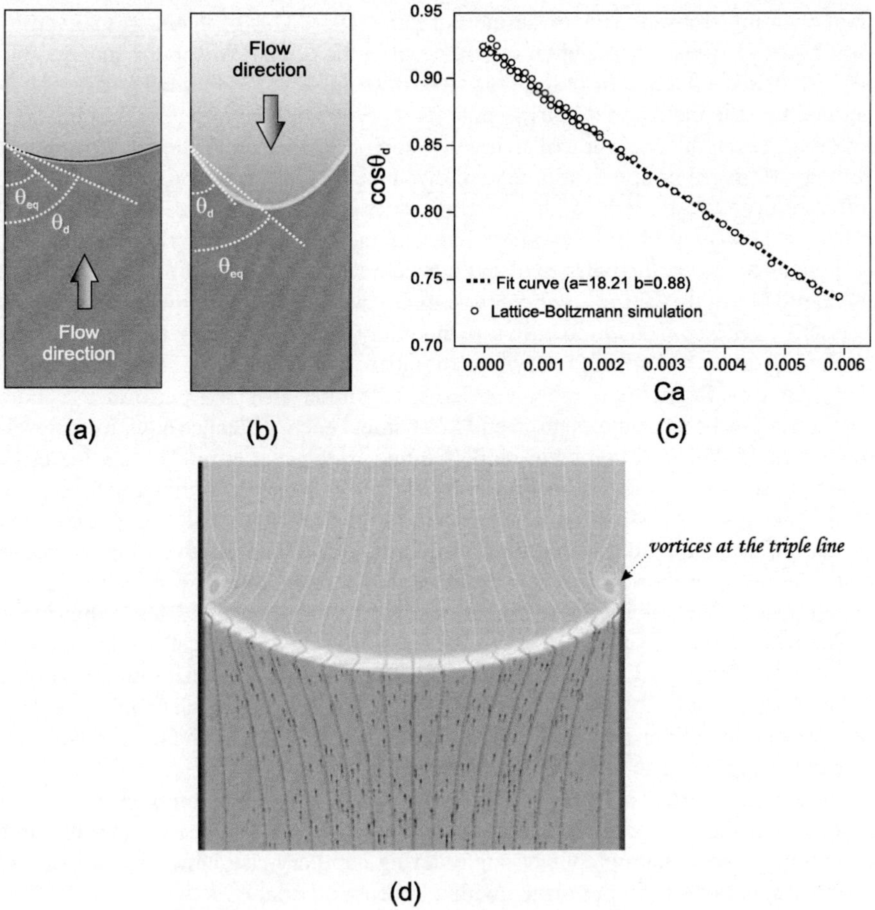

Fig. 6.12 Sketch of the variation of the dynamic contact angle in accordance with the flow direction (**a**, **b**); dependence of the contact angle with the capillary number (**c**, **d**) streamlines showing the formation of vortices near the triple line [43]

and enhanced oil recovery. Lucas[10] and Washburn[11] published the first attempts to describe the dynamic process involved in capillary rise. They verified that the process can be described using a mechanical balance between the capillary, viscous, and gravitational forces, leading to the well-known Lucas–Washburn equation. Their model has been verified experimentally even for very thin capillaries and is generally accepted as a valid approach under a laminar flow regime [2]. However, the Lucas–Washburn equation fails in the prediction of the first stages of the meniscus rising, predicting that the initial velocity is infinite at time zero, a non-physical solu-

[10] R. Lucas, Kolloid. Zh. 23 (1918) 15–22.

[11] E. W. Washburn, Phys. Rev. 17 (1921) 273–283.

tion, resulting from the lack of an inertial term [193]. This problem was solved by Bosanquet [18] who included an inertial term in the Lucas–Washburn equation but neglected any influence from the fluid flow in the reservoir. For such a case, it was verified that the inclusion of the inertial effects avoided the non-physical initial condition, allowing the definition of an initial velocity at time zero, called the Bosanquet velocity. High-resolution experiments showed that the observed initial velocities are different from those predicted by the Bosanquet equation [90]. Szekely et al. [168] carried out an analysis of the capillary rise with inertial and entrance effects by using arguments based on the balance of the mechanical energy equation. They modified the inertial term by adding an *apparent mass* related to the fluid flow from the reservoir and a *vena contracta* effect due to the inlet flow. Considering that the analysis by Szekely et al. was for high Reynolds number flow, while most cases of capillary rising are low Reynolds number problems, Levine et al. [104] performed a more detailed analysis of the momentum and mechanical energy balance equations, developing one of the most complete models found in the literature. The Levine et al. model presents different factors for the inertial terms and an additional term, related to the inlet viscous dissipation. Dreyer et al. [44] and Stange et al. [161] carried out similar analysis, including some improvements, and obtained governing equations that are similar to those found by Levine et al. These equations were confirmed by experiments under microgravity conditions. As mentioned above, the capillary rise problem has been the subject of several discussions related to the short-time aspects of meniscus formation and the resulting flow velocity. Its complete understanding continues to be, nevertheless, an open subject, particularly due to the complex physicochemical interactions between the liquid and the solid surface and the fundamental role of the triple line in the liquid displacement.

Figure 6.13, extracted from Wolf et al. [187] shows how complex can be the first steps of the meniscus formation in capillary rise, and Fig. 6.14 shows the time evolution of the dynamic contact angle during capillary rise between two parallel plates Simulations were performed with a lattice-Boltzmann model.

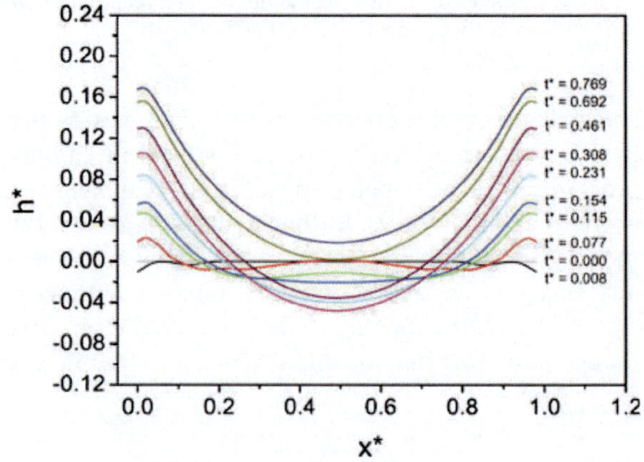

Fig. 6.13 First steps of the meniscus formation in capillary rise between parallel plates simulated with a lattice-Boltzmann model. $h^* = \frac{h}{2R}, x^* = \frac{x}{2R}, t^* = \frac{t}{t_0}$ are dimensionless variables [187]

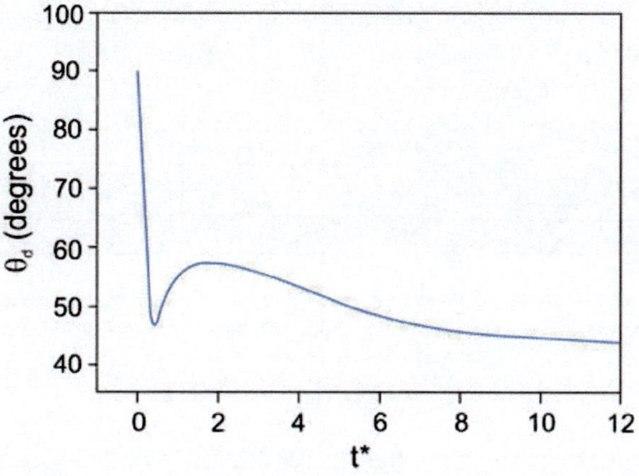

Fig. 6.14 Time evolution of the dynamic contact angle during the capillary rise between two parallel plates when the equilibrium contact angle is 44° [187]

In the next section, we present Cahn's theory [24] for predicting the surface energies γ_{fs} related to the intermolecular interaction between a fluid f—in its liquid or vapor physical state—and a solid surface s. Cahn's theory was conceived to retrieve the surface energies $\gamma_{\ell s}$ and γ_{vs} in terms of the equilibrium contact angle θ_{eq} for a system liquid–vapor–solid in equilibrium.

In dynamical problems, these surface energies are to be used instead of the measured values of the equilibrium contact angle.

6.7 Surface Energies: Cahn's Theory

When gas molecules are close to solid surfaces, they are subjected to the attraction fields from the solid molecules and form an adsorbed layer.

De Boer and Zwicker[12] explained the adsorption of non-polar molecules on adsorbents by assuming that the uppermost layer of the adsorbent induces dipoles in the first layer of adsorbed molecules, which in turn induce dipoles in the next layer and so on until several layers are built up (Fig. 6.15).

Physical adsorption is a subject of fundamental interest and received outstanding contributions since Langmuir in 1918.[13] References [21, 38] are suggested for readers interested in historical papers on this subject. Nevertheless, since our scale of interest is much larger than the molecular dimensions corresponding to the thickness of the

[12] De Boer and Zwicker, Z. physik. Chem., B3, 407 (1929) [38].

[13] Langmuir I (1918) The adsorption of gases on plane surfaces of glass, mica, and platinum. J Am Chem Soc 40:1361–1403. https://doi.org/10.1021/ja02242a004 [98].

Fig. 6.15 The electrostatic field produced by a solid surface induces polarization on non-polar molecules. These polarized molecules induce dipoles in the next layers until several layers are built up

adsorbed layer, it can be supposed that the semi-infinite fluid interacts with this layer and not with the solid surface itself.

When a drop of liquid is deposed on a solid surface, the only available experimental information is the contact angle θ after the liquid drop reaches an equilibrium configuration. From the Young–Laplace–Dupré equation,

$$\cos\theta = \frac{\gamma_{vs} - \gamma_{\ell s}}{\gamma_{\ell v}}, \tag{6.21}$$

we know that this contact angle is dependent on the difference between the Helmholtz energy in excess in the vapor–solid, γ_{vs}, and in the liquid–solid, $\gamma_{\ell s}$, transition layers. Since both the liquid and vapor phases are, in fact, interacting with an adsorbed layer of fluid, we can suppose that this interaction is ruled by a single potential which is only dependent on the amount of adsorbed molecules.

Cahn [24] assumed that the interactions between the surface and the fluid are sufficiently short-range in such a manner that the contribution to the free energy of a unit area of this layer is $\Phi(\rho_s)$, where ρ_s is the limiting density of the fluid at $x = 0$. So the contribution of the adsorbed layer to the free energy is of an additive constant.

Therefore, following the same reasoning of Sect. 6.5.4, the question we seek to solve is to get the density profile and excess free energy due to the adsorbed layer by finding the function which minimizes the excess free energy of a unit area of surface,

$$\Psi_{exc} = \Phi(\rho_s) + \int_0^\infty \left(\psi_{exc}(\rho, T) + \frac{\kappa}{2} \left(\frac{d\rho}{dx} \right)^2 \right) dx, \tag{6.22}$$

where

$$\psi_{exc}(\rho, T) = \psi(\rho, T) - \psi_0(\rho_\infty, T),$$

is here interpreted as the free energy required to create a unit volume of uniform fluid of composition from a large reservoir with density ρ_∞, $\frac{\kappa}{2}\left(\frac{d\rho}{dx}\right)^2$ is the van der Waals contribution due to the presence of a gradient and ρ_∞ is the bulk density of the fluid, far from the solid surface.

As it was seen in Sect. 6.5.4, in equilibrium conditions the solution to the minimization problem is given by,

$$\psi_{exc}(\rho, T) = \frac{\kappa}{2}\left(\frac{d\rho}{dx}\right)^2. \tag{6.23}$$

Cahn supposed that the free energy at the surface is responsible for the density gradient at $x = 0$,

$$\Phi(\rho_s) = \rho_s \kappa \left.\frac{d\rho}{dx}\right)_{x=0}.$$

Therefore, a *surface property* related to the wettability can be defined as

$$\phi = -\kappa \left.\frac{d\rho}{dx}\right)_{x=0},$$

considered as a macroscopic property related to the net balance of the intermolecular forces between the fluid molecules and the solid and between the fluid molecules close to the solid and the fluid phase.

When the surface is hydrophilic, $\phi > 0$ and the fluid density decreases as $x \to \infty$. When $\phi < 0$ the surface is hydrophobic and the fluid density increases as $x \to \infty$ (Fig. 6.16).

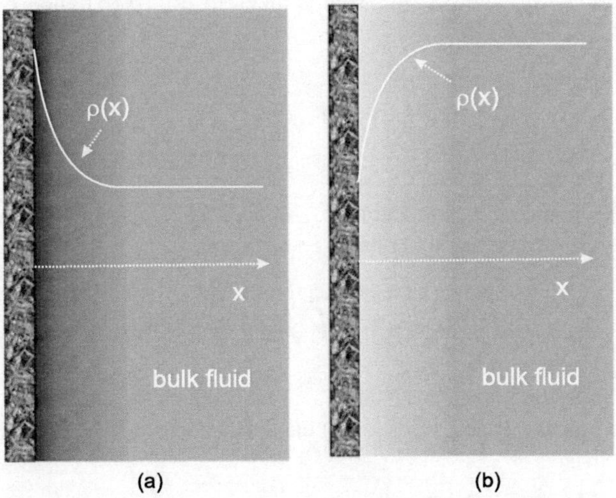

Fig. 6.16 Density distribution at the fluid–solid interface: **a** hydrophilic surface, $\phi > 0$, **b** hydrophobic surface, $\phi < 0$

From Eq. (6.23)

$$\frac{d\rho}{dx} = \pm\sqrt{\frac{2\psi_{exc}(\rho, T)}{\kappa}}. \tag{6.24}$$

So

$$\phi = \pm\sqrt{2\kappa\psi_{exc}(\rho, T)}. \tag{6.25}$$

If ϕ is assumed to be positive by taking the plus sign in the above equation, this means that $\left.\frac{d\rho}{dx}\right)_{x=0} < 0$, or that $\rho_s > \rho > \rho_\infty$. If it is assumed to be negative by taking the minus sign in the above equation, this means that $\left.\frac{d\rho}{dx}\right)_{x=0} > 0$, or that $\rho_s < \rho < \rho_\infty$.

The minimum value of the excess free energy ψ_{exc} is the interface surface tension γ_{fs}. So, by replacing Eq. (6.23) into Eq. (6.22) we get

$$\begin{aligned}\gamma_{fs} &= \Phi(\rho_s) + \int_0^\infty \kappa\left(\frac{d\rho}{dx}\right)^2 dx \\ &= \Phi(\rho_s) + \int_{\rho_s}^{\rho_\infty} \kappa\frac{d\rho}{dx}d\rho.\end{aligned} \tag{6.26}$$

The last term on the r.h.s. of the above equation is always positive. So if ϕ is positive (*hydrophilic surfaces*) and the plus sign is taken in Eq. (6.25), then Eq. (6.26) will be written as

$$\gamma_{fs} = \Phi(\rho_s) + \int_{\rho_\infty}^{\rho_s} \sqrt{2\kappa\psi_{exc}(\rho, T)}d\rho. \tag{6.27}$$

On the other hand, if ϕ is negative (*hydrophobic surfaces*) and the minus sign is taken in Eq. (6.25), then Eq. (6.26) will be written as

$$\gamma_{fs} = \Phi(\rho_s) + \int_{\rho_s}^{\rho_\infty} \sqrt{2\kappa\psi_{exc}(n, T)}d\rho. \tag{6.28}$$

In dimensionless variables Eq. (6.25) can be written as

$$\phi^* = \frac{\phi}{\sqrt{\kappa P_c}} = \pm\sqrt{2\psi^*_{exc}(\rho, T)}. \tag{6.29}$$

Considering the van der Waals equation of state for the fluid, the Helmholtz's excess energy is given by

$$\psi_{exc}^{*} = 3\rho^{*}\rho_{\infty}^{*}\left(1-\frac{\rho^{*}}{\rho_{\infty}^{*}}\right)+\frac{8}{3}\rho^{*}T^{*}\ln\frac{\rho^{*}\left(3-\rho_{\infty}^{*}\right)}{\rho_{\infty}^{*}\left(3-\rho^{*}\right)} \\ +\left(\frac{8\rho_{\infty}^{*}T^{*}}{\left(3-\rho_{\infty}^{*}\right)}-3n_{\infty}^{*2}\right)\left(1-\frac{\rho^{*}}{\rho_{\infty}^{*}}\right). \tag{6.30}$$

6.7.1 Hydrophilic Surfaces

Figure 6.17 shows how to use Eq. (6.29) for finding the fluid density n_s^* at the solid surface for a hydrophilic surface. Equation (6.29) has 2 solutions for the vapor–solid interaction and 2 other solutions for the liquid–solid interaction. Since in hydrophilic surfaces, $\rho_s^* > \rho_{\infty}^*$, where ρ_{∞}^* is the density of the bulk phase (*vapor or liquid*), the two solutions where $\rho_s^* < \rho_{\infty}^*$ are to be discarded.

The solution of the non-linear equation

$$\phi^{*} = \sqrt{2\psi_{exc}^{*}\left(\rho, T\right)}, \tag{6.31}$$

can be performed by using a numerical method such as the Newton–Raphson method and leads to the two solutions for the vapor and liquid densities at the solid surface. So, when $\phi^* = 0.4$ as in Fig. 6.17, the reduced densities of the fluid at the solid surface will be $\rho_s^* = 0.49337$ for the vapor phase and $\rho_s^* = 2.14093$ for the liquid phase.

Equation (6.27) can now be used for the calculation of the interface tensions γ_{vs} and $\gamma_{\ell s}$,

$$\Phi\left(\rho_s\right) = -\rho_s\phi = -\rho_s^*\rho_c\phi^*\sqrt{\kappa P_c}.$$

The water critical properties are $P_c = 221.2 \times 10^5$ Pa, $\rho_c = 322$ kg/m^3. For the van der Waals equation of state (see Table 6.2) at $T^* = 0.8$. $\rho_{vs}^* = 0.23962$, $\rho_{\ell s}^* = 1.93279$, $\kappa = 1.37541 \times 10^{-15}\ \frac{\mathrm{N}}{\left(\mathrm{kg/m^3}\right)^2}$. So, for the vapor phase

$$\Phi\left(\rho_s\right) = -6.5038 \times 10^{-3}\ \frac{\mathrm{N}}{\mathrm{m}},$$

and

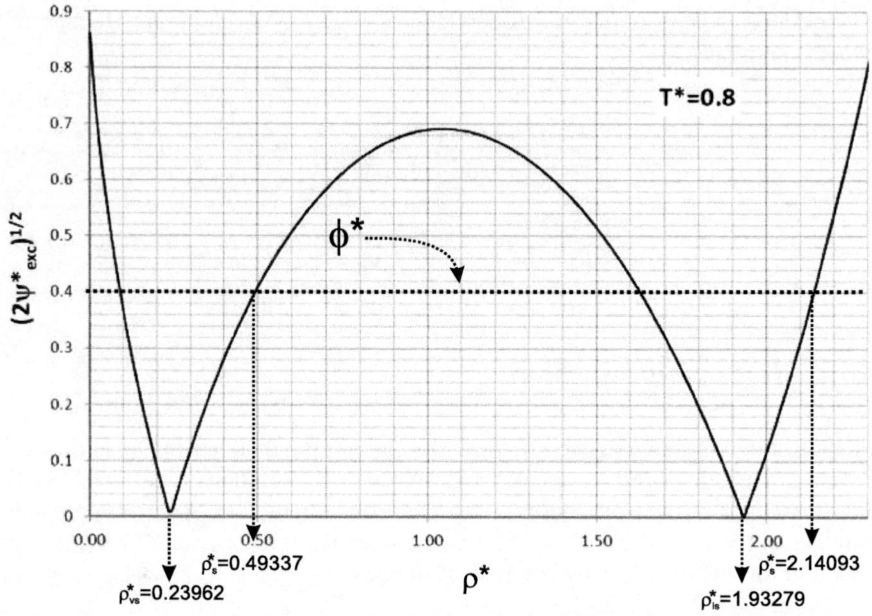

Fig. 6.17 Dimensionless excess function for water at T* = 0.8 calculated using the van der Waals EOS plotted against the reduced density. The information on the solid–fluid interaction is given by the dimensionless specific free energy ϕ^* considered in this case as $\phi^* = 0.4$ (*hydrophilic surface*). In accordance with Cahn's theory, the water reduced density at the solid surface will be $\rho_s^* = 0.49337$ and $\rho_s^* = 2.14093$ corresponding, respectively, to the vapor and liquid in equilibrium with the solid

$$\gamma_{vs} = -6.5038 \times 10^{-3} + \rho_c\sqrt{2\kappa P_c} \int_{0.23962}^{0.49337} \sqrt{\psi_{exc}^*} d\rho^*$$

$$= -0.0046598 \frac{\text{N}}{\text{m}}.$$

For the liquid phase,

$$\Phi(\rho_s) = -0.028223 \frac{\text{N}}{\text{m}},$$

and

$$\gamma_{\ell s} = -0.028223 + \rho_c\sqrt{2\kappa P_c} \int_{1.93279}^{2.14093} \sqrt{\psi_{exc}^*} d\rho^*$$

$$= -0.026859 \frac{\text{N}}{\text{m}}.$$

Table 6.2 Surface properties of water with the van der Waals equation of state

T^*	$\rho^*_{v\infty}$	$\rho^*_{\ell\infty}$	$\kappa\left(\frac{\text{N}}{(\text{kg/m}^3)^2}\right)$
0.4220	0.00725	2.56345	4.68314×10^{-16}
0.4622	0.01336	2.51190	5.18317×10^{-16}
0.5023	0.02238	2.45787	5.71457×10^{-16}
0.6027	0.06127	2.30967	7.25571×10^{-16}
0.7016	0.12958	2.13966	9.35490×10^{-16}
0.8005	0.23962	1.93279	1.37541×10^{-15}
0.9025	0.43234	1.65075	2.13247×10^{-15}
0.9426	0.55217	1.49691	2.88014×10^{-15}

The water surface tension at this temperature is

$$\gamma_{\ell v} = 0.0264 \, \frac{\text{N}}{\text{m}}.$$

Finally, the contact angle can be calculated using the Young–Laplace–Dupré equation

$$\cos\theta = \left(\frac{\gamma_{vs} - \gamma_{\ell s}}{\gamma_{\ell v}}\right),$$

or $\theta = 32.77°$.

In the above calculations, we notice that the surface free energies γ_{vs} and $\gamma_{\ell s}$ are both negative when the surface is hydrophilic and that $\gamma_{\ell s}$ is more negative than γ_{vs}. Indeed, the dominant term in Eq. (6.27) is the surface free energy $\Phi\,(n_s)$ which is negative when the surface is hydrophilic, indicating that work is needed for freeing the adsorbed fluid layer from the solid surface. The reason for $\gamma_{\ell s}$ be more negative than γ_{vs} is that, while the specific free energy at the surface, ϕ, was written to be only dependent on the fluid–solid molecular interaction, the amount of adsorbed water in the liquid–solid interface is, in the present case, around 6 times the amount of adsorbed water in the vapor–solid interface.

Figure 6.17 also shows that when $\phi^* = 0$, both $\gamma_{\ell s}$ and γ_{vs} are null, meaning that there is no interaction between the fluid and the solid surface, or that $\theta = 90°$. On the other hand the maximum value of ϕ^* for which $\theta = 0°$ is the one satisfying

$$\frac{\gamma_{vs} - \gamma_{\ell s}}{\gamma_{\ell v}} = 1.$$

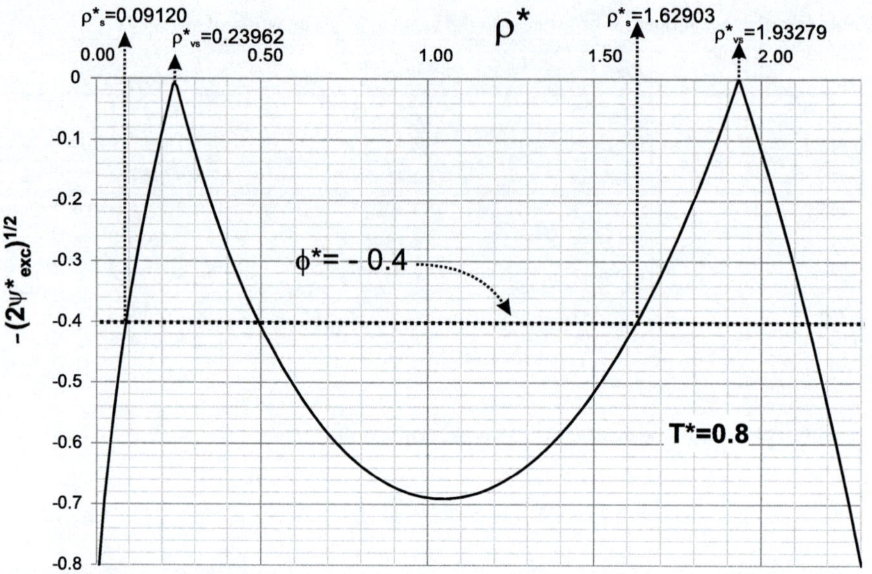

Fig. 6.18 Hydrophobic surface

6.7.2 Hydrophobic Surfaces

Figure 6.17 shows how to use Eq. (6.29) for finding the fluid density n_s^* at the solid surface for a hydrophilic surface. Equation (6.29) has 2 solutions for the vapor–solid interaction and 2 other solutions for the liquid–solid interaction. Since in hydrophobic surfaces, $n_s^* < n_\infty^*$, where n_∞^* is the density of the bulk phase (*vapor or liquid*), the two solutions where $n_s^* > n_\infty^*$ are to be discarded.

So, when $\phi^* = -0.4$ as in Fig. 6.18, the reduced densities of the fluid at the solid surface will be $n_s^* = 0.09120$ for the vapor phase and $n_s^* = 1.62903$ for the liquid phase.

6.7.3 Contact Angle and Adhesion Energy

Table 6.3 shows the reduced densities at the solid surface for the vapor ρ_{s-v}^* and liquid $\rho_{s-\ell}^*$ phases of water for a reduced temperature $T^* = 0.8$. The van der Waals equation of state, Eq. (6.30) was used in the solution of the non-linear equation, Eq. (6.31), by a Newton–Raphson method. The interface tensions γ_{sv} and $\gamma_{s\ell}$ are also shown in this table. They were calculated for each surface parameter ϕ^* using Eq. (6.27) for hydrophilic surfaces and Eq. (6.28) for hydrophobic surfaces. Finally,

Table 6.3 Reduced densities at the solid surface, interface tensions and equilibrium contact angle for several solid surface parameters ϕ^* when $T^* = 0.8$ and the van der Waals equation of state

ϕ^*	ρ^*_{s-v}	$\rho^*_{s-\ell}$	$\gamma_{sv} \times 10^3$	$\gamma_{s\ell} \times 10^2$	θ
Reduced values			N/m	N/m	°
0.478	0.56798	2.17530	−6.0199	−3.2404	1.97
0.475	0.56483	2.17401	−5.9640	−3.2189	6.59
0.470	0.55965	2.17185	−5.8714	−3.1832	10.47
0.460	0.54947	2.16751	−5.6888	−3.1117	15.60
0.440	0.52988	2.15876	−5.3334	−2.9692	22.68
0.400	0.49337	2.14093	−4.6598	−2.6859	32.77
0.300	0.41412	2.09429	−3.1695	−1.9883	50.72
0.20	0.35100	2.04500	−2.7092	−1.3889	64.95
0	0.23976	1.93255	0	0	90.00
−0.20	0.15600	1.79600	1.2934	1.2364	114.79
−0.30	0.12171	1.72002	1.7501	1.8174	128.47
−0.40	0.09120	1.62903	2.0996	2.3703	144.92
−0.44	0.08009	1.58795	2.2123	2.5826	153.44
−0.47	0.07215	1.55480	2.2875	2.7382	161.90
−0.475	0.07086	1.54905	2.2993	2.7638	163.70
−0.48	0.06958	1.54323	2.3108	2.7893	165.70
−0.49	0.06705	1.53138	2.3333	2.8400	170.89
−0.495	0.06579	1.52533	2.3443	2.8652	175.21

Reduced densities ρ^*_{s-v} and $\rho^*_{s-\ell}$ were found from the non-linear equation, Eq. (6.31), using a Newton–Raphson method

the Young–Laplace–Dupré equation was used for calculating the equilibrium contact angle θ.

The adhesion energy is defined as

$$E_{adh} = \gamma_{vs} - \gamma_{\ell s},$$

and is a measure of the interaction energy between the liquid and solid surface.

Figure 6.19 shows the dependence of the adhesion energy on the solid surface specific energy ϕ^*. A linear fit appears to be an excellent approximation for the adhesion energy for almost the entire range of ϕ^*. The figure also shows the dependence of the equilibrium contact angle on the surface parameter ϕ^* for hydrophilic and hydrophobic interactions.

Figure 6.20 shows the reduced fluid densities at the solid surface in terms of the equilibrium contact angle for vapor–solid and liquid–solid interactions. The knowledge of the liquid and vapor densities at the solid surface in terms of the equilibrium contact angle is of crucial importance because they are the only parameters that are to be used in dynamic simulations of the liquid–vapor interface displacement in microcavities. This is the case of capillary rising. As we have seen in Sect. 6.6.2 it is not possible to consider the contact angle as a constant when a liquid column rises

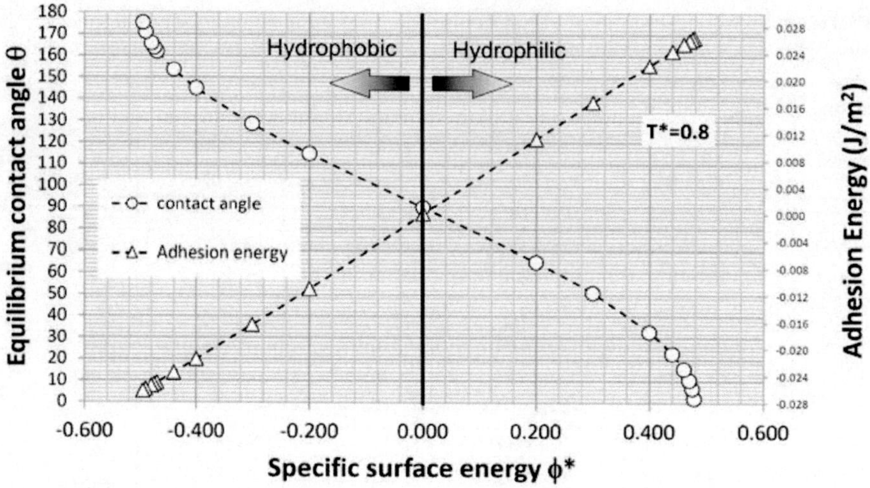

Fig. 6.19 Adhesion energy and contact angle in terms of parameter ϕ^*

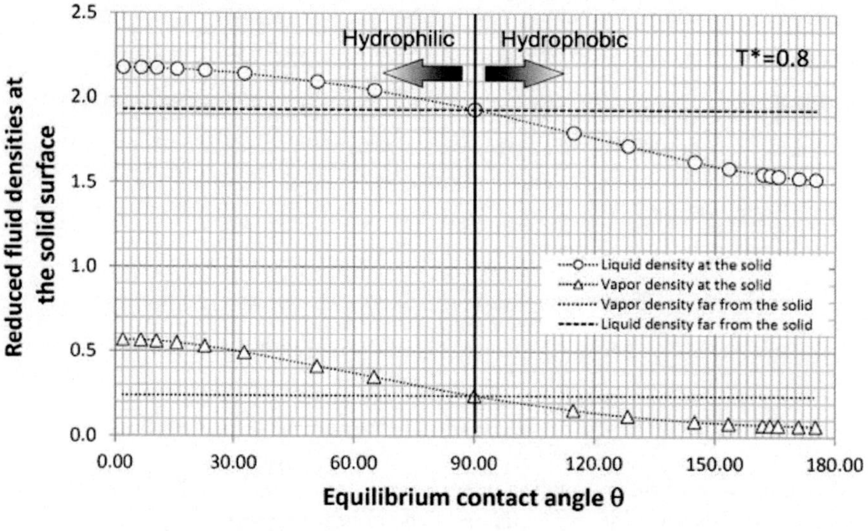

Fig. 6.20 Vapor ρ^*_{s-v} and liquid $\rho^*_{s-\ell}$ densities at the solid surface in terms of the equilibrium contact angle θ for water at $T^* = 0.8$, when a van der Waals equation of state is used

inside a capillary as it was assumed in Bosanquet model [18]. In accordance with Fig. 6.14, the contact angle has a strong variation at the beginning of the rising and only approaches its equilibrium value at the end of ascension. During this process, both the meniscus shape and the contact angle are dependent on the capillary, gravity, viscous, and inertial forces.

6.7.4 *Exercise*

The equilibrium contact angle between water and glass was measured as 13° at $T = 300$ K. Suppose that water follows a van der Waals equation of state and find parameter ϕ^* and the surface energies $\gamma_{\ell s}$ and γ_{vs}. In the following, calculate parameter ϕ^* corresponding to the contact angles $\theta = 0°$ and $\theta = 180°$.

6.8 Emulsions

Emulsions in one form or another, have been used for centuries and early societies learned to take advantage of this natural phenomenon. The ancient Egyptians used eggs to emulsify berry extracts with oils to form crude emulsified paints. In medicine, emulsions are delivered in the form of medical ointments. Cream and butter are literally the same thing. To make butter, you simply mix cream until the emulsion reverses; that is, it transforms from an oil-in-water emulsion into a water-in-oil emulsion. But despite this being the only difference between cream and butter, the effect on taste and texture are significant. Many natural and processed foods consist either partly or wholly as emulsions or have been in an emulsified state at some time during their production; such foods include milk, cream, butter, margarine, fruit beverages, soups, cake batters, mayonnaise, cream liqueurs, sauces, desserts, salad cream, and ice cream.

Oil and water are immiscible liquids and do not mix. The reason behind this behavior must be searched in the molecular scale. The intermolecular attraction forces between water molecules are much stronger than the ones between oil molecules. Water is a polar molecule and the distribution of its unpaired electrons enforces this force in the form of hydrogen bonds. Oil has apolar molecules that attract themselves with weak London forces. So, each water molecule will push other water molecules in its neighborhood, segregating the oil molecules that are found in this neighborhood. Suppose now that you add, e.g., oil to water inside a container. Since oil is less

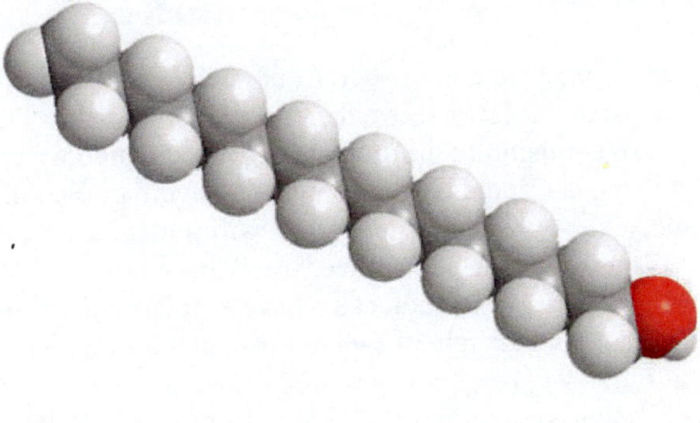

Fig. 6.21 The molecule of cetyl alcohol $CH_3(CH_2)_{15}OH$

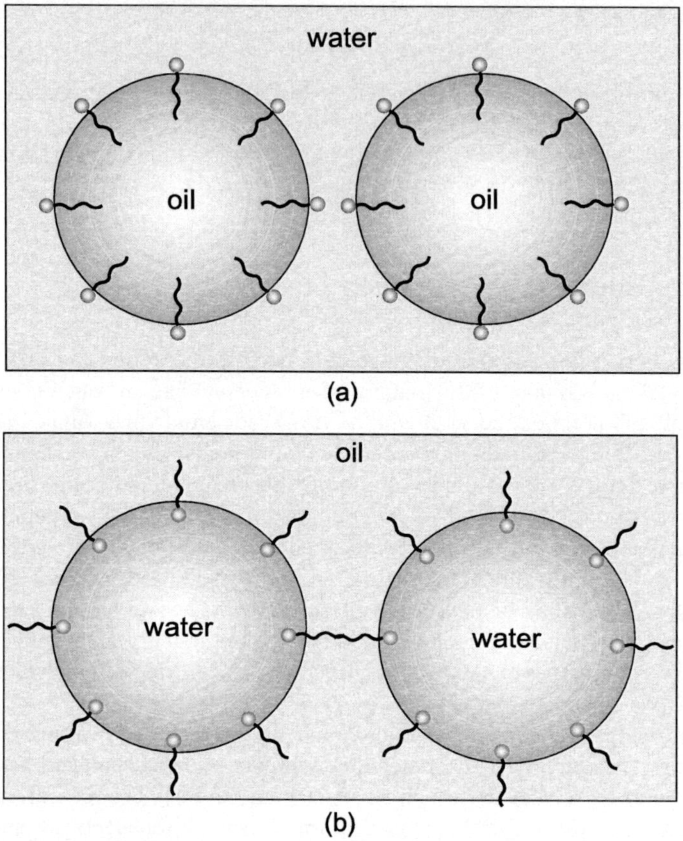

Fig. 6.22 Due to its amphiphilic nature, when surfactants are added to an oil–water system their molecules migrate to the interface. **a** When oil is fragmented and dispersed in a continuous water phase the polar heads of the emulsifier at the interface attract the water molecules in their neighborhoods preventing the oil drops from coalescing. **b** When water is dispersed in oil, the long tails of the emulsifier prevent their polar heads to attract the water molecules in the contiguous water drop, because they are beyond the interaction length of these intermolecular polar forces

dense than water, their segregated molecules will move up in the container forming a separate oil layer above the water level.

An emulsion of oil droplets dispersed inside water can be formed by vigorously shaking the liquids in the container. Nevertheless, this emulsion is drastically unstable, and, very soon, water and oil will segregate again forming two separated layers.

Emulsions can be stabilized by adding emulsifiers to it. Emulsifiers are amphiphilic substances, whose molecules have a polar hydrophilic head and a long hydrophobic tail. An example of such a substance is cetyl-alcohol, whose molecule is shown in Fig. 6.21. Due to its amphiphilic nature, when these substances are added to an oil–water system their molecules migrate to the interface. When oil is fragmented

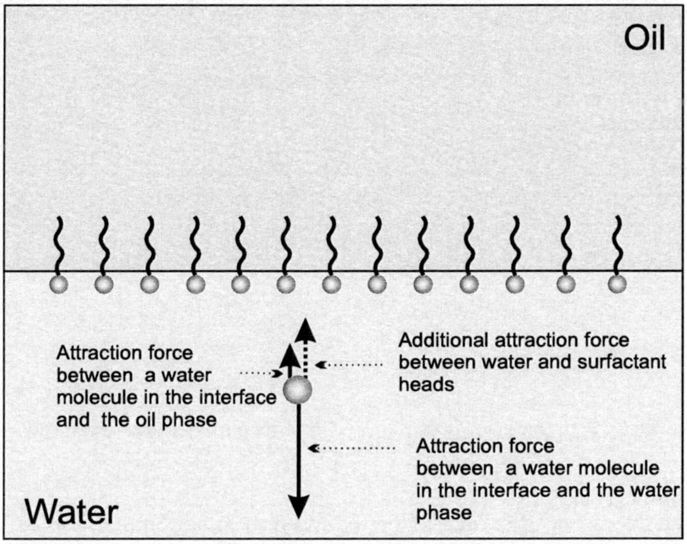

Fig. 6.23 When a surfactant is added to an oil–water system, the polar heads in the interface will create an additional force from the surface that will reduce the force imbalance responsible for the interface tension

and dispersed in a continuous water phase as in Fig. 6.22a, the polar heads of the emulsifier at the interface will attract the water molecules in their neighborhoods with a force that is much stronger than the one it attracts the oil molecules. So a water layer will be stabilized by these segregating forces between any two oil drops, preventing the oil drops from coalescing.

When water is dispersed in oil as in Fig. 6.22b the long tails of the emulsifier will prevent their polar heads to attract the water molecules in the contiguous water drop, because they will be beyond the interaction length of these intermolecular polar forces.

Emulsifiers are found in milk, mayonnaise, butter, ice cream, chocolate, and several other food products. Proteins and phospholipids prevent the milk fat dispersed in water from coalescing. Other emulsifiers are the lecithin present in the egg yolk and responsible for stabilizing the mayonnaise. Cetyl alcohol is used as an emulsifier in the cosmetic industry in the production of skin creams and lotions.

Surfactants are responsible for the reduction in the interface tension between two immiscible fluids. In fact, the interface tension may be thought as the result of the imbalance between the forces that attract a fluid molecule in the interface to its bulk phase and the forces from the surface that attract this molecule to its contiguous phase. So, when a surfactant is added to an oil–water system as in Fig. 6.23, the polar heads in the interface will create an additional force from the surface that will reduce this force imbalance.

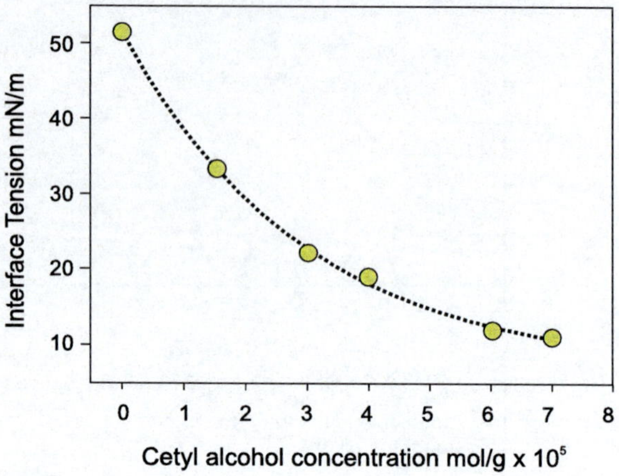

Fig. 6.24 Effect of the addition of cetyl alcohol on the interface tension of a water–nujol system at 30 °C. Experimental results from Wolf [186]

Figure 6.24 shows the effect of the addition of small amounts of cetyl alcohol on the interface tension of a water–nujol system at 30 °C [186]. Nujol is a mixture of alkanes C_nH_{2n+2} with a very large value of n and is often used for simulating water–oil displacements in petroleum rocks. The water–nujol interface tension at 30 °C was measured as 51.6 mN/m.

Part III
Introduction to Non-equilibrium Thermodynamics

Chapter 7
Non-equilibrium States

Abstract Non-equilibrium thermodynamics provides us with a general framework for the macroscopic description of thermodynamic systems in non-equilibrium states, where the state variables are treated as field variables, i.e., as continuous functions of space coordinates and time. In this field, the balance equation for entropy plays a central role, expressing the fact that the entropy of a volume element changes with time for two reasons. First, it changes because entropy flows into the volume element, and second, because there is an entropy source due to irreversible phenomena inside the volume element. In this chapter, the differential equations for density, momentum, and energy are presented when these thermodynamic variables are considered field variables. An application of the theory is given considering the heating and dissipation processes when a cavity filled with an ideal gas is heated. The chapter ends with the entropy balance equation in its differential and integral forms.

7.1 Introduction

In accordance with de Groot and Mazur [39], thermodynamic considerations were first applied to the treatment of irreversible processes by W. Thomson in 1854,[1] in the study of the thermoelectric phenomena. Duhem,[2] Natanson,[3] Jaumann[4] and later Eckart[5] attempted to obtain expressions for the rate of change of the local entropy

---

[1] Thomson, W. (1882) *On the Dynamical Theory of Heat, with numerical results deduced from Mr Joule's equivalent of a Thermal Unit, and M. Regnault's Observations on Steam*, Mathematical and Physical Papers, 174–332 [174].

[2] Duhem, P. (1911) *Traité d' énergétique ou de thermodynamique générale* (2 vols.). Gauthier-Villars [47].

[3] Natanson, L. (1896). *Uber die Gezetze nicht umkehrbarer Vorgange*. Z. Phys. Chem., 193 [121].

[4] Jaumann, G. (1905). *Die Grundlagen der Bewegungslehre*. Barth Verlag [75]. Jaumann, G. (1911). *Geschlossenes System physikalischer und chemischer Diffrerentialgesetze*. K. k. Hof- und Staats-druckerei [76].

[5] Eckart, C. (1940a). *The Thermodynamics of Irreversible Processes. I. The Simple Fluid*. Physical Review, 58(3), 267–269 [49]. Eckart, C. (1940b). *The Thermodynamics of Irreversible Processes. II. Fluid Mixtures*. Physical Review, 58(3), 269–275 [50].

P. C. Philippi, *Thermodynamics*, https://doi.org/10.1007/978-3-031-49357-7_7

in nonuniform systems by combining the second law of thermodynamics with the macroscopic laws of conservation of mass, momentum, and energy.

The field of non-equilibrium thermodynamics provides us with a general framework for the macroscopic description of irreversible processes. As such it is a branch of macroscopic physics, which has connections with other macroscopic approaches such as fluid dynamics and heat transfer, insofar as the latter disciplines are also concerned with non-equilibrium situations.

Thus non-equilibrium thermodynamics should be set up from the start as a continuum theory, treating the state variables of the theory as field variables, i.e., as continuous functions of space coordinates and time. Moreover one would like to formulate the basic equations of the theory in such a way that they only contain quantities referring to a single point in space at one time, i.e. in the form of local equations. In equilibrium thermodynamics such a local formulation is generally not needed, since the state variables are usually independent of the space coordinates.

In non-equilibrium thermodynamics, the so-called balance equation for entropy plays a central role. This equation expresses the fact that the entropy of a volume element changes with time for two reasons. First, it changes because entropy flows into the volume element, and second because there is an entropy source due to irreversible phenomena inside the volume element.

As we have seen in Sect. 2.2 the entropy source is always a non-negative quantity. It only vanishes when the process is reversible. This is the local formulation of the second law of thermodynamics. The main aim is to relate the entropy source explicitly to the various irreversible processes that occur in a system.

To this end, one needs the macroscopic conservation laws of mass, momentum, and energy, in local, i.e. differential form. These conservation laws contain a number of quantities such as the diffusion flows, the heat flow and the pressure tensor, which are related to the transport of mass, energy, and momentum. It turns out that the entropy source has a very simple appearance: it is a sum of terms each one being a product of a flux characterizing an irreversible process, and a quantity, called thermodynamic force, which is related to the non-uniformity of the system. e.g., the gradient of the temperature or to the deviations of some internal state variables from their equilibrium values. The entropy source strength can thus serve as a basis for the systematic description of the irreversible processes occurring in a system.

This set of equations contains the irreversible fluxes as unknown parameters and can therefore not be solved with given initial and boundary conditions for the state of the system. At this point, we must therefore supplement our equations with an additional set of equations that relate the irreversible fluxes and the thermodynamic forces appearing in the entropy source strength. In the first approximation, the fluxes are linear functions of the thermodynamic forces. Fick's law of diffusion, Fourier's law of heat conduction, and the stress–strain relationship in fluid mechanics, for instance, belong to this class of linear phenomenological laws.

In this sense, an outstanding contribution to elucidate the relationships between fluxes and forces in complex systems was made by Onsager[6] establishing the *reciprocity theorem*, connecting the coefficients, which occur in the linear phenomenological laws that describe the irreversible processes.

Finally, Prigogine[7] formulated a consistent phenomenological theory of irreversible processes, incorporating both Onsager's reciprocity theorem and the explicit calculation for a certain number of physical situations of the so-called entropy source strength. In this way, a new field of *thermodynamics of irreversible processes* was born.

The problem of justifying the principles of non-equilibrium thermodynamics can alternatively be approached from the viewpoint of the kinetic theory. The irreversibility itself is already contained in the fundamental equation of the kinetic theory, the Boltzmann integro-differential equation. One may then justify the use of thermodynamic relations for systems outside equilibrium, and derive the Onsager reciprocal relations (see, e.g., Philippi and Brun [129]).

In this part, we limit ourselves to presenting the transport equations for non-equilibrium systems and, in Chaps. 8 and 9 to use the entropy balance equation as a tool for finding the fundamental balance equations in multiphase, multicomponent systems, by using the Gibbs's formal relations. The book of de Groot and Mazur [39] is suggested to interested readers aiming to improve their understanding of irreversible phenomena, including chemical reactions, cross-effects in multicomponent systems and electromagnetic theory.

7.2 The Local Approach

Consider a thermodynamic system with a single component. When this system is in equilibrium, the energy of its molecules, or the internal energy U, is a homogeneous function of the entropy S, the volume V and the mass m,

$$U = U\left(S, V, m\right).$$

Since the temperature T can be defined as $T = (\partial U/\partial S)_{V,m}$, the pressure as $P = -\left(\partial U/\partial V\right)_{S,m}$ and the chemical potential as $\mu = (\partial U/\partial m)_{S,V}$ the following differential equation can be derived

$$dU = TdS - PdV + \mu dm,$$

[6] L. Onsager (1931), *Reciprocal Relations in Irreversible Processes. I.* Phys. Rev. 37, 405–426 [122]. L. Onsager (1931), *Reciprocal Relations in Irreversible Processes. II.* Phys. Rev. 38, 2265–2279 [123].

[7] I. Prigogine, Etude thermodynamique des phenomenes irreversibles (Dunod, Paris and Desoer, Liege, 1947).

or, in terms of intensive variables, as

$$du = Tds - Pdv, \tag{7.1}$$

because in homogeneous systems,

$$u = Ts - Pv + \mu,$$

and

$$d\mu = -sdT + vdP, \tag{7.2}$$

in accordance with the well-known Gibbs–Duhem equation.

In this section, we will prefer to work with the mass density $\rho = 1/v$ instead of the specific volume v as an independent variable. So Eq. (7.1) will be written as

$$\rho du = \rho T ds + \frac{P}{\rho} d\rho. \tag{7.3}$$

Equation (7.3) is strictly valid for homogeneous systems in thermodynamic equilibrium. Non-equilibrium systems can, nevertheless, be described with this equation using the separation of scales hypothesis in a continuum system. With this assumption, each point of the space is considered to have a characteristic length much greater than the molecular dimensions and, at the same time, much smaller than the characteristic length of the whole system. So, each point can be thought as an elementary system in local thermodynamic equilibrium, although its local thermodynamic properties differ from the ones of its neighbor points.

In fluid dynamics, within the framework of continuum mechanics, a *fluid element* or *fluid particle* is a very small amount of fluid, identifiable throughout its dynamic history while moving with the fluid flow. As it moves, the mass of a fluid element remains constant, while—in a compressible flow—its volume may change. The shape of this fluid particle may also change due to the distortion by the flow.

Therefore by assuming the continuum hypothesis and, consequently, by supposing each point to be an elementary representative volume in local thermodynamic equilibrium, the governing differential equations for this fluid in non-equilibrium states will be established.

Let us first rewrite Eq. (7.3) in the form

$$\rho \frac{ds}{dt} = \frac{\rho}{T}\frac{du}{dt} - \frac{P}{\rho T}\frac{d\rho}{dt}, \tag{7.4}$$

where d/dt refers to the material time derivative, $d/dt = \partial_t + \mathbf{v} \cdot \nabla$, i.e., to the changes in the course of time that an elementary material volume will suffer in its thermodynamic properties when flowing along its trajectory.

A trajectory or path line of a fluid particle can be described by a function $\mathbf{x} = \mathbf{x}(\mathbf{x}_0, t)$ where $\mathbf{x}$ is the position of the particle at time t and $\mathbf{x}_0$ is the position this particle had at time $t = 0$. The initial coordinates $\mathbf{x}_0$ of a particle is referred to as the material coordinates of the particle. The term *Lagrangian coordinates* is also used. The spatial coordinates $\mathbf{x}$ of the particle can be referred to as its position or place. Associated with the spatial and Lagrangian descriptions are two derivatives with respect to time. We denote them by

$$\partial_t = \left(\frac{\partial}{\partial t}\right)_{\mathbf{x}} = \textit{local time derivative}$$

$$\frac{d}{dt} = \left(\frac{\partial}{\partial t}\right)_{\mathbf{x}_0} = \textit{material time derivative}$$

Thus $\partial_t F$ is the rate of change of F as observed at a fixed point $\mathbf{x}$, whereas $\frac{dF}{dt}$ is the rate of change as observed when moving with the particle. In particular, the material derivative of the position of a particle is its velocity.

This allows us to establish a connection between the two derivatives,

$$\frac{dF}{dt} = \frac{\partial}{\partial t}\left(F\left(\mathbf{x}\left(\mathbf{x}_0, t\right), t\right)\right)_{\mathbf{x}_0} = \frac{\partial}{\partial t}(F)_{\mathbf{x}} + \frac{\partial F}{\partial \mathbf{x}} \cdot p\left(\frac{\partial \mathbf{x}}{\partial t}\right)_{\mathbf{x}_0} = \partial_t F + \mathbf{v} \cdot \nabla F.$$

Therefore, the material derivative of the entropy in Eq. (7.4) is dependent on the material derivatives of the internal energy and the density. These derivatives will be given in the next two sections in relation to the mass conservation and the internal energy balance equations.

7.3 Mass Conservation

The law of mass conservation is attributed to Antoine Lavoisier.[8]

In its modern integral form, the mass conservation equation is related to the invariance of mass in a material or Lagrangian volume $V(t)$ as depicted in Fig. 7.1

[8] It is generally accepted that Lavoisier's great accomplishments in chemistry largely stem from his changing the science from a qualitative to a quantitative one. Lavoisier is most noted for his discovery of the role oxygen plays in combustion. He recognized and named oxygen (1778) and hydrogen (1783) and opposed the phlogiston theory. Lavoisier helped construct the metric system, wrote the first extensive list of elements, and helped to reform chemical nomenclature. He predicted the existence of silicon (1787) and was also the first to establish that sulfur was an element (1777) rather than a compound. He discovered that, although matter may change its form or shape, its mass always remains the same [101].

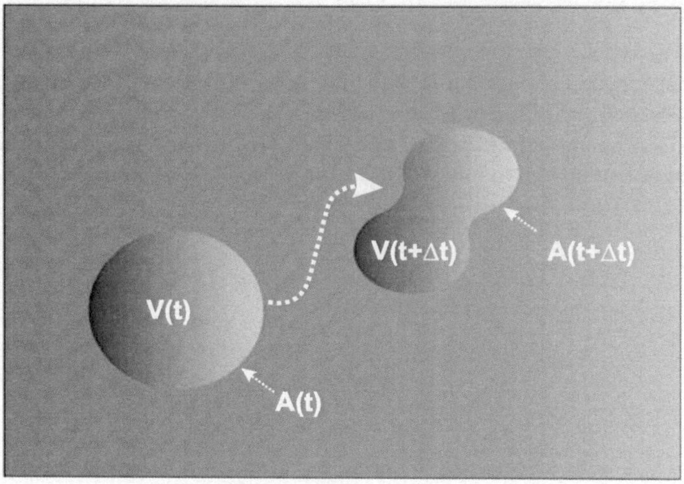

Fig. 7.1 Material or Lagrangian volume $V(t)$ moving inside a fluid has its volume and shape affected by pressure and viscous forces, but the mass m is always the same

$$\frac{dm}{dt} = \frac{d}{dt} \int_{V(t)} \rho dV = 0.$$

By using the Reynolds transport theorem[9] we obtain

$$\frac{d}{dt} \int_{V(t)} \rho dV = \int_{V(t)} \partial_t \rho dV + \int_{A(t)} \rho \mathbf{v} \cdot \mathbf{n} dA = 0. \tag{7.5}$$

The divergence theorem

$$\int_{A(t)} \rho \mathbf{v} \cdot \mathbf{n} dA = \int_{V(t)} \nabla \cdot p\,(\rho \mathbf{v})\, dV,$$

can, thus, be used for the surface integral resulting in the mass conservation equation in its differential form

$$\partial_t \rho + \nabla \cdot (\rho \mathbf{v}) = 0 \tag{7.6}$$

or, in terms of the material derivative of the density

$$\frac{d\rho}{dt} = \partial_t \rho + \mathbf{v} \cdot p \nabla \rho,$$

[9] The Reynolds transport theorem is derived in fluid mechanics textbooks, see, e.g., [7, 93, 127].

as

$$\frac{d\rho}{dt} = -\rho\nabla \cdot \mathbf{v} \tag{7.7}$$

The local form of the mass conservation equation, Eq. (7.6) can be generalized to any arbitrary thermodynamic property F. So, let f be the value of this property per unit mass. The time variation of this property in a material volume $V(t)$, is given by

$$\frac{d}{dt}\int_{V(t)} \rho f dV = -\int_{A(t)} \mathbf{j}(f) \cdot \mathbf{n} dA + \int_{V(t)} \mathscr{P}(f) dV. \tag{7.8}$$

The above equation is a general conservation law and means that the time variation of the thermodynamic property F in a material volume $V(t)$ is due to the diffusive flux $\mathbf{j}(f)$ across its surface boundary and to a source term $\mathscr{P}(f)$ that generates the property F inside this material volume.

By applying the Reynolds transport equation, Eq. (7.5) and the divergence theorem, Eq. (7.8) can be written in its differential form

$$\partial_t(\rho f) + \nabla \cdot (\rho f \mathbf{v} + \mathbf{j}) = \mathscr{P}(f), \tag{7.9}$$

where $\rho f \mathbf{v}$ is the advective flux of f.

7.3.1 Exercises

1. Integrate Eq. (7.6) in an Eulerian volume V_0, invariant with time and show that

$$\frac{d}{dt}\int_{V_0} \rho dV = -\int_{A_0} \rho\mathbf{v} \cdot \mathbf{n} dA,$$

where A_0 denotes the external area of the Eulerian volume. What is the physical meaning of the r.h.s. of the above equation?

2. Consider now Eq. (7.5). What is the physical meaning of

$$\int_{A(t)} \rho\mathbf{v} \cdot \mathbf{n} dA,$$

when $V(t)$ is a Lagrangian volume with a constant mass?

3. Consider a mixture of two *non-reacting* fluids A and B and $\mathbf{J}_i$, $i = A, B$ to be the total flux of each one of these fluids. The flux $\mathbf{J}_A$ has two components: the advective flux $\rho_A \mathbf{v}$ and the diffusive flux $\mathbf{j}_A = \rho_A(\mathbf{v}_A - \mathbf{v})$.

(a) What would be the form of the differential equation, Eq. (7.9), for ρ_A and ρ_B?
(b) Write these equations in accordance with an integral Lagrangian form.
(c) Show that $\mathbf{j}_A = -\mathbf{j}_B$, i.e., that the two diffusive flux must cancel themselves.

7.4 Momentum Balance Equation

The momentum per unit mass is represented by $f = \mathbf{v}$. It is a vectorial quantity and is transported in accordance with Eq. (7.9), where $\rho f \mathbf{v} = \rho \mathbf{v}\mathbf{v}$ is the momentum per unit volume $\rho \mathbf{v}$ that is advected with the velocity $\mathbf{v}$. In this case a external force field $\mathbf{g}$—the gravity field, for instance—is a source of momentum at each point of the physical domain and Eq. (7.9) is written as

$$\partial_t (\rho \mathbf{v}) + \nabla \cdot (\rho \mathbf{v}\mathbf{v} + \mathbf{P}) = \rho \mathbf{g}. \tag{7.10}$$

Pressure tensor $\mathbf{P}$ represents the diffusive flux of momentum and has two components, an elastic component $P\boldsymbol{\delta}$ and a viscous component, the viscous stress tensor $\boldsymbol{\tau}$,

$$\mathbf{P} = P\boldsymbol{\delta} + \boldsymbol{\tau},$$

where $\boldsymbol{\delta}$ is the Kronecker-delta tensor and P is the local thermodynamic pressure.

In components notation, Eq. (7.10) is also written as

$$\partial_t (\rho v_\alpha) + \partial_\beta \left(\rho v_\alpha v_\beta + P\delta_{\alpha\beta} + \tau_{\alpha\beta}\right) = \rho g_\alpha. \tag{7.11}$$

By using the mass conservation equation, Eq. (7.7), the above equation can be written as

$$\rho \frac{dv_\alpha}{dt} = \rho g_\alpha - \partial_\alpha P - \partial_\beta \tau_{\alpha\beta}, \tag{7.12}$$

which is the usual form of the momentum balance equation that is found in most textbooks on fluid mechanics.

For Newtonian fluids that satisfy the Stokes hypothesis[10] the relationship between the viscous stress and strain rate tensors is given by

[10] The Stokes hypothesis [162] supposes that the trace of the viscous stress tensor is null

$$\tau_{11} + \tau_{22} + \tau_{33} = 0.$$

This hypothesis is confirmed by the kinetic theory for systems of particles limited to translational degrees of freedom. The paper of Buresti [22] is suggested for further reading.

$$\tau_{\alpha\beta} = -2\eta S_{\alpha\beta} + \frac{2}{3}\eta\nabla \cdot p\mathbf{v}, \tag{7.13}$$

where η is the dynamical viscosity and

$$S_{\alpha\beta} = \frac{1}{2}\left(\partial_\alpha v_\beta + \partial_\beta v_\alpha\right),$$

is the strain rate tensor.

The Lagrangian integral form of Eq. (7.10) gives, perhaps, a better picture of the transport of momentum in a fluid,

$$\frac{d}{dt}\int\limits_{V(t)} \rho\mathbf{v}dV = -\int\limits_{A(t)} P\mathbf{n}dA - \int\limits_{A(t)} \boldsymbol{\tau}\cdot\mathbf{n}dA + \int\limits_{V(t)} \rho\mathbf{g}dV, \tag{7.14}$$

which can be interpreted as giving the forces that affect the momentum of a material volume $V(t)$:

- the normal pressure forces $P\mathbf{n}$ and the viscous forces $\boldsymbol{\tau}\cdot\mathbf{n}$ acting on the external surface $A(t)$ of the material volume;
- the external forces $\rho\mathbf{g}$ acting in all the points of $V(t)$.

7.4.1 Exercises

1. With the help of Eq. (7.13) show that for incompressible flow with constant viscosity,

$$\rho\frac{dv_\alpha}{dt} = \rho g_\alpha - \partial_\alpha P + \eta\nabla^2 v_\alpha. \tag{7.15}$$

2. Neglecting the gravity, consider the incompressible, two-dimensional, and full-developed, Poiseuille channel flow as depicted in Fig. 7.2. In this case

$$\nabla\cdot\mathbf{v} = 0,$$

$$v_1 = v_1(x_2),$$

and

$$v_2 = 0.$$

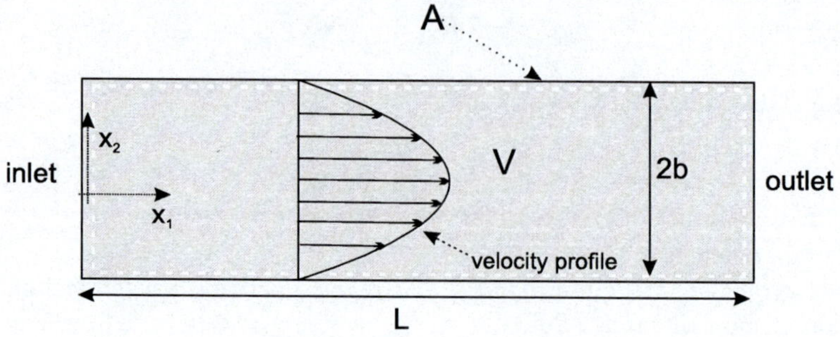

Fig. 7.2 A two-dimensional channel flow

(a) Simplify Eq. (7.15) for the components 1 and 2 of the momentum and show that the pressure P is only dependent on the horizontal coordinate, $P = P(x_1)$.
(b) Show that $\partial^2 v_1/\partial x_2^2$ is a constant and related to dP/dx_1.
(c) Find the velocity profile $v_1 = v_1(x_2)$.
(d) In the Eulerian volume V shown in the figure, integrate the resulting equation for the horizontal velocity component momentum equation and show that the pressure difference between the inlet and the outlet must counterbalance the friction forces between the fluid and the channel wall, for maintaining the flow.

7.5 Energy Conservation

The principle of energy conservation was enunciated by Hermann Ludwig Ferdinand von Helmholtz (Fig. 7.3). Based on the works of Sadi Carnot, Émile Clapeyron and James Prescott Joule, he postulated a relationship between mechanics, heat, light, electricity and magnetism by treating them all as manifestations of a single force (*or energy*). He published his theories in his book 'Über die Erhaltung der Kraft' (*On the conservation of force*) in 1847 [68].

In a fluid, the total energy can be written as the sum of its internal (U), kinetic (E_c), and potential $\left(E_p\right)$ energies

$$E = U + E_c + E_p.$$

In this section, we derive the balance equation for the internal energy from the balance equations for the kinetic and potential energy and from the conservation equation for the total energy.

Fig. 7.3 Helmholtz in 1848

Balance Equation for the Kinetic Energy
The kinetic energy balance equation can be retrieved by making the inner product of the momentum balance equation, Eq. (7.11) by the local velocity **v**. Therefore, for the time derivative term

$$v_\alpha \partial_t (\rho v_\alpha) = \partial_t \left(\rho v^2\right) - \underbrace{\rho v_\alpha \partial_t (v_\alpha)}_{= v_\alpha \partial_t (\rho v_\alpha) - v^2 \partial_t \rho} .$$

So

$$v_\alpha \partial_t (\rho v_\alpha) = \partial_t \left(\frac{1}{2}\rho v^2\right) + \frac{v^2}{2}\partial_t \rho. \tag{7.16}$$

On the other hand, for the momentum advection flux

$$v_\alpha \partial_\beta \left(\rho v_\alpha v_\beta\right) = \partial_\beta \left(\rho v^2 v_\beta\right) - \underbrace{\rho v_\alpha v_\beta \partial_\beta v_\alpha}_{= v_\alpha \partial_\beta (\rho v_\alpha v_\beta) - v^2 \partial_\beta (\rho v_\beta)} .$$

So

$$v_\alpha \partial_\beta \left(\rho v_\alpha v_\beta\right) = \partial_\beta \left(\frac{1}{2}\rho v^2 v_\beta\right) + \frac{v^2}{2}\partial_\beta \left(\rho v_\beta\right). \tag{7.17}$$

Summing Eqs. (7.16) and (7.17), we have,

$$v_\alpha \partial_t (\rho v_\alpha) + v_\alpha \partial_\beta (\rho v_\alpha v_\beta) = \partial_t \left(\frac{1}{2}\rho v^2\right) + \partial_\beta \left(\frac{1}{2}\rho v^2 v_\beta\right) + \frac{v^2}{2}\underbrace{(\partial_t \rho + \partial_\beta (\rho v_\beta))}_{=0}. \tag{7.18}$$

For the remaining terms

$$v_\alpha \partial_\beta (P\delta_{\alpha\beta}) = v_\alpha \partial_\alpha P = \partial_\alpha (P v_\alpha) - P\nabla \cdot \mathbf{v}, \tag{7.19}$$

and

$$v_\alpha \partial_\beta \tau_{\alpha\beta} = \partial_\beta (\tau_{\alpha\beta} v_\alpha) - \tau_{\alpha\beta}\partial_\beta v_\alpha. \tag{7.20}$$

From Eqs. (7.18), (7.19), and (7.20) results

$$\partial_t \left(\frac{1}{2}\rho v^2\right) + \partial_\beta \left(\frac{1}{2}\rho v^2 v_\beta + P v_\alpha + \tau_{\alpha\beta} v_\alpha\right) = \rho g_\alpha v_\alpha + P\nabla \cdot \mathbf{v} + \tau_{\alpha\beta}\partial_\beta v_\alpha,$$

or in vector form

$$\partial_t \left(\frac{1}{2}\rho v^2\right) + \nabla \cdot \left(\frac{1}{2}\rho v^2 \mathbf{v} + P\mathbf{v} + \boldsymbol{\tau} \cdot \mathbf{v}\right) = \rho \mathbf{g} \cdot \mathbf{v} + \boldsymbol{\tau} : \nabla \mathbf{v} + P\nabla \cdot p\mathbf{v}. \tag{7.21}$$

In its Lagrangian integral form, the above equation reads as

$$\frac{d}{dt}\int\limits_{V(t)} \frac{1}{2}\rho v^2 dV = -\int\limits_{A(t)} P\mathbf{v} \cdot \mathbf{n} dA - \int\limits_{A(t)} \boldsymbol{\tau} \cdot \mathbf{v} \cdot \mathbf{n} dA + \int\limits_{V(t)} (\rho \mathbf{g} \cdot \mathbf{v} - \Phi + P\nabla \cdot \mathbf{v})\, dV.$$

In the above equation

- the first term on the r.h.s. represents the compression work exerted by normal forces on the boundary surface $A(t)$;
- the second term represents the change of kinetic energy due to shear forces between $V(t)$ and the external fluid.

There are three sources of kinetic energy inside the Lagrangian volume:

- the external body force $\mathbf{g}$;
- the viscous forces in the form of a dissipation term $\Phi == -\boldsymbol{\tau} : \nabla \mathbf{v} \geq \mathbf{0}$ that is responsible for the irreversible conversion of kinetic energy into heat;
- in compressible flows when $\nabla \cdot \mathbf{v} \neq 0$ the pressure also acts as a (*reversible*) source of kinetic energy.

Balance Equation for the Potential Energy
The balance equation for the potential energy is given in its Lagrangian form as

$$\frac{d}{dt} \int_{V(t)} \rho \Upsilon dV = - \int_{V(t)} \rho \mathbf{g} \cdot \mathbf{v} dV, \tag{7.22}$$

where Υ is the potential energy per unit mass. In fact, there is no diffusive flux for the potential energy, which is only transported by advection of the fluid and can be transformed into kinetic energy when the inner product between the external force $\mathbf{g}$ and the local velocity $\mathbf{v}$ is greater than zero. In its differential form, this balance equation is written as

$$\partial_t (\rho \Upsilon) + \nabla \cdot (\rho \Upsilon \mathbf{v}) = - \rho \mathbf{g} \cdot \mathbf{v}. \tag{7.23}$$

Energy Conservation
Consider now the total energy $E = U + E_c + E_p$. Due to the principle of energy conservation, there are no sources of total energy

$$\mathscr{P}(E) = \mathscr{P}(U) + \mathscr{P}(E_c) + \mathscr{P}\left(E_p\right) = 0,$$

and since

$$\mathscr{P}(E_c) + \mathscr{P}\left(E_p\right) = \boldsymbol{\tau} : \nabla \mathbf{v} + P \nabla \cdot \mathbf{v},$$

this result means that the source of thermodynamic internal energy is to be written as

$$\mathscr{P}(U) = \Phi - P \nabla \cdot \mathbf{v}.$$

Therefore, the balance equation for the internal energy ρu per unit volume, will be written as

$$\partial_t (\rho u) + \nabla \cdot (\rho u \mathbf{v} + \mathbf{q}) = \Phi - P \nabla \cdot \mathbf{v}, \tag{7.24}$$

where $\mathbf{q}$ is the Fourier diffusive flux of heat.

The Eulerian integral form of this equation is given by

$$\frac{d}{dt}\int_{V_0} \rho u dV = -\int_{A_0} (\rho u \mathbf{v} \cdot \mathbf{n} + \mathbf{q} \cdot \mathbf{n})\, dA + \int_{V_0} (\Phi - P\nabla \cdot \mathbf{v})\, dV. \tag{7.25}$$

In its differential form, the internal energy balance equation is more usually written as

$$\rho \frac{du}{dt} + \nabla \cdot \mathbf{q} = \Phi - P\nabla \cdot \mathbf{v}. \tag{7.26}$$

The total energy conservation equation can be written by summing Eqs. (7.21), (7.23) and (7.24),

$$\partial_t \left(\rho u + \frac{1}{2}\rho v^2 + \rho \Upsilon \right) + \nabla \cdot \begin{bmatrix} \left(\rho u + \frac{1}{2}\rho v^2 + \rho \Upsilon\right) \mathbf{v} \\ + \mathbf{q} + P\mathbf{v} + \boldsymbol{\tau} \cdot \mathbf{v} \end{bmatrix} = 0. \tag{7.27}$$

The meaning of the several terms that contribute to the transport of energy can be better appreciated when this equation is written in its Eulerian integral form. So, consider an Eulerian fixed volume V_0 delimited by an external surface with area A_0. Integrating Eq. (7.27) in the volume V_0 and using the Reynolds theorem, Eq. (7.8)

$$\begin{aligned} &\frac{d}{dt}\int_{V_0} \left(\rho u + \frac{1}{2}\rho v^2 + \rho \Upsilon \right) dV \\ &= -\int_{A_0} \left(\rho u + \frac{1}{2}\rho v^2 + \rho \Upsilon \right) \mathbf{v} \cdot \mathbf{n} dA - \int_{A_0} P\mathbf{v} \cdot \mathbf{n} dA \\ &- \int_{\mathbf{A_0}} \mathbf{q} \cdot \mathbf{n} dA - \int_{A_0} \boldsymbol{\tau} \cdot \mathbf{v} \cdot \mathbf{n} dA. \end{aligned} \tag{7.28}$$

The first term on the r.h.s. represents the net flux of total energy through the Eulerian volume V_0.

The total energy E of V_0 is also affected by:

- the pressure work $-\int_{A_0} P\mathbf{v} \cdot \mathbf{n} dA$ that is released to or from V_0 at, respectively, the inlet and outlet;
- the net heat flux $-\int_{\mathbf{A_0}} \mathbf{q} \cdot \mathbf{n} dA$ exchanged with the external environment;
- the friction work $-\int_{A_0} \boldsymbol{\tau} \cdot \mathbf{v} \cdot \mathbf{n} dA$ that is transferred through the boundary surface A_0 (Fig. 7.4).

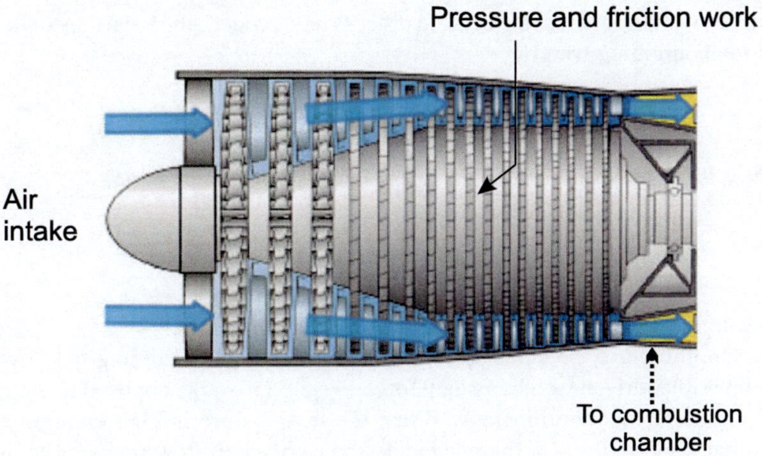

Fig. 7.4 Pressure work and friction are responsible for transferring the kinetic energy of the compressor blades to the air

Fig. 7.5 The heated cavity problem

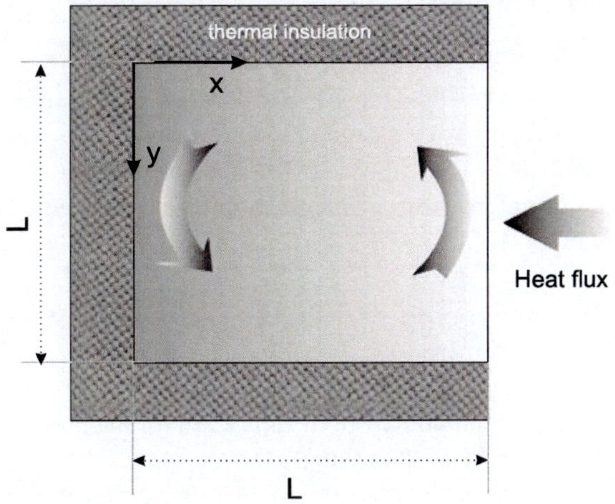

7.6 The Heated Cavity Problem

In this section, we apply the differential equations derived in the previous sections to formulate and find the temperature and velocity fields that are developed when a fluid inside a two-dimensional square cavity is heated from its right face (Fig. 7.5). We consider the fluid as an ideal gas submitted to a constant heat flux q_0 during a certain interval of time Δt. After this interval, the cavity's right face is insulated and we follow what happens with the temperature and velocity fields.

Further, we suppose that the heating process only reduces the local density $\rho\,(x, y)$ by an small amount given by

$$\Delta\rho = -\,\rho_0\beta\,(T - T_0)\,,$$

where ρ_0 is the unperturbed density at the initial temperature T_0 and

$$\beta = -\frac{1}{\rho_0}\left(\frac{\partial\rho}{\partial T}\right)_P,$$

is the volumetric expansion coefficient at a constant pressure.

We can, thus, use the Boussinesq's approximation[11] considering the flow to be incompressible, driven by a buoyancy force $-\,\rho_0 g\beta\,(T - T_0)$ opposed to the gravity force $\rho_0 g$ along the coordinate y. When the temperature is high enough in such a way that $\beta\,(T - T_0) > 1$ the gas inside the cavity will flow upwards by natural convection whereas in the colder regions, the gas moves downwards, closing the convection loop.

With these approximations, the potential energy inside the cavity is a constant and given by

$$-\int\limits_{V_0}(\rho_0 y)\,dV = -\,\rho_0\int\limits_0^L dx\int\limits_0^L ydy = -\,\rho_0\frac{L^2\times 1}{2}.$$

Therefore, from an integral point of view the equation to be used during the heating time is

$$\frac{d}{dt}\int\limits_{V_0}\left(\rho_0 u + \frac{1}{2}\rho_0 v^2\right)dV = q_0 L\times 1,$$

which is a suitable form of Eq. (7.28) for the problem since there is no flow through the boundary of the cavity.

It is, nevertheless, important to note that this invariance of the total potential energy is a consequence of the mathematical model used for describing the problem. A more realistic model would consider the local variations of the density with time due to heating and the incompressibility assumption would not apply to the problem.

The differential equations for the horizontal v_x and vertical v_y velocity components are,

$$\frac{\partial v_x}{\partial x} + \frac{\partial v_y}{\partial y} = 0,$$

---

[11] Boussinesq, Joseph (1897). Théorie de l'écoulement tourbillonnant et tumultueux des liquides dans les lits rectilignes a grande section. Vol. 1. Gauthier-Villars.

$$\frac{\partial v_x}{\partial t} + v_x \frac{\partial v_x}{\partial x} + v_y \frac{\partial v_x}{\partial y} = -\frac{1}{\rho_0} \frac{\partial P}{\partial x} + \nu \left(\frac{\partial^2 v_x}{\partial x^2} + \frac{\partial^2 v_x}{\partial y^2} \right), \tag{7.29}$$

$$\frac{\partial v_y}{\partial t} + v_x \frac{\partial v_y}{\partial x} + v_y \frac{\partial v_y}{\partial y} = - g\beta \left(T - T_0 \right) - \frac{1}{\rho_0} \frac{\partial \left(P - \rho_0 g y \right)}{\partial y} + \nu \left(\frac{\partial^2 v_y}{\partial x^2} + \frac{\partial^2 v_y}{\partial y^2} \right).$$

Since the flow was supposed to be incompressible the pressure P lost its meaning as a function of ρ and T given by the Clapeyron equation of state and became an additional field variable.

For closing the system given by Eq. (7.29), we use the Fourier law

$$\mathbf{q} = -\lambda \nabla T,$$

for heat conduction and the internal energy balance equation, Eq. (7.24) is written as

$$\rho_0 c_v \partial_t T + \rho_0 c_v \left[v_x \frac{\partial T}{\partial x} + v_y \frac{\partial T}{\partial y} \right] = \lambda \left(\frac{\partial^2 T}{\partial x^2} + \frac{\partial^2 T}{\partial y^2} \right) + \Phi, \tag{7.30}$$

where c_v is the specific heat, λ the thermal conductivity and η, the dynamical viscosity of the gas are supposed to be constants.

The last term on the r.h.s. of Eq. (7.30) corresponds to the irreversible transformation of kinetic energy into heat. For incompressible flows (see Sect. 7.7),

$$\Phi = 2\eta \mathbf{S} : \mathbf{S} = 2\eta \left(\left(\frac{\partial v_x}{\partial x} \right)^2 + \left(\frac{\partial v_y}{\partial y} \right)^2 \right) + \eta \left(\frac{\partial v_x}{\partial y} + \frac{\partial v_y}{\partial x} \right)^2.$$

Equations (7.29) and (7.30) can be written using dimensionless variables as

$$\frac{\partial \hat{v}_x}{\partial \hat{x}} + \frac{\partial \hat{v}_y}{\partial \hat{y}} = 0,$$

$$\frac{\partial \hat{v}_x}{\partial \hat{t}} + \hat{v}_x \frac{\partial \hat{v}_x}{\partial \hat{x}} + \hat{v}_y \frac{\partial \hat{v}_x}{\partial \hat{y}} = -\frac{\partial \hat{P}}{\partial \hat{x}} + \left(\frac{\partial^2 \hat{v}_x}{\partial \hat{x}^2} + \frac{\partial^2 \hat{v}_x}{\partial \hat{y}^2} \right),$$

$$\frac{\partial \hat{v}_y}{\partial \hat{t}} + \hat{v}_x \frac{\partial \hat{v}_y}{\partial \hat{x}} + \hat{v}_y \frac{\partial \hat{v}_y}{\partial \hat{y}} = -\frac{Ra}{Pr}\theta - \frac{\partial \hat{P}}{\partial \hat{y}} + \left(\frac{\partial^2 \hat{v}_y}{\partial \hat{x}^2} + \frac{\partial^2 \hat{v}_y}{\partial \hat{y}^2} \right), \tag{7.31}$$

$$\partial_{\hat{t}} \theta + \left[\hat{v}_x \frac{\partial \theta}{\partial \hat{x}} + \hat{v}_y \frac{\partial \theta}{\partial \hat{y}} \right] = \frac{1}{Pr} \left(\frac{\partial^2 \theta}{\partial \hat{x}^2} + \frac{\partial^2 \theta}{\partial \hat{y}^2} \right) + Ec\hat{\Phi},$$

where $\hat{t} = \nu t/L^2$, $\hat{x} = x/L$, $\hat{y} = y/L$, $\hat{v}_\alpha = Lv_\alpha/\nu$,

$$\theta = \frac{T - T_0}{\frac{q_0 L}{\lambda}}, \quad \hat{P} = \frac{P - \rho_0 g y}{\frac{\rho_0 \nu^2}{L^2}}.$$

The problem is driven by three non-dimensional parameters:

- the Rayleigh number related to the ratio between the buoyancy and the viscous forces $Ra = \frac{\rho_o \beta g c_v q_0 L^4}{\nu \lambda^2}$;
- the Prandtl number giving the ratio of momentum diffusivity to thermal diffusivity, $Pr = \frac{\rho_o c_v \nu}{\lambda}$;
- the Eckert number related to the dissipation of kinetic energy, $Ec = \frac{\nu^2 \lambda}{L^3 c_v q_0}$.

Equation (7.31) are highly non-linear and were solved by using a finite-volumes numerical method.

Figure 7.6 shows the temperature field and streamlines during the heating time of the cavity. The heated fluid near the right face moves upwards by natural convection driven by a buoyancy force and downwards in the colder regions. We can notice that the velocity and temperature gradients are very large close to the heated right face.

After the heating process is finished and the right face of the cavity is insulated, the convection loop is dissipated, and the kinetic energy of the flow being converted into internal energy (Fig. 7.7).

From Eq. (7.25), during the heating process, the internal energy increases with time in accordance with

$$\frac{d}{dt} \int_{V_0} \rho_0 c_v (T - T_0) \, dV = q_0 L \times 1 + \int_{V_0} \Phi dV,$$

meaning that the internal energy increase is due to the heat flow into the cavity and the viscous dissipation of kinetic energy. Remark that dissipation can be very strong in the heating time, due to the strong velocity gradients near the right face of the cavity.

During the dissipation time, after the heating is ceased and the right face of the cavity is insulated, the kinetic energy is entirely converted into internal energy by viscous dissipation,

$$\frac{d}{dt} \int_{V_0} \rho_0 c_v (T - T_0) \, dV = \int_{V_0} \Phi dV.$$

The numerical solution of the differential equations, Eq. (7.31) enables to find the temperature field $T(x, y, t)$ and, consequently, the total internal energy of the cavity as a function of time

$$U(t) = \int_{V_0} \rho_0 c_v (T(x, y, t) - T_0) \, dV.$$

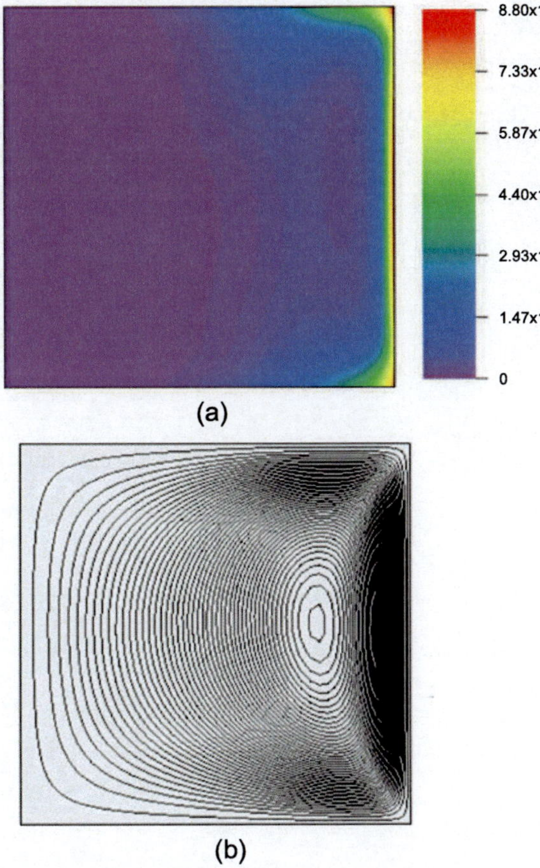

Fig. 7.6 **a** Temperature field and **b** streamlines during the heating time of the cavity ($\hat{t} = 0.008$). Dimensionless parameters are Pr = 1, Ra = 10^6, Ec = 10^{-4} [73]

Figure 7.8 shows the integrated result for the total internal energy $U(t)$ for the cavity problem. When the gas is heated both the total internal and total kinetic energy increase with time producing a convection loop that is further extinguished by viscous dissipation when heating is ceased.

Viscous dissipation was also measured using the simulation results for the velocity field. The results for

$$\hat{D} = Ec \int_{\hat{V}} \hat{\Phi} d\hat{V},$$

as a function of the dimensionless time $\hat{t}$ are presented in Fig. 7.9.

Viscous dissipation grows very fast during the heating period due to the strong velocity gradients close to the heated face. During the dissipation time, it is responsi-

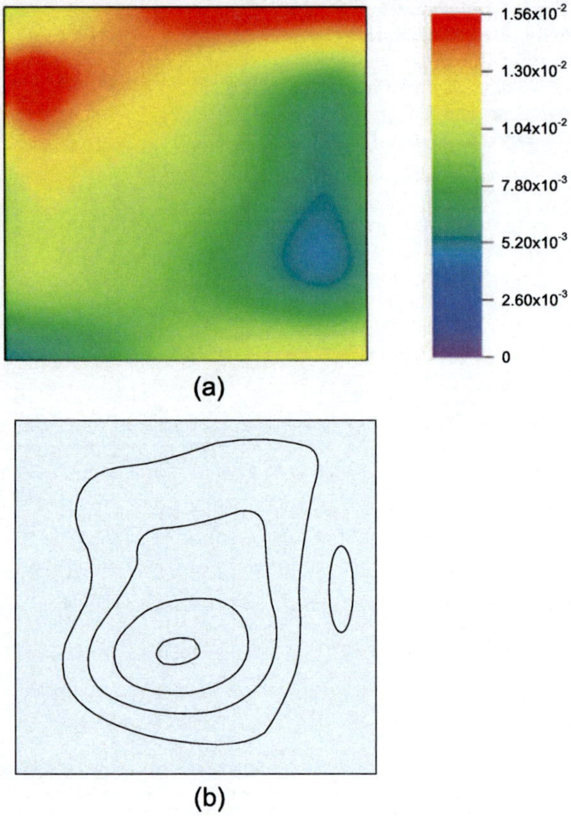

Fig. 7.7 **a** Temperature field and **b** streamlines at the dissipation period $\hat{t} = 0.026$. Dimensionless parameters are Pr = 1, Ra = 10^6, Ec = 10^{-4} [73]

ble for converting the kinetic energy of the flow into internal energy and for reducing these velocity gradients.

Oscillations in both Figs. 7.8 and 7.9 do not appear to have any physical meaning and are possibly due to numerical errors.

Equilibrium × Non-equilibrium Thermodynamics

In spite of the simplifications that were made for numerically solving the problem, the heated cavity is an important sample problem for distinguishing the main features of non-equilibrium thermodynamics when compared to equilibrium thermodynamics. The models and methods of *equilibrium thermodynamics* that were seen in Parts I and II of this book only apply to thermodynamic systems *in equilibrium*. The first law of thermodynamics

$$\delta Q + \delta W = dU,$$

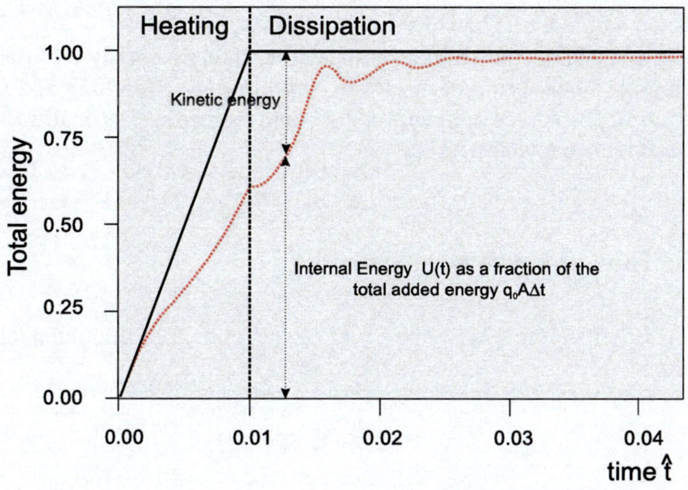

Fig. 7.8 A constant heat flow $\mathbf{q_0}$ into a cavity filled with an ideal gas increases its temperature and produces natural convection currents. After heating is ceased these currents are dissipated, and their kinetic energy is transformed into internal energy

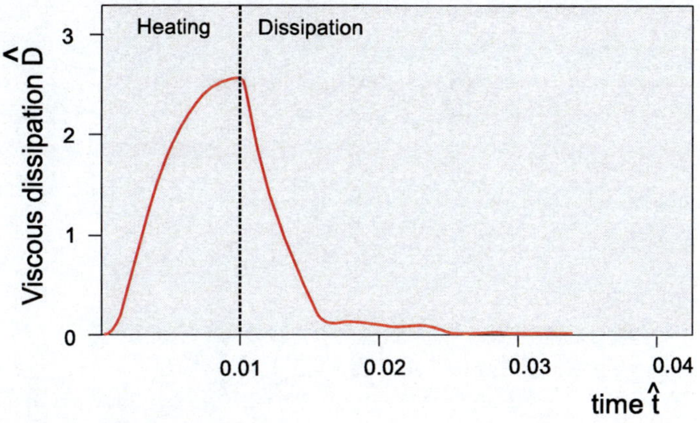

Fig. 7.9 Dimensionless viscous dissipation $\hat{\Phi}$ as a function of time [73]

for instance, can only be used when we are considering a thermodynamic process between any two *equilibrium states*. It cannot be used, for, e.g., finding the variation of the internal energy, when one or both states are non-equilibrium states.

Therefore, for the cavity problem, the variation of the internal energy between the initial 1 and final 2 *equilibrium states* is the total heat that was added to the cavity during the heating time Δt,

$$Q_{1-2} = q_0 L \times 1 \times \Delta t = U_2 - U_1,$$

in accordance with the first law of thermodynamics.

In any intermediary non-equilibrium state, the kinetic energy is different from zero, the influx of heat being used for creating the convection loop and increasing the temperature. The source of kinetic energy (and momentum) is, in this Boussinesq approximation, the buoyancy force.

7.7 The Law of Increasing Entropy

By inserting Eqs. (7.7) and (7.26) into Eq. (7.4), the following differential relationship can be derived,

$$\rho\frac{ds}{dt} = -\frac{1}{T}\left(\nabla\cdot\mathbf{q} - \Phi\right).$$

The first term on the r.h.s. of the above equation can be rewritten in the form

$$\frac{1}{T}\nabla\cdot\mathbf{q} = \nabla\cdot\left(\frac{\mathbf{q}}{T}\right) + \frac{1}{T^2}\mathbf{q}\cdot\nabla T, \tag{7.32}$$

so

$$\rho\frac{ds}{dt} = -\nabla\cdot\mathbf{q}_s + \mathscr{P}(s), \tag{7.33}$$

where

$$\mathbf{q}_s = \frac{\mathbf{q}}{T},$$

is the flux of entropy and

$$\mathscr{P}(s) = \left(-\frac{1}{T^2}\mathbf{q}\cdot\nabla T\right) + \frac{\Phi}{T} \geq 0,$$

is the irreversible source of entropy.

The Lagrangian integral of this equation is easily found as,

$$\underbrace{\frac{d}{dt}\int\limits_{V(t)}\rho s dV}_{\frac{dS}{dt}} = \underbrace{-\int\limits_{A(t)}\frac{\mathbf{q}\cdot\mathbf{n}}{T}dA}_{\frac{d_eS}{dt}} + \underbrace{\int\limits_{V(t)}\left(-\frac{1}{T^2}\mathbf{q}\cdot\nabla T + \frac{\Phi}{T}\right)dV}_{\frac{d_iS}{dt}}. \tag{7.34}$$

This equation, to be attributed to Ilya Prigogine [140], states that the material derivative of the entropy $\frac{dS}{dt}$ is due to the net influx of heat through the external surface of the material volume and to the existence of internal non-equilibrium in the temperature and velocity fields. By the second law of thermodynamics, the internal production of entropy $\frac{d_i S}{dt}$ must be non-negative

$$\frac{d_i S}{dt} \geq 0.$$

Let us consider a Newtonian fluid that satisfies the Stokes hypothesis for which the relationship between the viscous stress $\tau_{\alpha\beta}$ and the rate of strain $S_{\alpha\beta} = \frac{1}{2}\left(\partial_\alpha v_\beta + \partial_\beta v_\alpha\right)$ tensors is given by

$$\tau_{\alpha\beta} = -\,2\eta S_{\alpha\beta} + \frac{2}{3}\eta \nabla \cdot \mathbf{v}\delta_{\alpha\beta}.$$

Using the symmetry with respect to a change of index, the viscous dissipation term can be written as

$$\Phi = -\,\tau_{\alpha\beta} : \partial_\alpha v_\beta = -\,\tau_{\alpha\beta}\frac{1}{2}\left(\partial_\alpha v_\beta + \partial_\beta v_\alpha\right) = -\,\tau_{\alpha\beta} S_{\alpha\beta},$$

or

$$\Phi = \left(2\eta S_{\alpha\beta} S_{\alpha\beta} - \frac{2}{3}\eta \left(\nabla \cdot \mathbf{v}\right)^2\right),$$

because

$$S_{\alpha\beta}\delta_{\alpha\beta} = \frac{1}{2}\left(2\partial_x v_x + 2\partial_y v_y + 2\partial_z v_z\right) = \nabla \cdot \mathbf{v}.$$

Therefore

$$\Phi = 2\eta \left(S_{\alpha\beta} - \frac{1}{3}\nabla \cdot \mathbf{v}\right)^2,$$

because

$$\begin{aligned}\left(S_{\alpha\beta} - \frac{1}{3}\nabla \cdot \mathbf{v}\right)^2 &= S_{\alpha\beta} S_{\alpha\beta} - \frac{2}{3} S_{\alpha\beta}\delta_{\alpha\beta}\nabla \cdot \mathbf{v} + \frac{1}{9}\left(\nabla \cdot \mathbf{v}\right)^2 \delta_{\alpha\beta}\delta_{\alpha\beta} \\ &= S_{\alpha\beta} S_{\alpha\beta} - \frac{2}{3}\left(\nabla \cdot \mathbf{v}\right)^2 + \frac{1}{3}\left(\nabla \cdot \mathbf{v}\right)^2,\end{aligned}$$

since, in three dimensions

$$\delta_{\alpha\beta}\delta_{\alpha\beta} = 3.$$

On the other hand, by restricting ourselves to isotropic fluids that satisfy the Fourier's law for heat conduction

$$\mathbf{q} = -\lambda \nabla T,$$

the heat conduction contribution to the source of entropy can be written as

$$-\frac{\mathbf{q} \cdot \nabla T}{T^2} = \lambda \left(\frac{\nabla T}{T}\right)^2,$$

in such a way that

$$\frac{d_i S}{dt} = \int_{V(t)} \left[\lambda \left(\frac{\nabla T}{T}\right)^2 + \frac{2\eta}{T}\left(S_{\alpha\beta} - \frac{1}{3}\nabla \cdot \mathbf{v}\right)^2\right] dV \geq 0, \tag{7.35}$$

and we are led to the conclusion that for a Newtonian and isotropic fluid both the dynamical viscosity η and the thermal conductivity λ must be positive, otherwise the second law of thermodynamics would be broken. This appears to be an obvious conclusion since we always expect that heat will flow from higher to lower temperatures and that momentum will flow from higher to smaller velocities, but the important meaning of Eq. (7.35) is that dissipation will always happen every time we have a gradient in our thermodynamic system. In other words, gradients are the source of irreversibility and will always increase the entropy.

When a thermodynamic system is subjected to a process, starting from an equilibrium state 1 at $t = 0$ to another equilibrium state 2 at $t = t_\infty$, where t_∞ is the time required by the system for reaching the equilibrium state 2, Eq. (7.34) can be integrated between $t = 0$ and $t = t_\infty$ giving,

$$S_2 - S_1 = (S_2 - S_1)_e + (S_2 - S_1)_i$$

For concreteness let us reconsider the abrupt isothermal expansion sample problem of Sect. 2.5.1 and shown in Fig. 7.10. Numerical simulation of the velocity and temperature field during the abrupt gas expansion is a difficult task for this problem since it involves a compressible flow with moving boundaries and subjected to friction forces between the piston and the cylinder walls. As commented in Sect. 2.1.2 the friction work is absorbed by the thermal pond and will not contribute to defining the final equilibrium state. Nevertheless, these forces have an important contribution to the dynamics of gas expansion, or to the time the system will take for reaching the final equilibrium state.

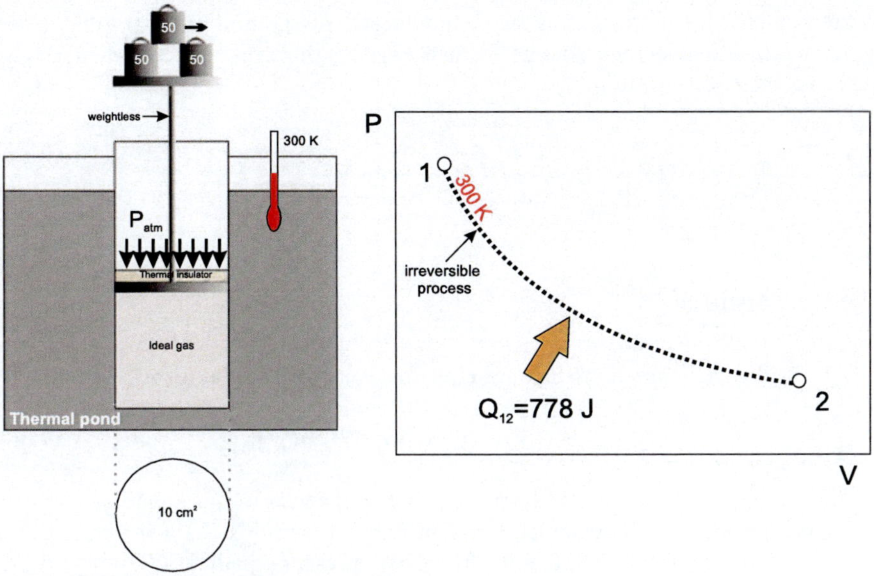

Fig. 7.10 Isothermal expansion of an ideal gas

We can easily find the total variation of entropy as

$$S_2 - S_1 = R \ln \frac{V_2}{V_1} = 3.10 \, \frac{\mathrm{J}}{\mathrm{K}}.$$

On the other hand, since the boundary surface is maintained at a constant temperature during the whole process, the entropy variation due to the heat transfer from the thermal pond is

$$(S_2 - S_1)_e = -\int_0^{t_\infty} \int_{A(t)} \frac{\mathbf{q} \cdot \boldsymbol{n}}{T} dA = \frac{1}{T} Q_{1-2},$$

giving

$$(S_2 - S_1)_e = 2.5933 \, \frac{\mathrm{J}}{\mathrm{K}}.$$

Therefore the amount

$$(S_2 - S_1)_i = 0.5067 \, \frac{\mathrm{J}}{\mathrm{K}},$$

corresponds to the entropy increase by the internal dissipation processes due to the temperature and velocity gradients inside the system that are produced when the gas has an abrupt expansion

$$(S_2 - S_1)_i = \int_0^{t_\infty} \int_{V(t)} \left(-\frac{1}{T^2} \mathbf{q} \cdot \nabla T + \frac{\Phi}{T} \right) dV.$$

7.8 Exercises

1. Consider the momentum balance equation in its differential form

$$\partial_t (\rho \mathbf{v}) + \nabla \cdot (\rho \mathbf{v}\mathbf{v} + \mathbf{P}) = \rho \mathbf{g}. \tag{7.36}$$

   (a) Write the Eulerian and Lagrangian integral forms of this equation.
   (b) Consider the 2D channel drawn in Fig. 7.11. In this channel a fluid flows along a horizontal direction, driven by a pressure gradient $(P_0 - P_L)/L$. Use the integral Eulerian form of the momentum balance equation supposing the flow to be incompressible and considering the control volume inside the dashed line shown in the figure. What is the physical meaning of the several terms that appear in the formulation? Simplify the equations for a stationary flow.

2. An ideal gas in chamber A at 20 °C is put in thermal contact with a liquid bath at 90 °C (Fig. 7.12). In consequence, a Bénard cellular convection process will take place in the gas phase (*consider that the liquid bath remains with* $T = 90$ °C *during the whole process*). Using the Boussinesq approximation shows that the momentum balance equation for the gas can be written as[12]

$$\rho_0 \frac{d\mathbf{v}}{dt} = -\nabla \cdot \boldsymbol{\tau} - \nabla P + \underbrace{\rho_0 (1 - \beta (T - T_0))\, \mathbf{g}}_{\text{source of momentum}} \tag{7.37}$$

   where $\mathbf{g}$ is the gravity, ρ_0 is the fluid density at $t = 0$ and β is the fluid volumetric expansion coefficient

$$\beta = -\frac{1}{\rho} \left(\frac{\partial \rho}{\partial T} \right)_P$$

[12] In the Boussinesq approximation the flow is supposed to be incompressible with $\nabla \cdot \mathbf{v} = 0$. So the density variation produced by heating is supposed to be small.

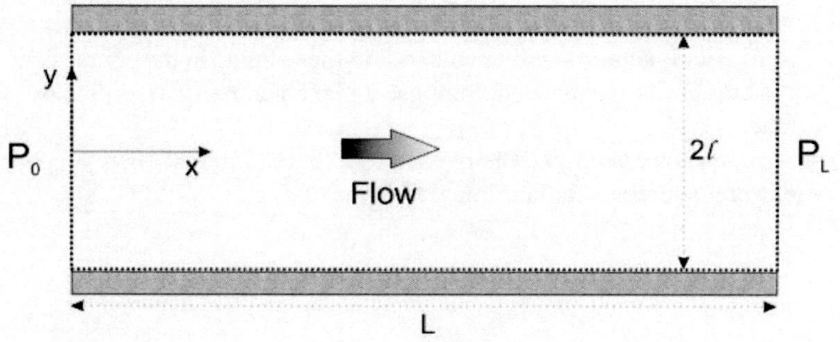

Fig. 7.11 Channel flow

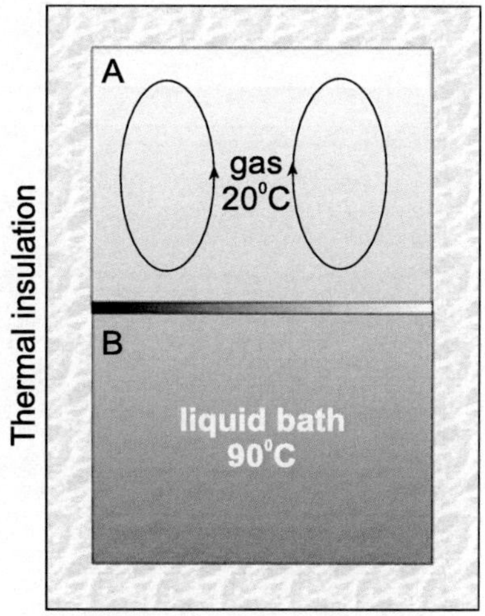

Fig. 7.12 Bénard cellular convection

Therefore, the buoyancy force $\rho_0 (1 - \beta (T - T_0))\, \mathbf{g}$ is a source of momentum contrary to gravity, and responsible for the convection process.

(a) Multiply Eq. (7.37) by $\mathbf{v}$ and find the balance equation for the advection kinetic energy of the fluid.
(b) Write the balance equation for the potential energy.
(c) Find the balance equation for the total energy in its differential and integral form.
(d) Find the balance equation for the internal energy in its differential and integral form.
(e) Find the balance equation for the entropy in its differential and integral form. Finally, try to identify and explore the possible symmetries of the problem and to write/identify a closed system of differential equations and

their boundary conditions, for the velocity, temperature, and entropy whose (numerical) solution will enable to find these fields in the course of time till the equilibrium is reached. Suppose a plane geometry described by coordinates x and z.

Also, try to explain the role of each term in the internal energy balance and entropy equations in their integral forms.

Chapter 8
Multiphase Systems

Abstract Two-phase liquid–vapor systems in non-equilibrium conditions are often found in nature and devices such as cooling systems of power plants, as sprays in the combustion chamber of internal combustion engines, and in boiling and industrial devices. The liquid–vapor interface has a thickness of a few nanometers and the fluid density has strong variation along the interface thickness. In numerical simulations, the way the interface position is determined plays a key role in modeling two-phase systems and we can classify the several methods that were developed for two-phase systems in two families. The first one is based on the singular interface approach when the interface is considered as a boundary separating the two phases. The second one is based on the diffuse interface approach when the interface is considered a diffused zone. These two families of methods are presented in this chapter. For illustrating purposes, the lattice-Boltzmann simulation results for the coalescence process between two liquid droplets and vapor condensation are presented at the end of the chapter.

8.1 Introduction

Many processes in nature and industry are dependent on the correct understanding of multiphase systems. These include omnipresent phenomena and applications, such as rain clouds or fuel injection into combustion chambers, rocket engines, or spray-based production processes and petroleum exploitation.

Droplets have been the focus of scientific interest for a long time: the generation of droplets from liquid jets under the influence of surface tension was studied by Plateau[1] and Lord Rayleigh [143] in the nineteenth century. The Plateau–Rayleigh instability describes how a liquid jet breaks up into droplets. In Fig. 8.1, we can notice the waviness in the jet before the breakdown. These undulations are nothing but small disturbances, which grow with the distance from the jet exit. In accordance with Laplace's law, the tiniest curvature in the jet causes an imbalance in the liquid pressure due to surface tension. Liquid pressure is higher in the jet regions with

[1] Plateau JAF (1873) Statique expérimentale et théorique des liquides soumis aux seules forces moléculaires, vol 2. Gauthier-Villars.

P. C. Philippi, *Thermodynamics*, https://doi.org/10.1007/978-3-031-49357-7_8

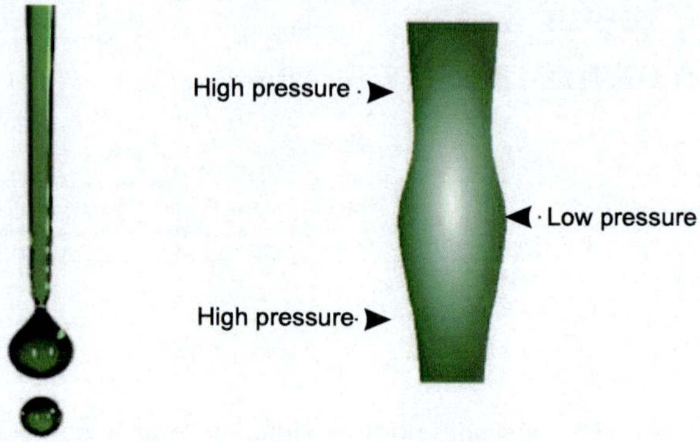

Fig. 8.1 The Plateau–Rayleigh instability

smallest curvature radius and the liquid flows from the high to the small pressures. Considering the inset in Fig. 8.1, the downward movement of the liquid is assisted by the pressure difference between the superior and central regions. In contrast, it is resisted by the pressure difference between the inferior and central regions. This promotes the growth of an undulation, which eventually transforms the central region into a droplet. Because the system is unstable, any small changes in the pressure due to the undulations will become larger, ultimately resulting in the jet breaking into droplets.

As pointed out by Lefebvre and McDonnel in their seminal book [102], in the process of atomization, a liquid jet is disintegrated by the kinetic energy of the liquid, by the exposure to high-velocity air or gas, or by mechanical energy applied externally through a rotating or vibrating device. Combustion of liquid fuels in diesel engines, spark-ignition engines, gas turbines, rocket engines, and industrial furnaces are dependent on effective atomization to increase the specific surface area of the fuel and thereby achieve high rates of mixing and evaporation.

Another subject of great scientific interest is the liquid–vapor spinodal decomposition (Fig. 8.2). As was pointed out in Chap. 4, at a given temperature, a thermodynamic system that is suddenly put into a state where the density is between the vapor and liquid saturated densities, cannot subsist in this state as a single-phase system and any small perturbation will break up this system into a two-phase liquid–vapor system. Phase separation is very fast in the beginning, however, it becomes slower due to the large number of liquid droplets competing for those molecules that are in the vapor phase. Smaller drops evaporate faster than larger drops. This is due to the effects of surface tension on the interface with the greatest curvature, which leads to a greater pressure difference in smaller droplets, leading them to evaporate earlier, while larger drops tend to grow, by condensation or coalescence between neighbor droplets.

Spinodal decomposition is not restricted to liquid–vapor systems, but to liquid–liquid multicomponent systems and metallic alloys.

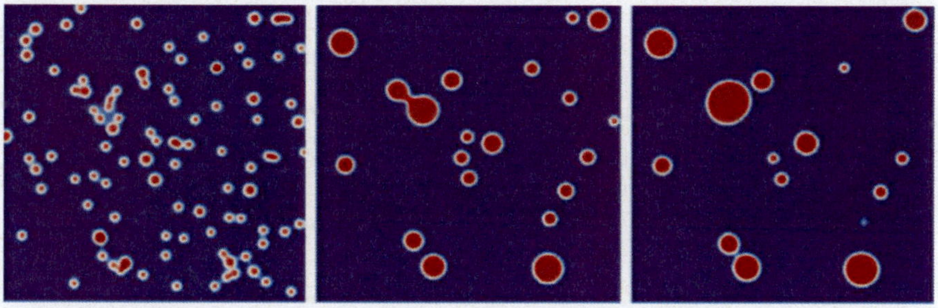

Fig. 8.2 The first stages of spinodal decomposition in liquid–vapor systems

Typically, a material interface separating two different phases is on the order of few nanometers, and, in an engineering context, can be considered as a sharp discontinuity. However, the way the interface is treated and how its position is determined plays a key part in multi-species, multiphase flow modeling, and differentiates the numerous methods introduced over the last three decades. Following a physical rigor principle, one can consider two families of methods, one that rigorously does not allow interface mixing, namely, the singular interface methods, and the other one that considers the interface as a diffused zone, namely, the diffuse interface methods.

In this chapter, these two approaches are exposed, leading to the macroscopic equations for the density, momentum, and internal energy.

8.2 The Singular Interface Approach

In the singular surface approach, the transport equations are solved in each phase and the interface is considered as a boundary, where the field variables from the two phases are linked. In the following, these boundary conditions are derived.

8.2.1 Mass Conservation

Consider a material volume $\Omega = \Omega_a \cup \Omega_b$ as shown in Fig. 8.3. This material volume involves the interface Σ between phases a and b and the interface is supposed to be moving with a velocity $\mathbf{W}$.

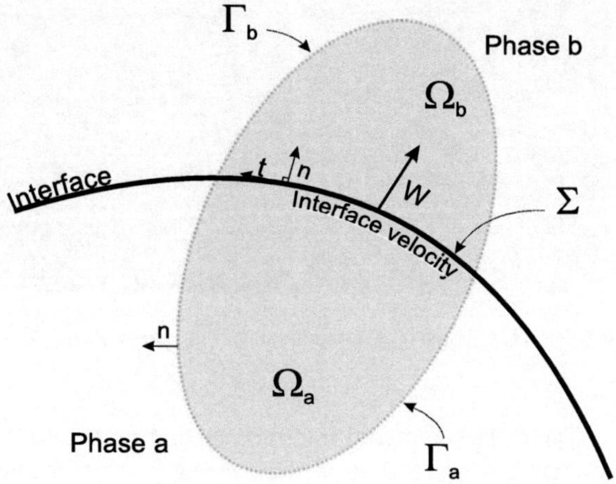

Fig. 8.3 The singular interface approach

Since the mass m inside Ω is constant

$$\frac{d}{dt}\left(\int\limits_{\Omega_a} \rho_a dV + \int\limits_{\Omega_b} \rho_b dV\right) = 0. \tag{8.1}$$

But from the Reynolds transport theorem

$$\begin{aligned}\frac{d}{dt}\int\limits_{\Omega_a} \rho_a dV &= \int\limits_{\Omega_a} \frac{\partial \rho_a}{\partial t} dV + \int\limits_{\Gamma_\alpha} \rho_a \mathbf{v}_a \cdot \mathbf{n} dA + \int\limits_{\Sigma} \rho_a \mathbf{W} \cdot \mathbf{n} dA \\ &= \int\limits_{\Omega_a} \frac{\partial \rho_a}{\partial t} dV + \int\limits_{\Gamma_a} \rho_a \mathbf{v}_a \cdot \mathbf{n} dA + \int\limits_{\Sigma} \rho_a \mathbf{v}_a \cdot \mathbf{n} dA \\ &\quad - \int\limits_{\Sigma} \rho_a \mathbf{v}_a \cdot \mathbf{n} dA + \int\limits_{\Sigma} \rho_a \mathbf{W} \cdot \mathbf{n} dA,\end{aligned}$$

and so

$$\frac{d}{dt}\int\limits_{\Omega_a} \rho_a dV = \int\limits_{\Omega_a} \left(\frac{\partial \rho_a}{\partial t} + \nabla \cdot (\rho_a \mathbf{v}_a)\right) dV - \int\limits_{\Sigma} \rho_a (\mathbf{v}_a - \mathbf{W}) \cdot \mathbf{n} dA.$$

In the same way,

$$\frac{d}{dt}\int\limits_{\Omega_b} \rho_b dV = \int\limits_{\Omega_b} \frac{\partial \rho_b}{\partial t} dV + \int\limits_{\Gamma_b} \rho_b \mathbf{v}_b \cdot \mathbf{n} dA + \int\limits_{\Sigma} \rho_b \mathbf{W} \cdot (-\mathbf{n})\, dA$$
$$= \int\limits_{\Omega_b} \frac{\partial \rho_b}{\partial t} dV + \int\limits_{\Gamma_\beta} \rho_b \mathbf{v}_b \cdot \mathbf{n} dA - \int\limits_{\Sigma} \rho_b \mathbf{v}_b \cdot (-\mathbf{n})\, dA$$
$$+ \int\limits_{\Sigma} \rho_b \mathbf{v}_b \cdot (-\mathbf{n})\, dA - \int\limits_{\Sigma} \rho_b \mathbf{W} \cdot (-\mathbf{n})\, dA,$$

because the normal $\mathbf{n}$ is oriented from phase a to phase b in Σ. So,

$$\frac{d}{dt}\int\limits_{\Omega_b} \rho_b dV = \int\limits_{\Omega_b} \left(\frac{\partial \rho_b}{\partial t} + \nabla \cdot (\rho_b \mathbf{v}_b) \right) dV + \int\limits_{\Sigma} \rho_b (\mathbf{v}_b - \mathbf{W}) \cdot \mathbf{n} dA.$$

Replacing the above equations in Eq. (8.1),

$$\underbrace{\int\limits_{\Omega} \left(\frac{\partial \rho_a}{\partial t} + \nabla \cdot (\rho_a \mathbf{v}_a) + \frac{\partial \rho_b}{\partial t} + \nabla \cdot (\rho_b \mathbf{v}_b) \right) dV}_{=0}$$
$$+ \int\limits_{\Sigma} \rho_b (\mathbf{v}_b - \mathbf{W}) \cdot \mathbf{n} dA - \int\limits_{\Sigma} \rho_a (\mathbf{v}_a - \mathbf{W}) \cdot \mathbf{n} dA$$
$$= 0,$$

or

$$\int\limits_{\Sigma} \rho_a (\mathbf{v}_a - \mathbf{W}) \cdot \mathbf{n} dA = \int\limits_{\Sigma} \rho_b (\mathbf{v}_b - \mathbf{W}) \cdot \mathbf{n} dA.$$

Since the interface Σ is arbitrary and can be made as small as possible, this equation can be written in its differential form as

$$\rho_a (\mathbf{v}_a - \mathbf{W}) \cdot \mathbf{n} = \rho_b (\mathbf{v}_b - \mathbf{W}) \cdot \mathbf{n} = \dot{m}. \tag{8.2}$$

The above equation means that the mass $\dot{m}$ (*per unit area*) that exits phase a, per unit time, is the same as the one that enters into phase b.

When phases a and b are immiscible, $\dot{m} = 0$ and

$$\mathbf{v}_a \cdot \mathbf{n} = \mathbf{v}_b \cdot \mathbf{n} = \mathbf{W} \cdot \mathbf{n}. \tag{8.3}$$

If, in addition, it is possible to suppose that there is no slip between the two phases, i.e., the tangential velocities of the two phases are the same in the interface

$$\mathbf{v}_a = \mathbf{v}_b, \quad \text{in } \Sigma. \tag{8.4}$$

8.2.2 Momentum Balance Equation

Neglecting the body forces in the volume element Ω

$$\begin{aligned} \frac{d}{dt}\left(\int\limits_{\Omega_a} \rho_a \mathbf{v}_a dV + \int\limits_{\Omega_b} \rho_b \mathbf{v}_b dV\right) \\ = -\int\limits_{\Gamma_a} \mathbf{P}_a \cdot \mathbf{n} dA - \int\limits_{\Gamma_b} \mathbf{P}_b \cdot \mathbf{n} dA \\ + \underbrace{\int\limits_{\Sigma} (\nabla \gamma_{ab} \cdot \mathbf{t} + (\gamma_{ab} \nabla \cdot \mathbf{n})\, \mathbf{n})\, dA.}_{\text{surface forces}} \end{aligned} \tag{8.5}$$

In the above equation $\nabla \cdot \mathbf{n}$ is the curvature of the surface or the inverse of the curvature radius $\mathcal{R}$ and $(\gamma_{ab} \nabla \cdot \mathbf{n})\, \mathbf{n}$ is the normal or Laplace component of the surface forces whereas $\nabla \gamma_{ab} \cdot \mathbf{t}$ corresponds to the tangential or Marangoni component of these forces,[2]

$$\mathbf{P}_s = P_s \boldsymbol{\delta} + \boldsymbol{\tau}_s$$

[2] Marangoni forces were first identified in the so-called *tears of wine* by James Thomson in 1855. (James Thomson. *On certain curious Motions observable at the Surfaces of Wine and other Alcoholic Liquors*. The London, Edinburgh, and Dublin Philosophical Magazine and Journal of Science. 1855. pp. 330–333.) The general effect is named after Italian physicist Carlo Marangoni, who studied it for his doctoral dissertation at the University of Pavia and published his results in 1865. (Carlo Marangoni. *Sull'espansione delle goccie d'un liquido galleggianti sulla superficie di altro liquid*, Pavia, Italy: Fratelli Fusi. 1869.) A complete theoretical treatment of the subject was given by Gibbs [59, 60]. Since a liquid with a high surface tension pulls more strongly on the surrounding liquid than one with a low surface tension, the presence of a gradient in surface tension will naturally cause the liquid to flow away from regions of low surface tension. The surface tension gradient can be caused by a concentration gradient or by a temperature gradient. The *tears of wine* effect is a consequence of two factors: (i) alcohol has a lower surface tension and (ii) alcohol is more volatile than water. The water/alcohol solution rises up the surface of the glass as a film. Alcohol evaporates from the film leaving behind liquid with higher surface tension (more water, less alcohol). This region with a lower concentration of alcohol (greater surface tension) pulls on the surrounding fluid more strongly than the regions with a higher alcohol concentration (lower in the glass). The result is the liquid is pulled up until its own weight exceeds the force of the effect, and the liquid drips back down the vessel's walls.

for $s = a, b$, is the pressure tensor, P_s is the thermodynamic pressure and $\boldsymbol{\tau}_s$ is the viscous stress tensor in phase s.

But,

$$\frac{d}{dt}\int\limits_{\Omega_a} \rho_a \mathbf{v}_a dV$$

$$= \int\limits_{\Omega_a} \frac{\partial}{\partial t}(\rho_a \mathbf{v}_a)\, dV + \int\limits_{\Gamma_a} (\rho_a \mathbf{v}_a \mathbf{v}_a) \cdot \mathbf{n} dA$$

$$+ \int\limits_{\Sigma} (\rho_a \mathbf{v}_a \mathbf{W}) \cdot \mathbf{n} dA$$

$$+ \int\limits_{\Sigma} (\rho_a \mathbf{v}_a \mathbf{v}_a) \cdot \mathbf{n} dA - \int\limits_{\Sigma} (\rho_a \mathbf{v}_a \mathbf{v}_a) \cdot \mathbf{n} dA,$$

where $\rho_a \mathbf{v}_a$ is the momentum of phase a in the interface Σ and $\mathbf{W}$ is the interface velocity, which transports this momentum in Σ.

It results

$$\frac{d}{dt}\int\limits_{\Omega_a} \rho_a \mathbf{v}_a dV$$

$$= \int\limits_{\Omega_a} \left(\frac{\partial}{\partial t}(\rho_a \mathbf{v}_a)\, dV + \nabla \cdot (\rho_a \mathbf{v}_a \mathbf{v}_a)\right) dV$$

$$- \int\limits_{\Sigma} (\rho_a \mathbf{v}_a (\mathbf{v}_a - \mathbf{W})) \cdot \mathbf{n} dA,$$

or

$$\frac{d}{dt}\int\limits_{\Omega_a} \rho_a \mathbf{v}_a dV$$

$$= \int\limits_{\Omega_a} (-\nabla \cdot \mathbf{P}_a)\, dV - \int\limits_{\Sigma} (\rho_a \mathbf{v}_a (\mathbf{v}_a - \mathbf{W})) \cdot \mathbf{n} dA.$$

In the same way, for phase β,

$$\frac{d}{dt}\int_{\Omega_b} \rho_b \mathbf{v}_b dV$$
$$= \int_{\Omega_b} \left(\frac{\partial}{\partial t} \left(\rho_b \mathbf{v}_b \right) dV + \nabla \cdot \left(\rho_b \mathbf{v}_b \mathbf{v}_b \right) \right) dV$$
$$+ \int_{\Sigma} \left(\rho_b \mathbf{v}_b \left(\mathbf{v}_b - \mathbf{W} \right) \right) \cdot \mathbf{n} dA,$$

or

$$\frac{d}{dt}\int_{\Omega_b} \rho_b \mathbf{v}_b dV$$
$$= \int_{\Omega_b} \left(- \nabla \cdot \mathbf{P}_b \right) dV + \int_{\Sigma} \left(\rho_b \mathbf{v}_b \left(\mathbf{v}_b - \mathbf{W} \right) \right) \cdot \mathbf{n} dA.$$

So by replacing the above equations into Eq. (8.5)

$$\begin{aligned} &\int_{\Omega_a} \left(- \nabla \cdot \mathbf{P}_a \right) dV - \int_{\Sigma} \left(\rho_a \mathbf{v}_a \left(\mathbf{v}_a - \mathbf{W} \right) \right) \cdot \mathbf{n} dA \\ &\quad + \int_{\Omega_b} \left(- \nabla \cdot \mathbf{P}_b \right) dV + \int_{\Sigma} \left(\rho_b \mathbf{v}_b \left(\mathbf{v}_b - \mathbf{W} \right) \right) \cdot \mathbf{n} dA \\ &= - \int_{\Gamma_a} \mathbf{P}_a \cdot \mathbf{n} dA - \int_{\Gamma_b} \mathbf{P}_b \cdot \mathbf{n} dA \\ &\quad + \int_{\Sigma} \left(\nabla \gamma_{ab} \cdot \mathbf{t} + \left(\gamma_{ab} \nabla \cdot \mathbf{n} \right) \mathbf{n} \right) dA. \end{aligned} \tag{8.6}$$

By using the divergence theorem

$$\int_{\Omega_a} \left(- \nabla \cdot \mathbf{P}_a \right) dV = - \int_{\Gamma_a} \mathbf{P}_a \cdot \mathbf{n} dA - \int_{\Sigma} \mathbf{P}_a \cdot \mathbf{n} dA,$$
$$\int_{\Omega_b} \left(- \nabla \cdot \mathbf{P}_b \right) dV = - \int_{\Gamma_b} \mathbf{P}_b \cdot \mathbf{n} dA + \int_{\Sigma} \mathbf{P}_b \cdot \mathbf{n} dA.$$

So, in Eq. (8.6), the integrals along the surfaces Γ_α and Γ_β cancel themselves and we get

$$\int_\Sigma (\mathbf{P}_b - \mathbf{P}_a) \cdot \mathbf{n} dA + \int_\Sigma (\mathbf{v}_b \rho_b (\mathbf{v}_b - \mathbf{W})) \cdot \mathbf{n} dA - \int_\Sigma (\mathbf{v}_a \rho_a (\mathbf{v}_a - \mathbf{W})) \cdot \mathbf{n} dA = \int_\Sigma (\nabla \gamma_{ab} \cdot \mathbf{t} + (\gamma_{ab} \nabla \cdot \mathbf{n}) \mathbf{n}) \, dA,$$

or

$$(\mathbf{P}_b - \mathbf{P}_a) \cdot \mathbf{n} + (\mathbf{v}_b \rho_b (\mathbf{v}_b - \mathbf{W})) \cdot \mathbf{n} - (\mathbf{v}_a \rho_a (\mathbf{v}_a - \mathbf{W})) \cdot \mathbf{n} = (\nabla \gamma_{ab} \cdot \mathbf{t} + (\gamma_{ab} \nabla \cdot \mathbf{n}) \mathbf{n}) \cdot \mathbf{n}$$

Now, from Eq. (8.2), we know that $\rho_b (\mathbf{v}_b - \mathbf{W}) = \rho_a (\mathbf{v}_a - \mathbf{W}) \cdot \mathbf{n} = \dot{m}$. So,

$$(\mathbf{P}_b - \mathbf{P}_a) \cdot \mathbf{n} + (\mathbf{v}_b - \mathbf{v}_a) \dot{m} = (\nabla \gamma_{ab} \cdot \mathbf{t} + (\gamma_{ab} \nabla \cdot \mathbf{n}) \mathbf{n}) .$$

8.2.3 Particular Cases

When the phases are immiscible $\dot{m} = 0$, so

$$(\mathbf{P}_b - \mathbf{P}_a) \cdot \mathbf{n} = (\nabla \gamma_{ab} \cdot \mathbf{t} + (\gamma_{ab} \nabla \cdot \mathbf{n}) \mathbf{n}) .$$

When the phases are immiscible and at rest, $\boldsymbol{\tau}_a = \boldsymbol{\tau}_b = 0$ and so, $\mathbf{P}_b = P_b \boldsymbol{\delta}$ and $\mathbf{P}_a = P_a \boldsymbol{\delta}$, resulting

$$(P_b - P_a) \mathbf{n} = (\nabla \gamma_{ab} \cdot \mathbf{t} + (\gamma_{ab} \nabla \cdot \mathbf{n}) \mathbf{n}) .$$

Making the inner product by $\mathbf{n}$

$$P_b - P_a = \gamma_{ab} \nabla \cdot \mathbf{n} = \frac{\gamma_{ab}}{\mathcal{R}},$$

which is the Laplace's law.

8.2.4 Exercises

1. (a) Show that $\nabla \cdot \mathbf{n} = 2/R$ on the surface of a sphere of radius R (*Hint: use spherical coordinates*).
   (b) Using the same method that was presented in this section, derive the total and internal energy balance equations for the interface.
   (c) **Marangoni force**. Using the same reasoning as in Exercise 1 explain how a gradient in the interface tension along the interface will produce a tangential unbalanced force.
2. A typical Rayleigh–Taylor instability (RTI) occurs when a heavier fluid sits on top of a lighter fluid under the effect of gravity, characterizing an unstable interface. As the heavy fluid goes down, the RTI develops spikes with a characteristic mushroom shape along with the appearance of plumes flowing upwards. At the same time, the light fluid forms bubbles as it goes up to occupy the place of the heavy fluid (Fig. 8.4). It receives its name after Rayleigh [144], who was the first to study this phenomenon in the nineteenth century, and later Sir G. I. Taylor in the 1950s [170].
   By considering the heavier (h) and lighter (l) fluid as immiscible and neglecting the Marangoni forces what would be the flow equations for fluids h and l and the boundary conditions at the interface? You can suppose the problem to be two-dimensional and the lateral boundary conditions to be periodic. Try to explain the physical reasons for the mushroom shape of the interface.

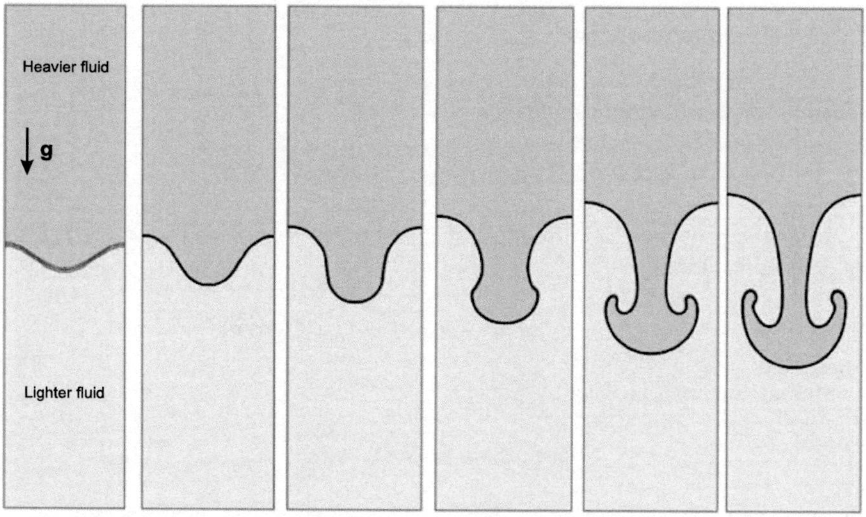

Fig. 8.4 The Rayleigh–Taylor instability

8.2.5 Numerical Methods for the Singular Interface Approach

Numerical methods based on Riemann solvers were developed by Fechter and collaborators [51–53], Rohde and Zeiler [146] and Schleper [151].

The level set method (LSM) developed by Osher and Sethian [125] and Sussman [166], uses a distance function from the interface, where the zero level marks the interface location, and the positive values correspond to one of the fluids and the negative values to the other one. In the original version, this method was affected by spurious oscillations that were corrected in the work of Fedkiw et al. [54]. In this technique, named *ghost fluid method*, a ghost cell storing the properties of the fluid b is defined in correspondence with each computational cell storing the properties of the fluid a. After each time step, the distance function, calculated with a level set equation, determines which set of properties has to be used to compute the solution, whether the one stored in the ghost or the computational cell [11]. This set of methods has the advantage of allowing a sharp definition of the interface for fluids with different equations of states. However, general drawbacks they share are the difficulty of enforcing mass conservation at the discontinuities.

Interested readers are suggested to read these papers.

8.3 The Diffuse Interface Approach

The singular interface approach is physically well-grounded as long as the interfaces are separated. However, it comes with a severe disadvantage when the interfaces meet, which is the case with the coalescence of droplets and droplet splitting. This renders the singular interface approach to fail. For such reasons, diffuse-interface models have been suggested to describe the dynamics of two-phase flow.

8.3.1 Surface Forces

The study of two-phase liquid–vapor systems was pioneered by Johannes Diderik van der Waals who in 1873 proposed a theoretical model[3] for condensable gases in his doctoral thesis.[4]

---

[3] J. D. van der Waals (1873), *Over de Continuiteit van den Gas en Vloeistoftoestand*, Ph.D. thesis, Leiden University [181].

[4] The main interest of van der Waals was in the field of thermodynamics. He was influenced by Rudolf Clausius' 1857 treatise entitled Über die Art der Bewegung, welche wir Wärme nennen (On the Kind of Motion which we Call Heat). Van der Waals was later greatly influenced by the writings of James Clerk Maxwell, Ludwig Boltzmann, and Willard Gibbs. Clausius' work led him to look for an explanation of Thomas Andrews' experiments that had revealed, in 1869, the existence of

Consider a system composed of a single fluid where the liquid (ℓ) and the vapor (v) phases coexist separated by an interface. Let this interface be a diffuse one, i.e., a region where the thermodynamic properties vary continuously from the liquid to the vapor phase and where it is possible to suppose local thermodynamic equilibrium at each point.

With these assumptions, the fundamental relationships of equilibrium thermodynamics can be used for each point in the space separating the two phases. So, consider the Helmholtz free energy per unit mass f in its differential form,

$$df = -sdT + \frac{P}{\rho^2}d\rho.$$

The Helmholtz energy can be rewritten in a volumetric basis as $\psi = \rho f$, in such a way that the above equation becomes

$$d\psi = -\rho sdT + \frac{\psi + P}{\rho}d\rho.$$

Since the chemical potential μ (*or the Gibbs energy per unit mass*, g) satisfies

$$\mu = g = \frac{\psi + P}{\rho}, \tag{8.7}$$

we can also write

$$d\psi = -\rho sdT + \mu d\rho,$$

and, so, the chemical potential can be calculated from the Helmholtz energy as

$$\mu = \left(\frac{\partial \psi}{\partial \rho}\right)_T.$$

Therefore, from Eq. (8.7), the thermodynamic pressure can be expressed in terms of the Helmholtz energy per unit volume.

---

critical temperatures in fluids. He managed to give a semi-quantitative description of the phenomena of condensation and critical temperatures in his 1873 thesis, entitled Over de Continuïteit van den Gasen Vloeistoftoestand (On the continuity of the gas and liquid state). This dissertation represented a hallmark in physics and was immediately recognized as such, e.g. by James Clerk Maxwell who reviewed it in Nature in a laudatory manner. In this thesis he derived the equation of state bearing his name. This work gave a model in which the liquid and the gas phases of a substance merge into each other in a continuous manner. It shows that the two phases are of the same nature. In deriving his equation of state van der Waals assumed not only the existence of molecules (the existence of atoms was disputed at the time), but also that they are of finite size and attract each other. Since he was one of the first to postulate an intermolecular force, however rudimentary, such a force is now sometimes called a van der Waals force.

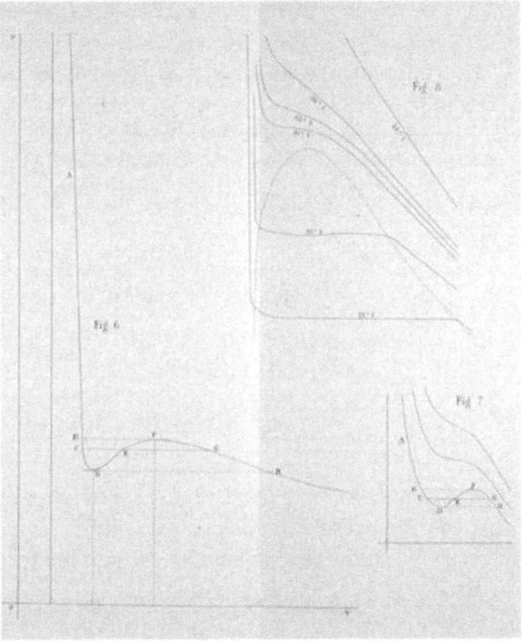

Fig. 8.5 Excerpt of van der Waals doctoral thesis [181]

$$P = \rho \left(\frac{\partial \psi}{\partial \rho} \right)_T - \psi. \tag{8.8}$$

In homogeneous systems the free energy ψ can be considered as a function of the density and temperature. Nevertheless, in the interface, the density gradients can be too strong in such a way that ψ can be supposed to be also dependent on the local density gradient, $\psi = \psi(\rho, T, \nabla \rho)$. Therefore, in the interface the Helmholtz potential has an excess due to the density gradients,

$$\psi = \psi_0(\rho, T) + \psi_{exc}, \tag{8.9}$$

where $\psi_0(\rho, T)$ is the free energy of an homogeneous system. The simplest form of this excess energy is given by,

$$\psi_{exc} = \frac{\kappa}{2} |\nabla \rho|^2,$$

firstly proposed by van der Waals (see Rowlinson and Widom [148]) and where κ can be shown to be a molecular parameter related to the attractive electrostatic forces between the fluid molecules [66, 156] (Fig. 8.5).

Therefore, in liquid–vapor interfaces, supposing that ρ, T and $\nabla \rho$ are independent variables

$$d\psi = -\rho s dT + \mu d\rho + \kappa \nabla\rho \cdot d(\nabla\rho).$$

The internal energy $u = f + Ts$ can be written in terms of the volumetric Helmholtz energy as

$$\rho u = \psi + \rho T s,$$

and so,

$$\rho du = (\mu + Ts - u) d\rho + \rho T ds + \kappa \nabla\rho \cdot d(\nabla\rho). \tag{8.10}$$

Since

$$\mu + Ts - u = \frac{\rho\mu - \psi}{\rho},$$

the difference $\rho\mu - \psi$ can be considered as a scalar pressure p, in accordance with Eq. (8.7). Nevertheless, this pressure differs from the thermodynamic pressure in the interface by terms that depend on the surface force parameter κ.

So, Eq. (8.10) can be rewritten as

$$\rho du = \frac{p}{\rho} d\rho + \rho T ds + \kappa \nabla\rho \cdot d(\nabla\rho),$$

or, in terms of material derivatives,

$$\rho \frac{du}{dt} = \frac{p}{\rho}\frac{d\rho}{dt} + \rho T \frac{ds}{dt} + \kappa \nabla\rho \cdot \frac{d(\nabla\rho)}{dt}, \tag{8.11}$$

giving for the material derivative of the entropy the following expression

$$\rho \frac{ds}{dt} = \frac{\rho}{T}\frac{du}{dt} - \frac{p}{\rho T}\frac{d\rho}{dt} - \frac{\kappa}{T}\nabla\rho \cdot \frac{d(\nabla\rho)}{dt}. \tag{8.12}$$

On the other hand, the entropy balance equation satisfies the general balance equation

$$\rho \frac{ds}{dt} = -\nabla \cdot \mathbf{q}_s + \mathscr{P}(s),$$

where $\mathbf{q}_s$ is the flux of entropy and $\mathscr{P}(s)$ is the irreversible source of entropy due to non-equilibrium.

So, from Eq. (8.12), the production of entropy in liquid–vapor phase-transition systems can be written as,

$$\mathscr{P}(s) = \frac{\rho}{T}\frac{du}{dt} + \nabla \cdot \mathbf{q}_s - \frac{p}{\rho T}\frac{d\rho}{dt} - \frac{\kappa}{T}\nabla\rho \cdot \frac{d(\nabla\rho)}{dt}. \tag{8.13}$$

For the last term in the above equation,

$$\nabla\rho\cdot\frac{d\left(\nabla\rho\right)}{dt}=\partial_{\beta}\left(\partial_{\beta}\rho\frac{d\rho}{dt}\right)-\partial_{\beta}\partial_{\beta}\rho\frac{d\rho}{dt}-\left(\partial_{\alpha}\rho\right)\left(\partial_{\beta}\rho\right)\left(\partial_{\beta}\mathrm{v}_{\alpha}\right),$$

or

$$\nabla\rho\cdot\frac{d\left(\nabla\rho\right)}{dt}=\nabla\cdot\left(\nabla\rho\frac{d\rho}{dt}\right)-\nabla^{2}\rho\frac{d\rho}{dt}-\nabla\rho\nabla\rho:\nabla\mathbf{v}. \tag{8.14}$$

From the mass conservation equation,

$$\frac{d\rho}{dt}=-\rho\nabla\cdot\mathbf{v} \tag{8.15}$$

On the other hand, the internal energy balance equation can be written in terms of the divergence of a heat flux vector and a source term given by the contraction of the pressure tensor **P**—the diffusive flux of momentum—and the gradient of the flow velocity,

$$\rho\frac{du}{dt}=-\nabla\cdot\mathbf{q}-\mathbf{P}:\nabla\mathbf{v}. \tag{8.16}$$

By replacing Eqs. (8.14), (8.15) and (8.16) into Eq. (8.13)

$$\begin{aligned}\mathscr{P}\left(s\right)=&-\frac{1}{T}\left[\mathbf{P}-\kappa\nabla\rho\nabla\rho-\left(p-\kappa\rho\nabla^{2}\rho\right)\boldsymbol{\delta}\right]:\nabla\mathbf{v}+\nabla\cdot\mathbf{q}_{s}\\&-\frac{1}{T}\nabla\cdot\mathbf{q}+\frac{1}{T}\kappa\nabla\cdot\left[\rho\left(\nabla\rho\right)\left(\nabla\cdot\mathbf{v}\right)\right].\end{aligned} \tag{8.17}$$

Now,

$$\frac{1}{T}\nabla\cdot\mathbf{q}=\nabla\cdot\left(\frac{\mathbf{q}}{T}\right)+\frac{1}{T^{2}}\mathbf{q}\cdot\nabla T, \tag{8.18}$$

and the last term on the r.h.s. of Eq. (8.17) may be written as

$$\frac{1}{T}\kappa\nabla\cdot\left[\rho\left(\nabla\rho\right)\left(\nabla\cdot\mathbf{v}\right)\right]=\nabla\cdot\left[\frac{\kappa}{T}\rho\left(\nabla\rho\right)\left(\nabla\cdot\mathbf{v}\right)\right]-\frac{1}{T^{2}}\kappa\left[\rho\left(\nabla\rho\right)\left(\nabla\cdot\mathbf{v}\right)\right]\cdot\nabla T.$$

Replacing the above equalities into Eq. (8.17),

$$\begin{aligned}\mathscr{P}(s) = &-\frac{1}{T}\left[\mathbf{P} - \kappa\nabla\rho\nabla\rho - \left(p - \kappa\rho\nabla^2\rho\right)\boldsymbol{\delta}\right] : \nabla\mathbf{v} \\ &+ \nabla\cdot\left[\mathbf{q}_s - \frac{1}{T}\left(\mathbf{q} - \kappa\rho\left(\nabla\rho\right)\left(\nabla\cdot\mathbf{v}\right)\right)\right] \\ &- \frac{1}{T^2}\left(\mathbf{q} - \kappa\rho\left(\nabla\rho\right)\left(\nabla\cdot\mathbf{v}\right)\right)\cdot\nabla T. \end{aligned} \tag{8.19}$$

The entropy production is always non-negative. In the above expression, this is assured by making

$$\mathbf{q} = -\lambda\nabla T + \kappa\rho\left(\nabla\rho\right)\left(\nabla\cdot\mathbf{v}\right), \tag{8.20}$$

reducing the flux of entropy to

$$\mathbf{q}_s = -\frac{\lambda\nabla T}{T},$$

and eliminating the divergence second term on the r.h.s. of Eq. (8.19).

Also, since $\boldsymbol{\tau} : \nabla\mathbf{v} \leq \mathbf{0}$, the entropy production is assured to be non-negative if, in addition to the above condition, we impose that

$$\mathbf{P} = \boldsymbol{\tau} + \kappa\nabla\rho\nabla\rho + \left(p - \kappa\rho\nabla^2\rho\right)\boldsymbol{\delta}. \tag{8.21}$$

Therefore, entropy production becomes

$$\mathscr{P}(s) = -\frac{1}{T}\boldsymbol{\tau} : \nabla\mathbf{v} + \frac{1}{T^2}\lambda\left(\nabla T\right)^2 \geq 0 \tag{8.22}$$

Equations (8.20) and (8.21) are sufficient conditions for $\mathscr{P}(s) \geq 0$, but not necessary. These conditions are, nevertheless, confirmed by kinetic theory.

In liquid–vapor interfaces the heat flux, Eq. (8.20), appears to be modified by an extra term related to the density gradient in the interface. From a molecular point of view, this term does not correspond to a flux of thermal energy, but is related to a work produced by intermolecular forces [48].

The term

$$\mathbf{S} = \kappa\nabla\rho\nabla\rho,$$

in Eq. (8.21) was postulated by Korteweg [91] and received his name, but it is also known as capillary tensor. This term is directly related to surface tension.

8.3.2 Momentum Balance Equation

Considering Eq. (8.21) for the pressure tensor, the momentum balance equation reads as

$$\partial_t (\rho \mathbf{v}) + \nabla \cdot (\rho \mathbf{v}\mathbf{v} + P_s \boldsymbol{\delta}) = \rho \mathbf{g} - \nabla \cdot \boldsymbol{\tau} - \nabla \cdot \mathbf{S}, \tag{8.23}$$

where P_s is a scalar pressure given by

$$P_s = p - \kappa \rho \nabla^2 \rho. \tag{8.24}$$

Pressure p can be found from Eq. (8.8)

$$p = \rho \left(\frac{\partial \psi}{\partial \rho} \right)_T - \psi, \tag{8.25}$$

where ψ is the Helmholtz free energy in the interface, given by

$$\psi = \psi_0 + \frac{\kappa}{2} (\nabla \rho)^2 .$$

So,

$$p = \underbrace{\rho \left(\frac{\partial \psi_0}{\partial \rho} \right)_T - \psi_0}_{=P} - \frac{\kappa}{2} (\nabla \rho)^2 ,$$

where the thermodynamic pressure is defined as in Eq. (8.8), but with the calculations performed with the homogeneous part of the Helmholtz free energy ψ_0,

$$P = \rho \left(\frac{\partial \psi_0}{\partial \rho} \right)_T - \psi_0.$$

Therefore the scalar pressure P_s in Eq. (8.24) becomes

$$P_s = P - \kappa \left(\frac{1}{2} (\nabla \rho)^2 + \rho \nabla^2 \rho \right),$$

in the liquid–vapor interface.

8.3.3 Summary

Table 8.1 summarizes the different forms of the momentum balance equation

Table 8.1 Single and two-phase flows

Kind of flow	Single phase	Two phases
Pressure	$P_s = P$	$P_s = P - \kappa\left(\frac{1}{2}(\nabla\rho)^2 + \rho\nabla^2\rho\right)$
Surface tensor	$\mathbf{S} = 0$	$\mathbf{S} = \kappa\nabla\rho\nabla\rho$

$$\partial_t(\rho\mathbf{v}) + \nabla\cdot(\rho\mathbf{v}\mathbf{v} + P_s\boldsymbol{\delta}) = \rho\mathbf{g} - \nabla\cdot\boldsymbol{\tau} - \nabla\cdot\mathbf{S},$$

when written for single and two-phase flows.

Therefore, under the assumption of diffuse interfaces, the momentum balance equation must take the liquid–vapor interface forces into account. The relationship between the surface parameter κ and the surface tension γ was deduced in Sect. 6.5.4 as

$$\kappa = \frac{\gamma^2}{2I^2 n_c^2 P_c}, \tag{8.26}$$

where γ is the experimental value of the surface tension evaluated at the local temperature, n_c is the critical value of the molecular density [number of molecules/unit volume], P_c is the critical pressure, and the dimensionless parameter I is given in terms of the excess Helmholtz energy in the interface as

$$I = \int_{n_{vs}^*}^{n_{\ell s}^*} \sqrt{\psi_{exc}^*}\, dn^*,$$

where $n^* = n/n_c$ and $\psi_{exc}^* = \psi_{exc}/P_c$.

8.3.4 Exercise

Consider a plane interface normal do the direction x. Suppose that this interface is in at rest ($\mathbf{v} = 0$). (a) Simplify Eq. (8.23) for this case and show that

$$P - \kappa n \frac{d^2 n}{dx^2} + \frac{\kappa}{2}\left(\frac{dn}{dx}\right)^2 = C,$$

where $P = P(x)$ is the thermodynamic pressure in the liquid–vapor interface. (b) Considering $x \to \pm\infty$ show that the constant C is the equilibrium pressure in the bulk phases P_∞, which is only dependent on the temperature, and must be the same in both phases since the interface curvature is null. (c) Define

$$H = \frac{1}{2}\kappa \left(\frac{dn}{dx}\right)^2,$$

and show that this function satisfies

$$n\frac{dH}{dn} - H = P - P_\infty.$$

(d) Show that

$$\frac{d}{dn}\left(\frac{H}{n}\right) = \frac{1}{n^2}\left(P - P_\infty\left(T\right)\right).$$

(e) Integrate the above equation between a point in the vapor phase and a point in the interface and show that the function H can be determined from the equation of state of the system

$$H\left(n, T\right) = n \int_{n_{v\infty}}^{n} \frac{1}{n'^2}\left(P\left(n', T\right) - P_\infty\left(T\right)\right) dn'.$$

(f) Using the last result and remembering that H is null in the bulk phases, show that

$$\int_{v_{\ell\infty}}^{v_{v\infty}} P dv' = P_\infty\left(v_{v\infty} - v_{\ell\infty}\right), \tag{8.27}$$

where $v = \frac{1}{n}$ is the molecular volume and Eq. (8.27) is the Maxwell area rule.

By going back to Sect. 6.5.4 it can be concluded that function H is, in fact, the Helmholtz energy in excess ψ_{exc} and that the surface tension is related to the excess energy we have in the liquid–vapor interface, due to the unbalanced intermolecular forces and following Eq. (8.26).

8.3.5 Internal Energy Balance Equation

The internal energy balance equation is given by Eq. (8.16), here reproduced,

$$\rho\frac{du}{dt} = -\nabla\cdot\mathbf{q} - \mathbf{P} : \nabla\mathbf{v}. \tag{8.28}$$

In the present case,

$$\rho u = \psi + \rho T s = \underbrace{\psi_0 + \rho T s}_{=\rho u_0} + \frac{\kappa}{2}\left|\nabla\rho\right|^2, \tag{8.29}$$

showing that the internal energy in the interface has two contributions: the internal energy per unit volume ρu_0 from an homogeneous system and a contribution from the surface forces $\frac{\kappa}{2} |\nabla\rho|^2$.

Therefore,

$$\rho\frac{du}{dt} = \rho\frac{du_0}{dt} + (u_0 - u)\frac{d\rho}{dt} + \kappa\nabla\rho\frac{d(\nabla\rho)}{dt}. \tag{8.30}$$

From Eq. (8.29)

$$(u_0 - u) = -\frac{\kappa}{2\rho}|\nabla\rho|^2,$$

and so, Eq. (8.30) becomes

$$\rho\frac{du}{dt} = \rho\frac{du_0}{dt} - \frac{\kappa}{2\rho}|\nabla\rho|^2\frac{d\rho}{dt} + \kappa\nabla\rho\frac{d(\nabla\rho)}{dt}. \tag{8.31}$$

The last term on the r.h.s. of the above equation is given by Eq. (8.14). Taking the mass conservation equation into account, it results

$$\begin{aligned}\rho\frac{du}{dt} = \rho\frac{du_0}{dt} &+ \frac{\kappa}{2}(\nabla\rho)^2\nabla\cdot\mathbf{v} - \kappa\nabla\cdot(\rho\nabla\rho\nabla\cdot\mathbf{v}) \\ &+ \kappa\rho\nabla^2\rho\nabla\cdot\mathbf{v} - \kappa\nabla\rho\nabla\rho\colon\nabla\mathbf{v}.\end{aligned}$$

On the other hand the heat flux vector in Eq. (8.28) was deduced in accordance with Eq. (8.20) in such a way that

$$-\nabla\cdot\mathbf{q} = \nabla\cdot(\lambda\nabla T) - \kappa\nabla\cdot(\rho\nabla\rho\nabla\cdot\mathbf{v}). \tag{8.32}$$

In the same way, using the deduced expression for the pressure tensor **P** in Eq. (8.21)

$$-\mathbf{P}:\nabla\mathbf{v} = -\boldsymbol{\tau}:\nabla\mathbf{v} - \kappa\nabla\rho\nabla\rho:\nabla\mathbf{v} - \left(p - \kappa\rho\nabla^2\rho\right)\nabla\cdot\mathbf{v}.$$

So, in its differential form the internal energy balance equation becomes

$$\rho\frac{du_0}{dt} = \nabla\cdot(\lambda\nabla T) - (\boldsymbol{\tau} + P\boldsymbol{\delta})\nabla\cdot\mathbf{v}, \tag{8.33}$$

in the interface.

The internal energy balance equation for non-homogeneous systems takes, thus, the same form as its related form for the bulk phases, where $\nabla\rho = 0$. Indeed, the surface force terms cancel themselves and do not have any direct effect on the temperature distributions. These effects are only indirect, since surface forces affect the density distributions in the interface and, consequently, the transport properties.

Nevertheless, the independence of the temperature distribution on the surface forces does not mean that the internal energy does not depend on the surface forces in the interface. This dependence can be easily seen in Eq. (8.29).

Explicit expressions for the internal energy equation, Eq. (8.33) in terms of ρ and T can be derived by using

$$\frac{du_0}{dt} = c_v \frac{dT}{dt} + \frac{1}{\rho^2}\left(P - T\left(\frac{\partial P}{\partial T}\right)_\rho\right)\frac{d\rho}{dt},$$

where c_v is the isovolumetric specific heat capacity.

8.3.6 Numerical Methods for the Diffuse Interface Approach

Two subclasses of numerical methods for the diffuse interface approach have emerged in literature, the first of which, also named phase-field method, considers a Cahn–Hilliard [25] visco-capillary structure for the mixture cells. The second subclass is to some extent an evolution of the volume of fluid method, where the classical system of conservation laws is extended with additional phase mass conservation and volume fraction advection equations. A very recent review of these methods is provided by Maltsev et al. [110].

The continuous development of more rigorous molecular or atomistic models of fluid systems, in conjunction with improvements in computational resources has resulted in a more in-depth understanding of the underlying physics of these systems. Nevertheless, molecular simulation models have excessively large computational requirements and are limited to extremely small time and/or space scales and simple geometries. In this context, mesoscopic models provide an alternative way for producing numerical tools that can improve the understanding of complex physical phenomena that are very difficult to describe at the hydrodynamic scale.

In the past two decades several numerical methods for liquid–vapor flows with phase transitions were proposed using mesoscopic modeling, especially in the framework of the Lattice Boltzmann Method (LBM), first introduced by Shan and Chen [152, 153], in the early 1990s. Although Shan and Chen (SC) isothermal model was not concerned with the first law of thermodynamics, there is a lack of thermodynamic consistency related to the failure of the model in predicting the liquid–vapor coexistence curves in agreement with the ones derived from the thermodynamic theory. Moreover, it was shown that the simulated coexistence curves depend on the relaxation time [72]. The use of this model is also limited to liquid–vapor density ratios of about 50 [190]. Another negative aspect of the SC model is that the interface tension and the attraction term of the equation of state are dependent on a single parameter G. Sbragaglia et al. [150] proposed a later modification of the Shan–Chen model by using two independent parameters on the pseudo-potential enabling to fix this drawback. Other SC model drawbacks are thoroughly discussed in Philippi et al. [133].

With the purpose of improving the Shan–Chen model, Sankaranarayanan et al. [149] and Yan and Schaeffer [190] proposed to write the effective mass of the SC model in terms of a given equation of state, incorporated into the model. In consequence, the interface tension becomes controlled, solely, by the temperature in the equation of state (EOS). It is observed that the maximum liquid to vapor density ratio is strongly dependent on the equation of state that is chosen: between 7 for the van der Walls EOS and superior to 1000 for the Carnahan Starling EOS. Nevertheless, as it was shown by Kupershtokh et al. [94], the agreement with the Maxwell liquid–vapor coexistence curves is only achieved by tuning up the relaxation time.

Swift et al. [167] developed a LBM model for multiphase flows based on the free energy excess related to a strong number density gradients inside the interface. Nevertheless, the model is not Galilean invariant [108]. As it was shown by He and Doolen [66], the imposed condition in Swift et al.'s model, equating the equilibrium distribution second order moments to the pressure tensor was the main factor leading to a break of Galilean invariance, with a spurious $O\left(u^2\right)$ term in the momentum balance equation. Ad-hoc corrections to the Swift and co-workers model were proposed and presented in Holdych et al. [70], Inamuro et al. [74] and Wagner [184].

Both pseudo-potential and free-energy models were based on ad-hoc assumptions. The successful establishment of the theoretical foundation of the LB method in the framework of kinetic theory has led to more rigorous ways of incorporating molecular interactions into the lattice-Boltzmann equation (LBE). The original Boltzmann equation leads to an ideal equation of state and is only suitable for the description of the non-equilibrium states of rarefied gases. The description of real fluids requires the attractive forces between the molecules to be considered. He and Doolen [66] considered that the integral term in the Boltzmann kinetic equation can be split into a short-range repulsion term and a long-range attraction term.

A major breakthrough in LB theory was the direct derivation of the LB equation from the continuous Boltzmann equation by He and Luo [67]. Shan et al. [154] and Philippi et al. [132] reopened the prospect by establishing a systematic link between the kinetic theory and the lattice Boltzmann method, determining the necessary conditions for the discretization of the velocity space. The lattices obtained through the method proposed by these authors, a *prescribed abscissas* quadrature, proved to be stable in flows over a wide range of parameters [158], by the use of high-order lattice Boltzmann schemes, leading to velocity sets which, when used in a discrete velocity kinetic scheme ensures accurate recovery of the high-order hydrodynamic moments and increasingly higher-order of isotropy of the lattice tensors.

The LB thermodynamic consistency for liquid vapor phase transitions was carefully investigated by Wagner [184] and Siebert et al. [156] leading to the conclusion that high-order approximations to the stream term are necessary for assuring the retrieval of the liquid–vapor coexistence curves. The essence is to derive LBEs from the discrete-velocity Boltzmann equation so that the resulting leading-order error terms are of higher-order with respect to the physical interface interaction terms. Such high-order thermodynamic consistent LBEs were applied for simulating two-phase liquid–vapor systems by Siebert et al. [156].

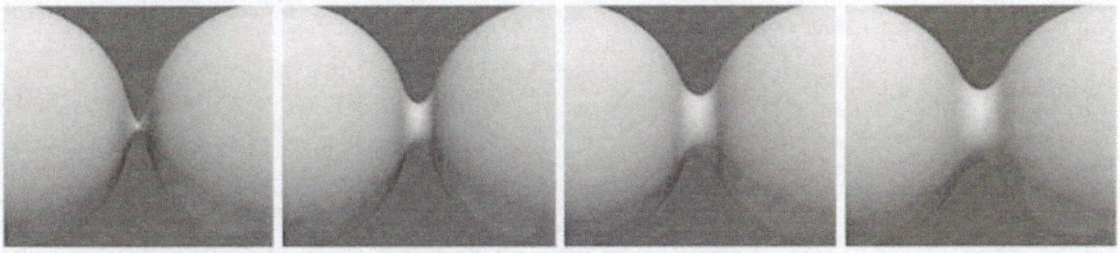

Fig. 8.6 Coalescence between two liquid droplets in a vapor atmosphere in different time steps δ [157]

Coalescence Between Two Liquid Droplets

Figure 8.6 shows the coalescence process between two liquid droplets in a vapor atmosphere. The numerical simulation was performed using a kinetic theory-based LB model developed by Siebert et al. [156]. When two liquid droplets are close enough, the molecules at the surface of each droplet are attracted by the intermolecular forces from the other droplet and form a liquid neck between them. The diameter of this neck increases with time, leading to coalescence. In accordance with Aarts et al. [1], the increase of the neck diameter D_0 with time is dependent on the Reynolds number Re given, in this case, by

$$Re = \frac{\rho \gamma D_0}{2\eta^2},$$

where the capillary velocity is given by $u_{cap} = \gamma/\eta$, γ is the surface tension and η the liquid dynamical viscosity. Experimental data indicate that for $Re < 1$ the capillary velocity is governed by viscous forces, and the neck diameter has a linear dependence with time, whereas for $Re > 1$ the inertial forces are dominant and the neck diameter has a $t^{1/2}$ dependence.

Lattice-Boltzmann simulations of the inertial regime showed to be in very good agreement with experimental data. Further details may be found in [157].

Vapor Condensation

Figure 8.7 shows what happens when a certain amount of vapor is suddenly cooled to a temperature in such a way that the volumetric average density $\langle \rho \rangle$ is between the saturated vapor density $\rho_{vs}(T)$ and the saturated liquid density $\rho_{\ell s}(T)$.

The simulation was performed in two dimensions with the lattice-Boltzmann method in the absence of gravity and considering periodic conditions at the boundaries. These simplifications avoid dealing with fluid–solid interactions and give a better picture of the dynamics of droplet formation by vapor condensation.

Being a mesoscopic method, LBM is based on the discretization of the Boltzmann equation, the attractive forces between molecules being modeled with molecular parameters a and κ of Sect. 6.5. Therefore, the molecular velocity ξ is discretized, in such a way that, at each node, the amount of molecules in this node is distributed

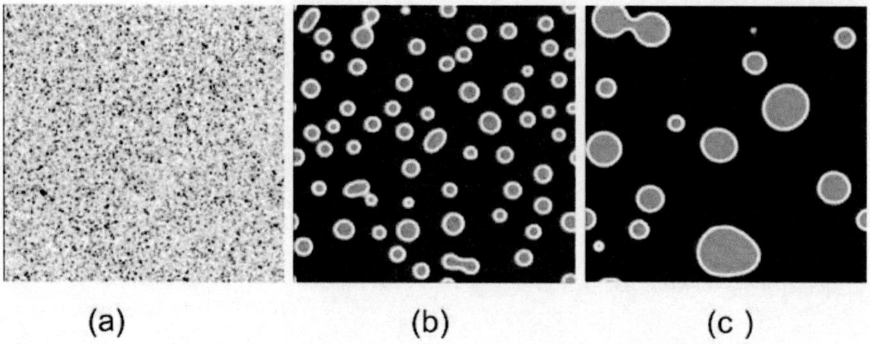

Fig. 8.7 Vapor condensation in the absence of gravity simulated with the lattice-Boltzmann method

among different populations with velocity ξ_i. The simulation starts by randomly distributing the fluid populations among the nodes (Fig. 8.7a). This gives rise to vapor condensation in the form of liquid droplets. Some liquid droplets grow by coalescence with their neighbors while others are unstable and reevaporate. Pressure waves are visible during simulation (Fig. 8.7b). Figure 8.7c shows a later stage of the simulation.

At the end of the simulation, a single liquid drop is visible with density $\rho_{\ell s}(T)$ inside a vapor bath with density $\rho_{vs}(T)$.

Chapter 9
Multicomponent Systems

Abstract Besides their technological importance in the chemical and petroleum industry, there is a great scientific interest in studying multicomponent systems due to the challenges to overcome when dealing with their dynamics and patterns in non-equilibrium states. Homogeneous non-ideal mixtures may segregate into two or even more phases when the temperature is decreased. The segregation mechanisms have a molecular origin and are due to the attractive forces between the molecules of the same species when these forces are dominant in comparison with diffusion and cross-interactions. The correct understanding of these complex systems requires, thus, a multiphysics approach and this complexity is a major difficulty for numerical simulations based on discrete forms of the macroscopic equations. On the other hand, the knowledge of the macroscopic equations that govern the highly non-linear physical processes when a non-ideal mixture changes from an equilibrium state to another is important for establishing a reference framework that every multiphysics model must satisfy. In this chapter, these macroscopic equations are derived based on the second law of thermodynamics. Particular attention is given to the molecular origin of segregation. Some lattice-Boltzmann simulation results are presented at the end of the chapter.

9.1 Introduction

From a molecular standpoint, the interface between two phases is a transition layer, where the molecules of each fluid are subjected to two opposite processes: the random thermal motion of the molecules, responsible for mixing, and the intermolecular attractive forces on each molecule from its own phase, promoting segregation.

These forces are of an electrical nature and are due to the permanent and temporarily asymmetric distribution of the electronic clouds around each molecule. Electrical neutral molecules have been addressed by several workers [138], mainly based on the ideas of Keesom, Debye, and London. The relative importance of these two competing effects is dependent on the thermodynamic state and on the molecular properties of the involved fluids. Indeed, miscibility is not an intrinsic property of a pair of liquids but is dependent on the thermodynamic state of the mixture. Two liquids that

P. C. Philippi, *Thermodynamics*, https://doi.org/10.1007/978-3-031-49357-7_9

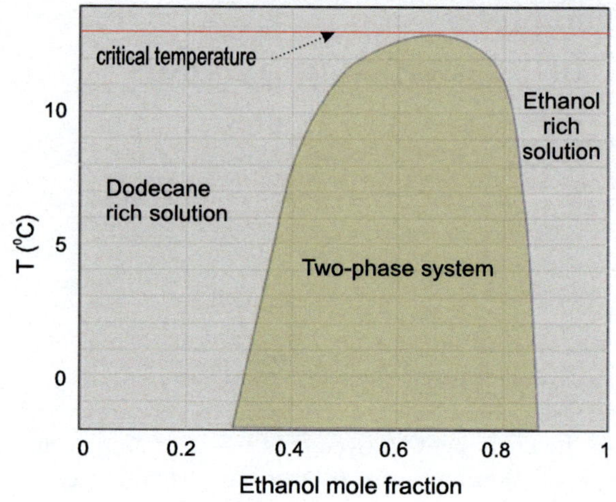

Fig. 9.1 Dodecane-ethanol phase diagram (from Dortmund Data Bank)

can mix in all proportions, forming a homogeneous solution at a given temperature can segregate when the temperature is decreased. The amount of solute that remains in the solution at this lower temperature is a measure of solubility and this property is dependent on the temperature and pressure.

Figure 9.1 shows an isobaric phase diagram of a thermodynamic system composed of dodecane and ethanol. When the temperature is above 13 °C the two fluids form a homogeneous solution in all proportions. Below this critical temperature, the dodecane-ethanol mixture is unstable for intermediate molar fractions of ethanol and suffers a spinodal decomposition breaking up into a two-phase system. This unstable region is shown as a bell in the phase diagram. On the left side of the bell, we have a single-phase system as a dodecane-rich solution and for thermodynamic states on the right side of the bell, we have an ethanol-rich solution (Fig. 9.1).

Considering their great technological importance, the equilibrium properties of real, non-ideal liquids and their mixtures were exhaustively studied in the last century, after van der Waals first attempt to understand the thermodynamic properties of vapors and phase transitions [117]. Equations of state and mixing rules have been heuristically proposed and successfully validated against pressure–temperature–volume (PVT) experiments. In the same manner, solubility and its temperature and pressure dependence have been measured for a large number of solutions.

In non-equilibrium conditions, the macroscopic equations for non-ideal fluids and their mixtures, in the miscibility range of these fluids, are usually written in the same form as in the ideal case, with an appropriate equation of state for the thermodynamic pressure. In a completely distinct framework, the flow of immiscible fluids has been numerically approached in classical computational fluid dynamics (CFD) by the introduction of an extra source term in the momentum balance equations, related to the surface forces and giving the Korteweg stress tensor. These macroscopic equations have not, however, been able to realistically describe the segregation processes that

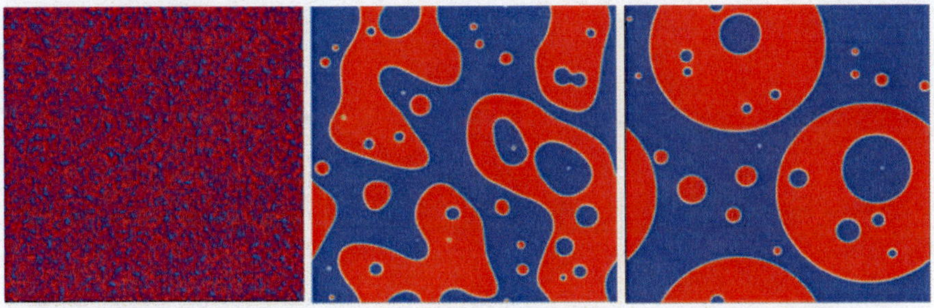

Fig. 9.2 Lattice Boltzmann simulation of the spinodal decomposition. The mixture forms a homogeneous solution at the beginning of the simulation (**a**). Intermolecular attraction forces between the molecules of the same species segregate the two components forming isolated islands of one fluid inside the other one (**b**), till the equilibrium is reached in (**c**)

occur when two partially immiscible fluids are mixed. This is a class of problems that cannot be fully understood in the Navier–Stokes (NS) scale, which is restricted to two extreme conditions: homogeneous non-ideal mixtures and immiscible flows.

In two-phase problems, the flow of individual drops and bubbles can be monitored using a Navier Stokes Euler–Lagrange approach, but heuristic assumptions must be used if it is necessary to describe a coalescence problem between two drops or to decide the time at which a liquid drop will snap off when it flows through a constriction. This does not reduce the value of continuum approaches in the study of two-phase flows, but knowledge of the underlying physical phenomena that control the complex two-phase macroscopic phenomena require downscaling toward molecular scales.

In this chapter, we establish the macroscopic transport equations for non-ideal mixtures, trying to put into evidence the role of the molecular attraction forces on segregation. Figure 9.2 shows the results of a lattice-Boltzmann simulation of a spinodal decomposition when two fluids p and q are mixed in equal proportions forming a homogeneous solution at the beginning of the simulation (Fig. 9.2a). Intermolecular attraction forces between the molecules of the same species segregate the two components forming isolated islands of one fluid inside the other one (Fig. 9.2b) till the equilibrium is reached in (Fig. 9.2c).

9.2 Mixtures Without Segregation Effects

In this section, we deal with mixtures where diffusion mechanisms, leading to mixing, are dominant with respect to the attractive forces between the molecules. So, these attractive forces do not lead to phase segregation and, consequently, to the formation of interfaces, corresponding to the regions outside the bell in Fig. 9.1.

For r-component systems in equilibrium, the Helmholtz free energy $F = U - TS$ satisfies

$$dF = -SdT - PdV + \sum_i \mu_i dm_i.$$

On the other hand, the Euler equation for homogeneous systems reads, in this case, as

$$F = -PV + \sum_i \mu_i m_i.$$

By dividing the last equation by the volume V

$$\psi = -P + \sum_i \mu_i \rho_i, \tag{9.1}$$

where ψ is the Helmholtz energy per unit volume.

Multicomponent systems in equilibrium form a homogeneous mixture and satisfy the homogeneity condition

$$\psi = \sum_i \psi_i, \quad P = \sum_i P_i,$$

and so

$$\psi_i = -P_i + \mu_i \rho_i.$$

By differentiating the last equation

$$d\psi_i = -dP_i + \mu_i d\rho_i + \rho_i d\mu_i. \tag{9.2}$$

An expression for $d\mu_i$ can be found from the Gibbs–Duhem equation

$$\sum_i m_i d\mu_i = -SdT + VdP. \tag{9.3}$$

Replacing

$$S = \sum_i m_i s_i, \quad V = \sum_i m_i v_i,$$

in Eq. (9.3), we get

$$d\mu_i = -s_i dT + v_i dP.$$

Now, v_i is the partial specific volume of component i in the mixture, or

$$v_i = \left(\frac{\partial V}{\partial m_i}\right)_{T,P,m_{j\neq i}} = \frac{1}{\left(\frac{\partial m_i}{\partial V}\right)_{T,P,m_{j\neq i}}} = \frac{1}{\rho_i}.$$

So

$$d\mu_i = -s_i dT + \frac{1}{\rho_i} dP.$$

Replacing the last equation into Eq. (9.2),

$$d\psi_i = d\left(P - P_i\right) + \mu_i d\rho_i - \rho_i s_i dT.$$

Summing the above equation over the components (*remember that P is an intensive variable and, so, independent of the position in homogeneous systems*)

$$d\psi = \sum_i \mu_i d\rho_i - \rho s dT. \tag{9.4}$$

Now, we know that $U = F + TS$, and so,

$$\rho u = \psi + \rho T s,$$

and so

$$\rho du = d\psi + \rho T ds + \rho s dT + T s d\rho - u d\rho.$$

By replacing $d\psi$ given by Eq. (9.4) in the above equation, we get

$$\rho du = \sum_i \mu_i d\rho_i + \rho T ds - \frac{\psi}{\rho} d\rho.$$

The Helmholtz energy is given by Eq. (9.1) and so,

$$\rho du = \sum_i \mu_i d\rho_i + \rho T ds + \frac{P}{\rho} d\rho - \sum_i \mu_i \omega_i d\rho.$$

On the other hand,

$$d\rho_i = d\left(\rho\omega_i\right) = \omega_i d\rho + \rho d\omega_i.$$

So

$$\rho du = \sum_i \mu_i \rho d\omega_i + \rho T ds + \frac{P}{\rho} d\rho.$$

The above equation can be written in terms of Lagrangian derivatives

$$\rho \frac{du}{dt} = \sum_i \mu_i \rho \frac{d\omega_i}{dt} + \rho T \frac{ds}{dt} + \frac{P}{\rho} \frac{d\rho}{dt}, \tag{9.5}$$

or

$$\rho \frac{ds}{dt} = \frac{\rho}{T} \frac{du}{dt} - \sum_i \frac{\mu_i}{T} \rho \frac{d\omega_i}{dt} - \frac{P}{\rho T} \frac{d\rho}{dt}. \tag{9.6}$$

Now, the entropy balance equation reads as

$$\rho \frac{ds}{dt} = -\nabla \cdot \mathbf{q_s} + \mathscr{P}(s)$$

where $\mathbf{q}_s$ is the diffusive flux of entropy and $\mathscr{P}(s)$ is the entropy production due to non-equilibrium. Replacing the entropy balance equation into Eq. (9.6),

$$\mathscr{P}(s) = \nabla \cdot \mathbf{q}_s + \frac{\rho}{T} \frac{du}{dt} - \sum_i \frac{\mu_i}{T} \rho \frac{d\omega_i}{dt} - \frac{P}{\rho T} \frac{d\rho}{dt}. \tag{9.7}$$

In the above equation, the material derivatives $\rho \frac{du}{dt}$, $\rho \frac{d\omega_i}{dt}$, $\frac{d\rho}{dt}$ can be replaced by using their respective balance equations. The advection–diffusion equation for component i reads as

$$\rho \frac{d\omega_i}{dt} = -\nabla \cdot \mathbf{j}_i, \tag{9.8}$$

where $\mathbf{j}_i$ is the diffusive flux of component i. The mass conservation equation is written as

$$\frac{d\rho}{dt} = -\rho \nabla \cdot \mathbf{v}, \tag{9.9}$$

and the internal energy balance equation, as

$$\rho \frac{du}{dt} = -\nabla \cdot \mathbf{q} - \mathbf{P} : \nabla \mathbf{v} \tag{9.10}$$

where $\mathbf{q}$ is the diffusive flux of heat and $\mathbf{P}$ is the pressure tensor.

Replacing Eqs. (9.8)–(9.10) into Eq. (9.7),

$$\mathscr{P}(s) = \nabla \cdot \mathbf{q}_s - \frac{1}{T}\nabla \cdot \mathbf{q} - \frac{1}{T}(\mathbf{P} - P\boldsymbol{\delta}) : \nabla \mathbf{v} + \sum_i \frac{\mu_i}{T}(\nabla \cdot \mathbf{j}_i). \tag{9.11}$$

Now,

$$\frac{1}{T}\nabla \cdot \mathbf{q} = \nabla \cdot \frac{\mathbf{q}}{T} + \frac{1}{T^2}\mathbf{q} \cdot \nabla T$$

$$\frac{\mu_i}{T}(\nabla \cdot \mathbf{j}_i) = \nabla \cdot \frac{\mu_i \mathbf{j}_i}{T} + \mu_i \frac{\mathbf{j}_i}{T^2} \cdot \nabla T - \frac{1}{T}\mathbf{j}_i \cdot \nabla \mu_i$$

Replacing the above equations into Eq. (9.11), it results

$$\mathscr{P}(s) = \nabla \cdot \left(\mathbf{q}_s - \frac{1}{T}\left(\mathbf{q} - \sum_i \mu_i \mathbf{j}_i\right)\right)$$
$$-\frac{1}{T}(\mathbf{P} - P\boldsymbol{\delta}) : \nabla \mathbf{v} - \frac{1}{T^2}\left(\mathbf{q} - \sum_i \mu_i \mathbf{j}_i\right) \cdot \nabla T - \sum_i \frac{1}{T}\mathbf{j}_i \cdot \nabla \mu_i \tag{9.12}$$

For the entropy production to be non-negative the divergence term must be null and so

$$\mathbf{q}_s - \frac{1}{T}\left(\mathbf{q} - \sum_i \mu_i \mathbf{j}_i\right) = 0,$$

and this is satisfied when

$$\mathbf{q} = \mathbf{q}_f + \sum_i \mu_i \mathbf{j}_i, \tag{9.13}$$

$$\mathbf{q}_s = \frac{\mathbf{q}_f}{T}. \tag{9.14}$$

The term $\mu_i \mathbf{j}_i$ is related to the transfer of energy due the diffusive flux of component i in addition to the Fourier diffusive flux of heat $\mathbf{q}_f$.

Also, to ensure that the entropy production is non-negative, the pressure tensor must satisfy

$$\mathbf{P} = P\boldsymbol{\delta} + \boldsymbol{\tau},$$

where $\boldsymbol{\tau}$ is the viscous stress tensor.

The entropy production will be thus given by

$$\mathscr{P}(s) = -\frac{1}{T}\boldsymbol{\tau} : \nabla \mathbf{v} - \frac{1}{T^2}\mathbf{q}_f \cdot \nabla T - \sum_i \frac{1}{T}\mathbf{j}_i \cdot \nabla \mu_i \geq 0, \tag{9.15}$$

as the sum of products between fluxes and the gradients that are the sources of these fluxes.

9.2.1 Transport Equations

We rewrite the macroscopic equations for locally homogeneous mixtures without segregation effects. The advection–diffusion equation for species i is given by Eq. (9.8),

$$\rho \frac{d\omega_i}{dt} = -\nabla \cdot \mathbf{j}_i, \tag{9.16}$$

where $\mathbf{j}_i$ is the diffusive flux of component i.

Since

$$\frac{d}{dt} = \frac{\partial}{\partial t} + \mathbf{v} \cdot \nabla,$$

where $\mathbf{v}$ is the local velocity at point $\mathbf{x}$, Eq. (9.16) can be rewritten in a more usual form as

$$\frac{\partial \rho_i}{\partial t} + \nabla \cdot (\rho_i \mathbf{v} + \mathbf{j}_i) = 0,$$

expressing the total flux of species i as the sum of an advective flux $\rho_i \mathbf{v}$ and a diffusive flux $\mathbf{j}_i$.

On the other hand, the momentum balance equation is given by

$$\frac{\partial}{\partial t}(\rho \mathbf{v}) + \nabla \cdot (\rho \mathbf{v}\mathbf{v} + \mathbf{P}) = \rho \mathbf{g},$$

where $\mathbf{P} = P\boldsymbol{\delta} + \boldsymbol{\tau}$ is the pressure tensor, P is the thermodynamic pressure and $\boldsymbol{\tau}$ is the viscous stress tensor.

The balance equation for the internal energy is given by Eq. (9.10), here rewritten as

$$\rho \frac{du}{dt} = -\nabla \cdot \mathbf{q} - P\nabla \cdot \mathbf{v} - \boldsymbol{\tau} : \nabla \mathbf{v}.$$

9.2.2 Exercises

1. In a single-phase, locally homogeneous mixture with r-components, in addition to its temperature dependence, the thermodynamic pressure is also dependent on the composition $P = P(\rho_1, \dots, \rho_r, T)$. In the same way the transport coefficients are dependent on the composition and on the temperature. Simplify the transport equations of Sect. 9.2.1 for a mixture of ideal gases.
2. Consider now a diluted solution of a component p in a solvent s. In this case $\omega_p \to 0$ and the viscosity η and thermal conductivity λ will be only dependent on the temperature. Simplify the transport equations of Sect. 9.2.1 for this case.

9.3 Segregation Effects in Non-ideal Mixtures

Similarly to single component systems, if $\mathbf{x}$ is a point belonging to an interface, the free energy at this point becomes also dependent on the local mass density gradients. Indeed, following van der Waals [181], the free energy has an excess in the interface given, in its simplest form, by

$$\psi_{exc} = \sum_{i,j} \frac{\kappa_{ij}}{2} \nabla \rho_i \cdot \nabla \rho_j \tag{9.17}$$

where κ_{ij} is related to the attraction forces between molecules of species i and j, and are at the origin of the interface tension [133]. In the above expression κ_{ii} represents the intermolecular attraction forces between identical molecules. These forces are responsible for segregating species i from the other species that are present in the multicomponent system. When $i \neq j$, κ_{ij} represent the attraction forces between different molecules. These forces have a surfactant effect, reducing the interfacial tension.

When segregating forces are to be taken into account, Eq. (9.4) becomes

$$d\psi = \sum_i \mu_i d\rho_i - \rho s dT + \sum_{i,j} \kappa_{ij} \nabla \rho_i \cdot d\left(\nabla \rho_j\right), \tag{9.18}$$

for all the points $\mathbf{x}$ where the mixture presents concentration gradients.

The last term in the above equation is symmetric with respect to the index i and j because $\kappa_{ij} = \kappa_{ji}$, in such a way that changing the index i and j has no effect on the surface term.

On the other hand, the internal energy $\rho u = \psi + \rho T s$ can be differentiated to give

$$\rho du = d\psi + \rho T ds + \rho s dT + T s d\rho - u d\rho. \tag{9.19}$$

By replacing $d\psi$ given by Eq. (9.18) in the above equation, we get

$$\rho du = \sum_i \mu_i d\rho_i + \rho T ds - \frac{\psi}{\rho} d\rho + \sum_{i,j} \kappa_{ij} \nabla \rho_i \cdot d\left(\nabla \rho_j\right). \tag{9.20}$$

The Helmholtz energy is given by Eq. (9.1)

$$\psi = -p + \sum_i \mu_i \rho_i,$$

where the pressure p differs from the thermodynamic pressure P by interfacial segregating forces.

In addition,

$$d\rho_i = d\left(\rho \omega_i\right) = \omega_i d\rho + \rho d\omega_i.$$

Therefore, writing Eq. (9.20) in terms of material derivatives

$$\rho \frac{du}{dt} = \frac{p}{\rho}\frac{d\rho}{dt} + \rho T \frac{ds}{dt} + \sum_{i,j} \kappa_{ij} \nabla \rho_i \cdot \frac{d\left(\nabla \rho_j\right)}{dt} + \sum_i \mu_i \rho \frac{d\omega_i}{dt}. \tag{9.21}$$

This expression can be compared with Eq. (8.11) for single-component systems

$$\rho \frac{du}{dt} = \frac{p}{\rho}\frac{d\rho}{dt} + \rho T \frac{ds}{dt} + \kappa \nabla \rho \cdot \frac{d\left(\nabla \rho\right)}{dt}.$$

The last term in the r.h.s. of Eq. (9.21) is related to the effect of the diffusion of species on the balance of energy in the interface. This term is also present in Eq. (9.5).

As in the previous sections, we rewrite Eq. (9.21) in the following form,

$$\rho \frac{ds}{dt} = \frac{\rho}{T}\frac{du}{dt} - \frac{p}{T\rho}\frac{d\rho}{dt} - \frac{1}{T}\sum_i \mu_i \rho \frac{d\omega_i}{dt} - \frac{1}{T}\sum_{i,j} \kappa_{ij} \nabla \rho_i \cdot \frac{d\left(\nabla \rho_j\right)}{dt}. \tag{9.22}$$

A general expression for the entropy balance equation is given by

$$\rho \frac{ds}{dt} = -\nabla \cdot \mathbf{q}_s + \mathscr{P}(s),$$

where $\mathbf{q}_s$ is the diffusive flux of entropy and $S(s)$ is the entropy production due to non-equilibrium. Replacing the entropy balance equation into Eq. (9.22), we find the entropy production as given by

$$\mathscr{P}(s) = \nabla \cdot \mathbf{q}_s + \frac{\rho}{T}\frac{du}{dt} - \frac{P}{\rho T}\frac{d\rho}{dt} - \frac{1}{T}\sum_i \mu_i \rho \frac{d\omega_i}{dt} - \frac{1}{T}\sum_{i,j} \nabla\rho_i \cdot \frac{d\left(\nabla\rho_j\right)}{dt} \tag{9.23}$$

For the last term in Eq. (9.23)

$$\nabla\rho_i \cdot \frac{d\left(\nabla\rho_j\right)}{dt} = \partial\rho_i \partial_\beta \left(\frac{d\rho_j}{dt}\right) - \left(\partial_\alpha\rho_j\right)\left(\partial_\beta\rho_i\right)\left(\partial_\beta \mathbf{v}_\alpha\right).$$

But

$$\left(\partial_\beta\rho_i\right)\partial_\beta\left(\frac{d\rho_j}{dt}\right) = \partial_\beta\left(\partial_\beta\rho_i\left(\frac{d\rho_j}{dt}\right)\right) - \left(\partial_\beta\partial_\beta\rho_i\right)\left(\frac{d\rho_j}{dt}\right).$$

So

$$\nabla\rho_i \cdot \frac{d\left(\nabla\rho_j\right)}{dt} = \partial_\beta\left(\partial_\beta\rho_i\left(\frac{d\rho_j}{dt}\right)\right) - \left(\partial_\beta\partial_\beta\rho_i\right)\left(\frac{d\rho_j}{dt}\right) - \left(\partial_\alpha\rho_j\right)\left(\partial_\beta\rho_i\right)\left(\partial_\beta \mathbf{v}_\alpha\right).$$

In vector notation

$$\nabla\rho_i \cdot \frac{d\left(\nabla\rho_j\right)}{dt} = \nabla \cdot \left(\nabla\rho_i\left(\frac{d\rho_j}{dt}\right)\right) - \nabla^2\rho_i\left(\frac{d\rho_j}{dt}\right) - \nabla\rho_i\nabla\rho_j : \nabla\mathbf{v}. \tag{9.24}$$

Since $\rho_j = \omega_j\rho$ and

$$\frac{d\rho_j}{dt} = \rho\frac{d\omega_j}{dt} + \omega_j\frac{d\rho}{dt},$$

Equation (9.24) can thus be rewritten as

$$\nabla\rho_i \cdot \frac{d\left(\nabla\rho_j\right)}{dt} = \nabla \cdot \left(\nabla\rho_i\left(\rho\frac{d\omega_j}{dt} + \omega_j\frac{d\rho}{dt}\right)\right) - \nabla^2\rho_i\left(\rho\frac{d\omega_j}{dt} + \omega_j\frac{d\rho}{dt}\right) - \nabla\rho_i\nabla\rho_j : \nabla\mathbf{v}. \tag{9.25}$$

The material derivatives $\rho\frac{du}{dt}, \rho\frac{d\omega_j}{dt}, \frac{d\rho}{dt}$ can be replaced by their respective balance equations. The advection–diffusion equation for component i reads as

$$\rho\frac{d\omega_j}{dt} = -\nabla \cdot \mathbf{j}_j, \tag{9.26}$$

where $\mathbf{j}_i$ is the diffusive flux of component i. The mass conservation equation is written as

$$\frac{d\rho}{dt} = -\rho\nabla\cdot\mathbf{v}, \tag{9.27}$$

and the internal energy balance equation, as

$$\rho\frac{du}{dt} = -\nabla\cdot\mathbf{q} - \mathbf{P} : \nabla\mathbf{v}, \tag{9.28}$$

where $\mathbf{q}$ is the diffusive flux of heat and $\mathbf{P}$ is the pressure tensor.

But,

$$\frac{1}{T}\nabla\cdot\mathbf{q} = \nabla\cdot\left(\frac{\mathbf{q}}{T}\right) + \frac{1}{T^2}\mathbf{q}\cdot\nabla T, \tag{9.29}$$

$$\frac{\mu_i}{T}\left(\nabla\cdot\mathbf{j}_i\right) = \nabla\cdot\left(\frac{\mu_i\mathbf{j}_i}{T}\right) + \mu_i\frac{\mathbf{j}_i}{T^2}\cdot\nabla T - \frac{1}{T}\mathbf{j}_i\cdot\nabla\mu_i. \tag{9.30}$$

Replacing the above relationships into Eq. (9.23) we finally arrive at the following expression for the entropy production

$$\begin{aligned}
\mathscr{P}(s) = {} & \nabla\cdot\left(\mathbf{q}_s - \frac{1}{T}\left(\begin{array}{c}\mathbf{q} - \sum_i \widetilde{\mu}_i\mathbf{j}_i \\ -\sum_{i,j}\kappa_{ij}\nabla\rho_j\left(\nabla\cdot\mathbf{j}_i + \rho_i\nabla\cdot\mathbf{v}\right)\end{array}\right)\right) \\
& - \frac{1}{T}\left(\mathbf{P} - \sum_{i,j}\kappa_{ij}\nabla\rho_i\nabla\rho_j - \left(p - \sum_{i,j}\kappa_{ij}\rho_i\nabla^2\rho_j\right)\boldsymbol{\delta}\right) : \nabla\mathbf{v} \\
& - \frac{1}{T^2}\left(\mathbf{q} - \sum_i\widetilde{\mu}_i\mathbf{j}_i - \sum_{i,j}\kappa_{ij}\nabla\rho_j\left(\nabla\cdot\mathbf{j}_i + \rho_i\nabla\cdot\mathbf{v}\right)\right)\cdot\nabla T \\
& - \sum_i\frac{1}{T}\mathbf{j}_i\cdot\nabla\widetilde{\mu}_i,
\end{aligned} \tag{9.31}$$

where

$$\widetilde{\mu}_i = \mu_i - \sum_j\kappa_{ij}\nabla^2\rho_j,$$

is a modified potential that takes the surface forces into account.

The entropy production is always non-negative. In the above expression, this is ensured by making

$$\mathbf{q} = -\lambda \nabla T + \sum_i \widetilde{\mu}_i \mathbf{j}_i + \sum_{i,j} \kappa_{ij} \nabla \rho_j \left(\nabla \cdot \mathbf{j}_i + \rho_i \nabla \cdot \mathbf{v} \right), \tag{9.32}$$

reducing the flux of entropy to

$$\mathbf{q}_s = -\frac{\lambda \nabla T}{T},$$

and eliminating the divergence term on the r.h.s. of Eq. (9.31).

The diffusive heat flux in a non-ideal mixture as expressed in Eq. (9.32) is considered to be due to the following contributions:

(i) The Fourier contribution $\mathbf{q}_f = -\lambda \nabla T$ due to temperature gradients;
(ii) The flux of energy due to the diffusion of the several species that are present in the interface $\sum_i \widetilde{\mu}_i \mathbf{j}_i$;
(iii) The work performed by the surface forces $\sum_{i,j} \kappa_{ij} \nabla \rho_j \left(\nabla \cdot \mathbf{j}_i + \rho_i \nabla \cdot \mathbf{v} \right)$.

Also, since $\tau : \nabla \mathbf{v} \leq \mathbf{0}$, the entropy production is ensured to be non-negative if, in addition to the condition given by Eq. (9.32), we impose that

$$\mathbf{P} = \tau : \nabla \mathbf{v} + \sum_{i,j} \kappa_{ij} \nabla \rho_i \nabla \rho_j + \left(p - \sum_{i,j} \kappa_{ij} \rho_i \nabla^2 \rho_j \right) \boldsymbol{\delta}. \tag{9.33}$$

The pressure p is given by

$$p = \rho \left(\frac{\partial \psi}{\partial \rho} \right)_T - \psi, \tag{9.34}$$

where ψ is the Helmholtz free energy in the interface,

$$\psi = \psi_0 + \sum_{i,j} \frac{\kappa_{ij}}{2} \nabla \rho_i \cdot \nabla \rho_j.$$

So,

$$p = \underbrace{\rho \left(\frac{\partial \psi_0}{\partial \rho} \right)_T - \psi_0}_{=P} - \sum_{i,j} \frac{\kappa_{ij}}{2} \nabla \rho_i \cdot \nabla \rho_j,$$

and the pressure tensor in Eq. (9.33) becomes

$$\mathbf{P} = \tau : \nabla \mathbf{v} + \mathbf{S} + P_s \delta, \tag{9.35}$$

where the scalar pressure P_s is now given by

$$P_s = P - \sum_{i,j} \frac{\kappa_{ij}}{2} \nabla \rho_i \cdot \nabla \rho_j - \sum_{i,j} \kappa_{ij} \rho_i \nabla^2 \rho_j. \tag{9.36}$$

This pressure reduces to the thermodynamic pressure P in the bulk phases. On the other hand, the Korteweg surface tensor is given by

$$\mathbf{S} = \sum_{i,j} \kappa_{ij} \nabla \rho_i \nabla \rho_j. \tag{9.37}$$

Considering the molecular parameters κ_{ij} as constants, the last term on the r.h.s. of Eq. (9.36) can be rewritten in a more symmetric form as

$$\sum_{i,j} \kappa_{ij} \rho_i \nabla^2 \rho_j = \frac{1}{2} \nabla^2 \left(\rho^2 \kappa \right) - \sum_{i,j} \kappa_{ij} \nabla \rho_i \nabla \rho_j,$$

where

$$\kappa = \sum_{i,j} \omega_i \omega_j \kappa_{ij},$$

is a force molecular parameter averaged by the mass fractions $\omega_i = \rho_i / \rho$. Therefore, Eq. (9.36) can also be rewritten as

$$P_s = P + \sum_{i,j} \frac{\kappa_{ij}}{2} \nabla \rho_i \cdot \nabla \rho_j - \frac{1}{2} \nabla^2 \left(\rho^2 \kappa \right). \tag{9.38}$$

Replacing Eqs. (9.32) and (9.33) into Eq. (9.31), the entropy production becomes

$$\mathscr{P}(s) = -\frac{1}{T} (\tau : \nabla \mathbf{v}) - \frac{1}{T^2} \mathbf{q}_f \cdot \nabla T - \sum_i \frac{1}{T} \mathbf{j}_i \cdot \nabla \widetilde{\mu}_i. \tag{9.39}$$

Equation (9.39) reduces to Eq. (9.15) in the absence of segregating forces.

Exercises

1. Consider Eq. (9.32) expressing the heat flux in a non-ideal mixture when the segregation forces are taken into account. (a) Show that Eq. (9.32) reduces to Eq. (9.13) expressing the heat flux in mixtures when the segregation forces can be discarded; (b) Show that Eq. (9.32) reduces to Eq. (8.20) for single component systems.
2. Show that the expression for the pressure tensor given by Eq. (9.35) reduces to Eq. (8.21) for single component systems.

9.3.1 Pressure Tensor and the Equation of State

In this section, we consider a non-ideal binary mixture, for simplicity. A molecule of species p in the interface will be subjected to an attractive potential energy $\Phi^{(ps)}$ related to the intermolecular interaction between this p-molecule and all the s-particles around it, $s = p, q$. This potential energy can be calculated at a point $\mathbf{x}$ as [131, 133],

$$\phi^{(ps)}(\mathbf{x}) = \int\limits_{\|\mathbf{r}-\mathbf{x}\|\geq\sigma_p} \phi_{ps}^{(1-2)}(\|\mathbf{r}-\mathbf{x}\|)\, n_s(\mathbf{r})\, d\mathbf{r}, \tag{9.40}$$

where $\phi_{ps}^{(1-2)}$ corresponds to the potential energy due to the binary interaction between a single molecule p and a single molecules s, and σ_p is related to the molecular diameter of the p-molecule. Since $\phi_{ps}^{(1-2)}$ corresponds to an attractive potential, $\phi^{(ps)} \leq 0$.

Let $n_s(\mathbf{r})$ be developed in a Taylor series around $n_s(\mathbf{x})$,

$$n_s(\mathbf{r}) = n_s(\mathbf{x}) + \eta_\alpha \partial_\alpha n_s + \frac{1}{2}\eta_\alpha \eta_\beta \partial_{\alpha\beta} n_s + \cdots,$$

where $\eta = \mathbf{r} - \mathbf{x}$. Inserting this expansion into Eq. (9.40), given that $\phi_{ps}^{(1-2)}(\|\eta\|)$ is an even function of η,

$$\phi^{(ps)}(\mathbf{x}) = -2a_{ps} n_s(\mathbf{x}) - \bar{\kappa}_{ps} \nabla^2 n_s + \cdots \tag{9.41}$$

where

$$a_{ps} = -\frac{1}{2} \int\limits_{\|\eta\|\geq\sigma_p} \phi_{ps}^{(1-2)}\, \|\mathbf{x}\|\, d\eta \geq 0,$$

$$\bar{\kappa}_{ps} = -\int\limits_{\|\eta\|\geq\sigma_p} \phi_{ps}^{(1-2)}(\|\eta\|)\, \frac{1}{6}\eta^2 d\eta \geq 0,$$

are molecular parameters that are only dependent on the interaction potential between the molecules p and s. Parameter $\bar{\kappa}_{ps}$ is the same force parameter as in the previous section, but it is here written on a molecular basis.

The force per unit mass acting on a single molecule p from all the molecules s around it, will be, thus, given by

$$\mathbf{g}^{(ps)} = \frac{2a_{ps}}{m_p}\nabla n_s + \frac{\bar{\kappa}_{ps}}{m_p}\nabla\nabla^2 n_s, \tag{9.42}$$

and the force per unit volume acting on n_p molecules in the interface can be written as

$$\rho_p \mathbf{g}^{(ps)} = 2a_{ps} n_p \nabla n_s + \bar{\kappa}_{ps} n_p \nabla \nabla^2 n_s.$$

We make use of the following relationship

$$2a_{ps} n_p \nabla n_s = 2a_{ps} \nabla \left(n_p n_s\right) - 2a_{ps} n_s \nabla n_p,$$

in such a way that

$$2a_{pp} n_p \nabla n_p = a_{pp} \nabla \left(n_p^2\right), \tag{9.43}$$

$$2a_{qq} n_q \nabla n_q = a_{qq} \nabla \left(n_q^2\right), \tag{9.44}$$

$$2a_{pq} n_p \nabla n_q + 2a_{pq} n_q \nabla n_p = 2a_{pq} \nabla \left(n_p n_q\right). \tag{9.45}$$

On the other hand,

$$\begin{aligned} n_p \nabla \nabla^2 n_s + n_s \nabla \nabla^2 n_p = {} & \left(\nabla \cdot \left(\nabla^2 \left(n_p n_s\right) - \nabla n_p \cdot \nabla n_s\right) \boldsymbol{\delta}\right) \\ & - 2\nabla \cdot \left(\nabla n_p \nabla n_s\right). \end{aligned}$$

So

$$\begin{aligned} \bar{\kappa}_{pp} n_p \nabla \nabla^2 n_p = {} & \frac{1}{2} \left(\nabla \cdot \left(\nabla^2 \left(\bar{\kappa}_{pp} n_p n_p\right) - \bar{\kappa}_{pp} \nabla n_p \cdot \nabla n_p\right) \boldsymbol{\delta}\right) \\ & - \nabla \cdot \left(\bar{\kappa}_{pp} \nabla n_p \nabla n_p\right), \end{aligned} \tag{9.46}$$

$$\begin{aligned} \bar{\kappa}_{qq} n_q \nabla \nabla^2 n_q = {} & \frac{1}{2} \left(\nabla \cdot \left(\nabla^2 \left(\bar{\kappa}_{qq} n_q n_q\right) - \bar{\kappa}_{qq} \nabla n_q \cdot \nabla n_q\right) \boldsymbol{\delta}\right) \\ & - \nabla \cdot \left(\bar{\kappa}_{qq} \nabla n_q \nabla n_q\right), \end{aligned} \tag{9.47}$$

$$\begin{aligned} \bar{\kappa}_{pq} n_p \nabla \nabla^2 n_q + \bar{\kappa}_{pq} n_q \nabla \nabla^2 n_p = {} & \left(\nabla \cdot \left(\nabla^2 \left(\bar{\kappa}_{pq} n_p n_q\right) - \bar{\kappa}_{pq} \nabla n_p \cdot \nabla n_q\right) \boldsymbol{\delta}\right) \\ & - \nabla \cdot \left(2\bar{\kappa}_{pq} \nabla n_p \nabla n_q\right). \end{aligned} \tag{9.48}$$

The total attractive force per unit volume at a given point $\mathbf{x}$ of the binary mixture is

$$\sum_{p,q} \rho_p \mathbf{g}^{(pq)} = \rho_p g^{(pp)} + \rho_p g^{(qp)} + \rho_q g^{(pq)} + \rho_q g^{(qq)}, \tag{9.49}$$

where we should remember that $g^{(pp)}$ is the force per unit mass acting on a single molecule p from all the molecules p around it, and that $g^{(pq)}$ is the force per unit mass acting on a single molecule p from all the molecules q around it. The same reasoning for molecules q. Replacing Eqs. (9.43)–(9.45) and Eqs. (9.46)–(9.48) into Eq. (9.49) we get the following relationship,

$$\sum_{p,q} \rho_p \mathbf{g}^{(pq)} = \nabla \cdot \left(\left(\begin{array}{c} an^2 \\ +\frac{1}{2}\nabla^2 \bar{\kappa} n^2 \\ -\frac{1}{2}\left(\bar{\kappa}_{pp}\left(\nabla n_p\right)^2 + 2\bar{\kappa}_{pq}\nabla n_p \cdot \nabla n_q + \bar{\kappa}_{qq}\left(\nabla n_q\right)^2\right) \end{array} \right) \boldsymbol{\delta} \right) - \nabla \cdot \mathbf{S}, \tag{9.50}$$

where

$$a = a_{pp}x_p^2 + 2a_{pq}x_p x_q + a_{qq}x_q^2, \tag{9.51}$$

$$\bar{\kappa} = \bar{\kappa}_{pp}x_p^2 + 2\bar{\kappa}_{pq}x_p x_q + \bar{\kappa}_{qq}x_q^2, \tag{9.52}$$

and

$$\mathbf{S} = \bar{\kappa}_{pp}\nabla n_p \nabla n_p + \bar{\kappa}_{qq}\nabla n_q \nabla n_q + 2\bar{\kappa}_{pq}\nabla n_p \nabla n_q. \tag{9.53}$$

In the above equations, we recognize parameter a and $\bar{\kappa}$ as the van der Waals force parameters for non-ideal mixtures. Both parameters are weighted by the molar fractions of species p and q in the mixture. We also recognize tensor **S** as the Korteweg surface tensor written on a molecular or molar basis, following the way we consider symbol n, as the number of molecules or the number of moles per unit volume.

Equation (9.50) can be easily generalized for an arbitrary multicomponent system, giving

$$\sum_{i,j} \rho_i \mathbf{g}^{(ij)} = \nabla \cdot \left(\left(\begin{array}{c} an^2 \\ +\frac{1}{2}\nabla^2 \bar{\kappa} n^2 \\ -\frac{1}{2}\sum_{i,j} \bar{\kappa}_{ij}\nabla n_i \cdot \nabla n_j \end{array} \right) \boldsymbol{\delta} \right) - \nabla \cdot \mathbf{S}. \tag{9.54}$$

It is now time to write the momentum balance equation for a mixture when it is subjected to segregation effects. We consider the thermodynamic pressure as composed of two parts,

$$P = P_T + P_{att},$$

where P_T is the pressure due to the random thermal motion of molecules corrected by the Enskog's volume exclusion term [30] and written on a molecular basis as

$$P_T = \frac{nkT}{1 - bn},$$

whereas P_{att} takes the attractive intermolecular forces into account. Therefore, we can write the momentum balance equation as

$$\frac{\partial}{\partial t}(\rho \mathbf{v}) + \nabla \cdot (\rho \mathbf{v}\mathbf{v} + \boldsymbol{\tau}) = \rho \mathbf{g} - \nabla \cdot (P_T \boldsymbol{\delta}) + \sum_{i,j} \rho_i \mathbf{g}^{(ij)}.$$

Replacing the last term of the above equation by Eq. (9.54),

$$\begin{aligned}\frac{\partial}{\partial t}(\rho \mathbf{v}) + \nabla \cdot (\rho \mathbf{v}\mathbf{v} + \boldsymbol{\tau}) \cdot = \rho \mathbf{g} &- \nabla \cdot \left(\left(P_T - an^2\right) \boldsymbol{\delta}\right) \\ &- \nabla \cdot \left(\frac{1}{2}\left(\sum_{i,j} \bar{\kappa}_{ij} \nabla n_i \cdot \nabla n_j - \nabla^2 \bar{\kappa} n^2\right) \boldsymbol{\delta}\right) \\ &- \nabla \cdot \mathbf{S},\end{aligned}$$

which can be rewritten in a more compact form as

$$\begin{aligned}\frac{\partial}{\partial t}(\rho \mathbf{v}) &+ \nabla \cdot (\rho \mathbf{v}\mathbf{v} + \boldsymbol{\tau}) \cdot \\ &= \rho \mathbf{g} - \nabla \cdot (P_s \boldsymbol{\delta}) - \nabla \cdot \mathbf{S},\end{aligned} \tag{9.55}$$

where the scalar pressure P_s is given by

$$P_s = P + \frac{1}{2}\left(\sum_{i,j} \bar{\kappa}_{ij} \nabla n_i \cdot \nabla n_j - \nabla^2 \bar{\kappa} n^2\right), \tag{9.56}$$

and the thermodynamic pressure P is written as

$$P = \frac{nkT}{1 - bn} - an^2.$$

Equation (9.56) has exactly the same form as Eq. (9.38). The first approach was based on pure thermodynamic concepts, or the second law of thermodynamics, while the second approach is based on the mechanical concept of force. Following this second approach, the resulting equation of state is the one first conceived by van der Waals.

9.3.2 Interface Physics

To get a better insight into the physics underlying a non-ideal mixture, we consider a binary mixture with components labeled p and q in isothermal conditions. When segregation forces are dominant over the random thermal motion of the molecules, fluids p and q will form separated phases and an interface will be formed between these two phases.

But,

(i) what are the physical mechanisms that sustain this interface, preventing fluids p and q to mix?;
(ii) how to model these physical mechanisms and quantify their effects on the interface formation?;
(iii) what are their contribution to the interface tension, or any other macroscopic mensurable property?

These are important questions we must answer to get an in-depth understanding of the macroscopic behavior of immiscible fluids. Nevertheless, their answers require downscaling to the molecular scale. Using the same approach of Sect. 6.5 we can, nevertheless, choose to work with a simplified model, considering the molecules as material points, and, in the case of neutral molecules, these material points as sources of electrostatic fields produced by permanent or temporary dipoles.

For a p, q binary mixture, in isothermal conditions, the diffusive flux of component p in the interface was found by Philippi et al. [133], using a kinetic theory approach as

$$\mathbf{j}^{(p)} = \frac{\tau_{pq}}{\chi}\left[\begin{array}{c} -nkT\nabla x_p + \rho\omega_p\omega_q\mathbf{g}^{(pp)} - \rho\omega_p\omega_q\mathbf{g}^{(qq)} \\ -\rho\omega_p\omega_q\frac{(m_p-m_q)}{m_p m_q}\left(m_p\mathbf{g}^{(pq)} - \frac{kT}{n}\nabla\left[n(1+bn\chi)\right]\right) \end{array}\right], \tag{9.57}$$

where τ_{pq} is a relaxation time for the collisions between molecules p and q related to the binary diffusivity $\mathscr{D}_{pq}$, χ is a dimensionless parameter that takes the volume b of the molecules into account

$$\chi = \frac{1}{1-bn},$$

m_s is the mass of molecule s, $s = p, q$, x_s is the molecular or molar fraction of component s and ω_s is the mass fraction of this component.

Equation (9.57) satisfies $\sum_s \mathbf{j}^{(s)} = 0$.
In the right-hand side of Eq. (9.57):

(i) the first term is the usual Fickean diffusion term, responsible for mixing the two fluids and proportional to the temperature;
(ii) the second and third terms are responsible for segregating these two fluids and are due to the intermolecular electrostatic forces among the molecules of the same species;
(iii) the fourth term is dependent on the molecular masses of the two involved species, being null when $m_p = m_q$. Since $\mathbf{g}^{(pq)}$ is related to the cross interactions between p and q molecules, the net effect of these cross-interactions in the interface is to be null, except when the p and q molecular masses are different. In this case, the force $\mathbf{g}^{(pq)}$ has a surfactant effect, reducing the interface tension. The second part of this last term is proportional to the density gradient and describes the well-known tendency of lighter molecules to be thrown away from the regions of higher number-density.

For simplicity, consider $m_p \approx m_q$ in such a manner that it is possible to neglect the last term on the r.h.s. of Eq. (9.57). Consider also that component q is apolar and that the force $\mathbf{g}^{(qq)} \approx 0$. When these conditions are satisfied

$$\mathbf{j}^{(p)} = \frac{\tau_{pq}}{\chi}\left[-nkT\nabla x_p + \rho\omega_p\omega_q\mathbf{g}^{(pp)}\right], \tag{9.58}$$

$$\mathbf{j}^{(q)} = \frac{\tau_{pq}}{\chi}\left[-nkT\nabla x_q - \rho\omega_p\omega_q\mathbf{g}^{(pp)}\right]. \tag{9.59}$$

Therefore, component q will be subjected to a diffusive flux promoted by the force $\mathbf{g}^{(pp)}$ between molecules p, even when velocity $\mathbf{v}_q$ is null. This result may appear somewhat strange, but we must remember that the local velocity at a given point in the interface is given by

$$\mathbf{v} = \omega_p\mathbf{v}_p + \omega_q\mathbf{v}_q = \omega_p\mathbf{v}_p,$$

and that

$$\mathbf{j}^{(q)} = \rho_q\left(\mathbf{v}_q - \mathbf{v}\right) = -\rho_q\mathbf{v},$$

and, so, for an observer traveling with the local velocity $\mathbf{v}$ the component q at rest will appear as moving upstream.

9.3.2.1 Interface Tension

Consider a plane interface orthogonal to the x-axis, separating two liquids p and q, and that this interface is at rest. In this condition, the interface tension can be obtained, using

$$\gamma^{(pq)} = \int_{-\infty}^{\infty} \left(P_{xx} - P_{yy}\right) dx. \tag{9.60}$$

The pressure tensor is given by Eq. (9.35) and since the spherical tensor $P_s\boldsymbol{\delta}$ is independent on the direction,

$$\gamma^{(pq)} = \int_{-\infty}^{\infty} \left(S_{xx} - S_{yy}\right) dx. \tag{9.61}$$

Using Eq. (9.37), the interface tension can be written as

$$\gamma^{(pq)} = \kappa^{(pp)} \int_{-\infty}^{\infty} \left(\frac{d\rho^{(p)}}{dx}\right)^2 dx + \kappa^{(qq)} \int_{-\infty}^{\infty} \left(\frac{d\rho^{(q)}}{dx}\right)^2 dx$$
$$+ \kappa^{(pq)} \int_{-\infty}^{\infty} \left(\frac{d\rho^{(p)}}{dx}\right)\left(\frac{d\rho^{(q)}}{dx}\right) dx. \tag{9.62}$$

In the above equation, the molecular parameters $\kappa^{(ss)}$, $s = p, q$, are related to the surface tensions $\gamma^{(ss)}$ for $s = p, q$, through

$$\kappa^{(ss)} = \frac{\gamma^{(ss)}}{\left(\int_{-\infty}^{\infty} \left(\frac{d\rho^{(s)}}{dx}\right)^2 dx\right)} \quad \text{for} \quad s = p, q, \tag{9.63}$$

and the cross-parameter $\kappa^{(pq)}$ can then be related to the interface tension $\gamma^{(pq)}$ by finding the profiles $\rho^{(p)}(x)$ and $\rho^{(q)}(x)$ inside the $p - q$ interface.

The result given by Eq. (9.62) can be further compared with the heuristic models for the interface tension that are widely used in the physical chemistry of surfaces [88],

$$\gamma^{(pq)} = \gamma^{(p)} + \gamma^{(q)} - \Upsilon^{(pq)}, \tag{9.64}$$

where the $\Upsilon^{(pq)}$ term is always positive, depends on the cross-interaction between the molecules from species p and q, and has different forms in accordance with the particular model that is being used.

9.3.3 Spinodal Decomposition in Multicomponent Systems

In this section, we present the numerical simulation results of a spinodal decomposition. Simulations are based on a lattice-Boltzmann model proposed by Philippi et al. [131, 133] for multicomponent systems. For simplifying reasons, a binary mixture of fluids p and q with identical molecular masses and critical temperatures is considered. In this case, the surfactant forces are null, as above explained. Therefore, from Eq. (9.57), the diffusion fluxes become

$$\mathbf{j}^{(p)} = \frac{\tau_{pq}}{\chi}\left[-nkT\nabla x_p + \rho\omega_p\omega_q\mathbf{g}^{(pp)} - \rho\omega_p\omega_q\mathbf{g}^{(qq)}\right], \tag{9.65}$$

and

$$\mathbf{j}^{(q)} = \frac{\tau_{pq}}{\chi}\left[-nkT\nabla x_q + \rho\omega_p\omega_q\mathbf{g}^{(qq)} - \rho\omega_p\omega_q\mathbf{g}^{(pp)}\right]. \tag{9.66}$$

The binary mixture is initially homogeneous and segregation is ruled by the intermolecular forces $\mathbf{g}^{(pp)}$ and $\mathbf{g}^{(qq)}$ between identical molecules p and q, respectively. In accordance with Eq. (9.42) these force are dependent on the molecular parameters a_{ss} and $\bar{\kappa}_{ss}$, $s = p, q$, and on the gradient of concentration. Force $\mathbf{g}^{(pp)}$ try to rescue molecules p from the mixture to phase p, whereas force $\mathbf{g}^{(qq)}$ have the same effect on molecules q. In the opposite sense, the Fickean diffuse forces try to mix the two components. These forces are proportional to the temperature in such a way that, above a certain temperature, they are dominant and there is no segregation.

Figure 9.3 shows the evolution over time of the interface length between the two components during the segregation process for four values of the reduced temperature T_r. The interface length is, in theory, infinite at time $t = 0$ and has an exponential

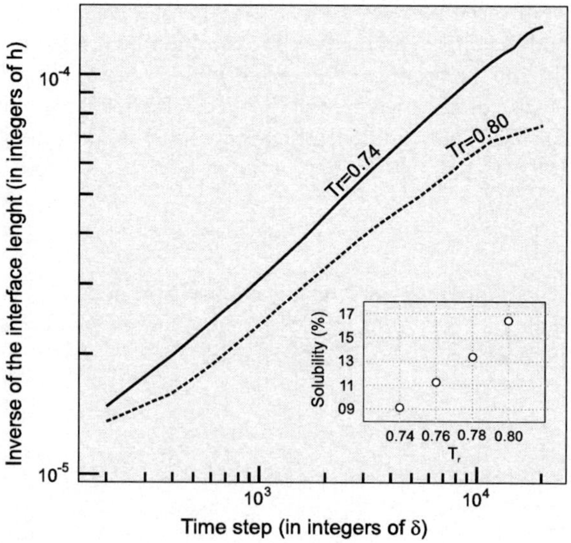

Fig. 9.3 Time evolution of the interface length [133]

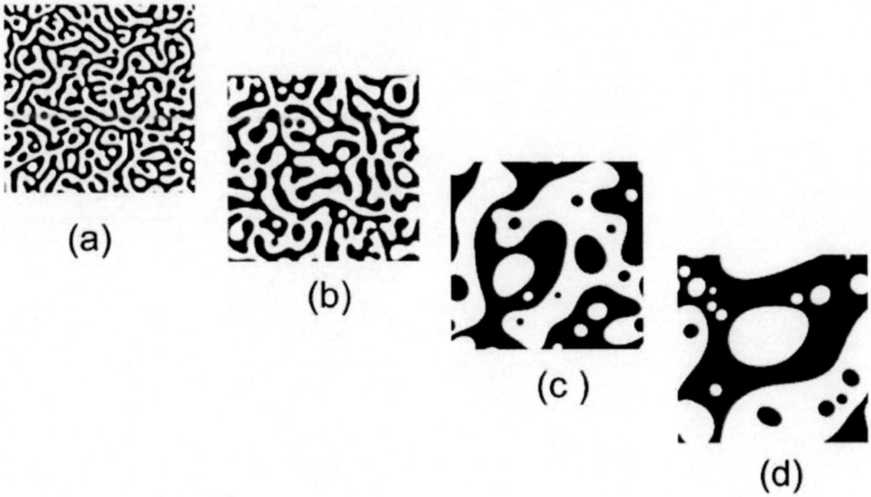

Fig. 9.4 When the temperature is below a critical value for a homogeneous mixture of two components, the segregating forces will lead to a two-phase system by spinodal decomposition. Snapshots are related to 500 (**a**), 1000 (**b**), 5000 (**c**), and 10,000 (**d**) time steps for a reduced temperature $T_r = 0.74$ [133]

decay with time. The exponent of decay in the first 600–10,000 time steps was found to be 0:51 for T_r = 0:74, and 0:43 for T_r = 0:80. After this initial period of time, the exponential decay is reduced and the segregation proceeds at a slower rate toward the equilibrium configuration. For reduced temperatures above T_r = 0:80 no segregation was noticed in the course of the simulation time and, in some cases, the simulations were destroyed by increasing high-frequency oscillations. For reduced temperatures below $T_r = 0.72$ the simulations were not stable for the parameters used in the simulations. The sequence of snapshots in Fig. 9.4 visually confirms that, in the absence of gravity, the segregation indeed proceeds through spinodal decomposition [25] with the formation of a very complex fine geometrical structure including isolated islands of one component inside the other. For better visualization, Fig. 9.4 presents black and white snapshots after the true gray-scale values were binarized.

This choice should not confuse the fact that a certain amount of component p will always be dissolved in phase q (and vice versa). Indeed, the inset in Fig. 9.3 shows the amount of component p (q), in terms of the temperature, that remains dissolved in phase q (p) after the equilibrium was reached. It is clearly seen that the simulation results are physically consistent, retrieving the tendency of solubility to increase with the temperature.

Chapter 10
Non-equilibrium Thermodynamics from a Kinetic Standpoint

Abstract The behavior of fluid, especially non-ideal fluids and their mixtures is largely conditioned by physical processes that require to downscale to the molecular level for a correct understanding of the underlying physics. Nevertheless, in numerical simulations of molecular dynamics, the round-off errors prevent the simulations from being performed beyond very limited spatial and time ranges. This difficulty increases for very large systems of particles because there is no means to ascertain the exact position and velocity of each particle at the initial time $t = 0$. The best it is possible to do is to try to prescribe the *probability* that this particle will have a certain velocity when it is in a position $\mathbf{x}$ at time t. This probabilistic approach led to the Boltzmann equation and the finding that a system of particles ruled by the deterministic Newton's laws of motion satisfies the second law of thermodynamics. In this chapter, we derive the Boltzmann equation starting from the Liouville equation for large systems of particles. The Boltzmann equation is restricted to material points considered sources of repulsion fields and collisions are binary involving, solely, a pair of particles. It is thus limited to rarefied gases. Considering their molecular nature and the great difficulty of the treatment of multiple collisions, the building up of kinetic equations considering only the information of the molecular scale is still a great scientific challenge. A kinetic model may be understood as a model that makes use of the information from the macroscopic scale to improve the simplifying assumptions that were required when deriving a model considering solely what is known from the molecular scale. Kinetic models are presented in this chapter for the collision term and non-ideal fluids and their mixtures.

10.1 Scales of Investigation

The behavior of fluids, especially non-ideal fluids and their mixtures is largely conditioned by the interaction processes between the molecules of the different involved fluids.

Restricting ourselves to neutral molecules, these interaction processes involve intermolecular forces of an electrical nature that result from the interaction between permanent dipoles (Keesom forces and hydrogen bonds), induced dipoles (Debye

P. C. Philippi, *Thermodynamics*, https://doi.org/10.1007/978-3-031-49357-7_10

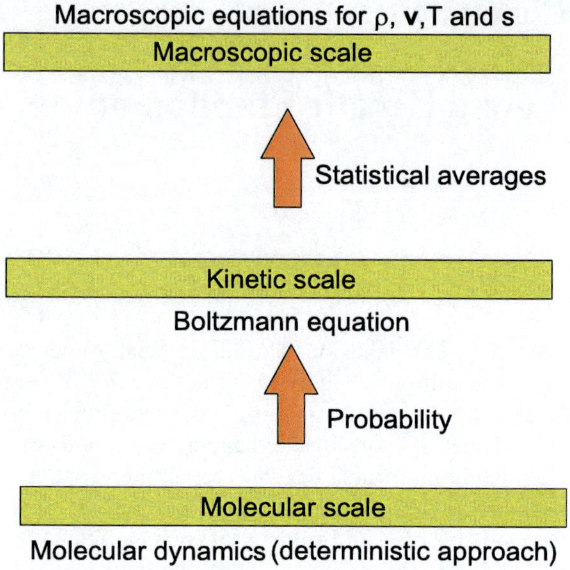

Fig. 10.1 The kinetic scale as a conceptual bridge between molecular physics and the macroscopic domain

forces), and temporary high-frequency fluctuations of the geometric center of the electronic clouds surrounding the molecules (London dispersive forces).

When these physical processes are important, the correct understanding of the behavior of thermodynamic systems in non-equilibrium conditions is not possible based only on the classical macroscopic equations, since the process is affected by physical phenomena that develop on the molecular scale. This is, e.g., the case of the displacement of the fluid-solid triple line, the segregation processes in multicomponent systems, and the formation of emulsions.

On the other hand, the description provided by molecular physics methods is restricted to a very small scale, measured in nanometers.

The kinetic scale may be understood as a conceptual bridge between molecular physics and the macroscopic domain, by means of certain simplifications resulting from the up-scaling processes of the detailed phenomena that develop at the molecular level. In this chapter, we deal with the kinetic approach to non-equilibrium thermodynamics (Fig. 10.1).

On a molecular scale, a fluid is a system of molecules whose thermodynamic properties result from molecular interactions. This is the molecular dynamics standpoint. In its simplest classical version, these molecules are considered as material points with an attraction-repulsion kernel ruled by Newton's laws of classical mechanics (Fig. 10.2).

Molecular dynamics in this classical context is, thus, a deterministic approach. The initial state of a system of N particles is given by the velocity $\boldsymbol{\xi}_i$ $(t = 0)$ and position $\mathbf{x}_i$ $(t = 0)$ of each particle and the next states are found by solving the Newton equations

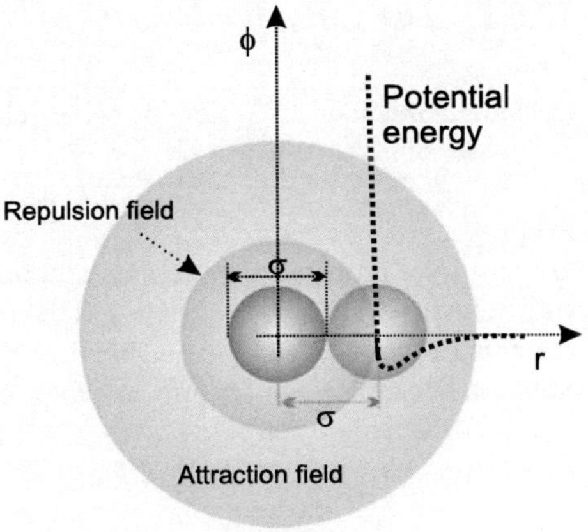

Fig. 10.2 Atraction-repulsion field around a single molecule

$$m\frac{d\boldsymbol{\xi}_i}{dt} = \sum_{j\neq i} \mathbf{F}_{i-j}, \quad i = 1, \ldots, N.$$

In numerical simulations of molecular dynamics, the round-off errors prevent the simulations from being performed beyond a very limited spatial and time ranges.

This difficulty increases for very large systems of particles, when $N \to \infty$, because there are no means to ascertain the exact position and velocity of each particle at the initial time $t = 0$. The best it is possible to do is to try to prescribe the *probability* that this particle will have a certain velocity when it is in a position $\mathbf{x}$ at time t.

As it was advanced in Sect. 2.6, this probabilistic standpoint was pursued by Clausius, Maxwell, and Boltzmann in their attempts to explain the full content of the second law of thermodynamics and the meaning of *entropy*, launching the foundations of the *The Mechanical Theory of Heat*.

This approach led to the Boltzmann equation [15, 16] and the outstanding finding that a system of particles ruled by the deterministic Newton's laws of motion satisfies the second law of thermodynamics.

10.2 The Boltzmann Equation

In this section, we show the main assumptions when the Boltzmann equation is derived, starting from the Liouville equation [107]. The books of Chapman and Cowling [30], Cercignani [29] and Kremer [92] are suggested for further reading and alternative derivations.

10.2.1 The Liouville Equation

Consider a mechanical system of N particles. Let

$$f^{(N)}\left(\mathbf{x}_1, \boldsymbol{\xi}_1, \dots \mathbf{x}_N, \boldsymbol{\xi}_N, t\right),$$

to be the joint probability of finding, at time t, dt the particle 1 at the position $\mathbf{x}_1$, $d\mathbf{x}_1$ with velocity $\boldsymbol{\xi}_1$, $d\boldsymbol{\xi}_1$, the particle 2 at the position $\mathbf{x}_2$, $d\mathbf{x}_2$ with velocity $\boldsymbol{\xi}_2$, $d\boldsymbol{\xi}_2$ and so on, until particle N at the position $\mathbf{x}_N$, $d\mathbf{x}_N$ with velocity $\boldsymbol{\xi}_N$, $d\boldsymbol{\xi}_N$. The Liouville equation describing the dynamical evolution of this system is given by

$$\partial_t f^{(N)} + \sum_i \boldsymbol{\xi}_i . \partial_{\mathbf{x}_i} f^{(N)} + \sum_i \mathbf{g}_i . \partial_{\boldsymbol{\xi}_i} f^{(N)} = 0, \tag{10.1}$$

where $\mathbf{g}_i$ is the acceleration due to the force acting on particle i,

$$\mathbf{g}_i = \mathbf{g}_i^{(e)} + \sum_{\substack{j=1 \\ j \neq i}}^{N} \mathbf{g}_{ij} .$$

Force $\mathbf{g}_i^{(e)}$ is related to the force on particle i due to an external field and $\mathbf{g}_{ij}$ is the force on particle i due to its interaction with particle j,

$$\mathbf{g}_{ij} = -\frac{1}{m} \frac{\partial \Phi\left(x_{ij}\right)}{\partial \mathbf{x}_i},$$

where $x_{ij} = \left|\mathbf{x}_i - \mathbf{x}_j\right|$ and Φ is the potential energy which will be supposed to be only dependent on the distance between particles i and j.

The joint probability $f^{(N)}$ gives a too much detailed description of the system, which is unnecessarily complex, since the dynamical evolution of an arbitrary, but, single particle, can be a reliable description of the whole mechanical system of particles, when these particles cannot be individually labeled. Therefore, probability $f^{(N)}$ can be integrated in the phase space $\mathbf{x}_2, \boldsymbol{\xi}_2, \dots \mathbf{x}_N, \boldsymbol{\xi}_N$ to give the marginal probability $f^{(1)}$ of finding, at time t, dt the particle 1 at the position $\mathbf{x}_1$, $d\mathbf{x}_1$ with velocity $\boldsymbol{\xi}_1$, $d\boldsymbol{\xi}_1$

$$f^{(1)}\left(\mathbf{x}_1, \boldsymbol{\xi}_1, t\right) = \int \cdots \int f^{(N)} d\mathbf{x}_2 \dots d\mathbf{x}_N d\boldsymbol{\xi}_2 \dots d\boldsymbol{\xi}_N .$$

After integrating the Liouville equation, considering the particles to be indistinguishable, Eq. (10.1) becomes

$$\begin{aligned}&\partial_t f^{(1)} + \boldsymbol{\xi}.\partial_{\mathbf{x}} f^{(1)} + \mathbf{g}^{(e)}.\partial_{\xi} f^{(1)} \\ &= -(N-1) \\ &\times \partial_{\xi_1} \int\int \mathbf{g}_{12} f^{(2)} \left(\mathbf{x}_1, \boldsymbol{\xi}_1, \mathbf{x}_2, \boldsymbol{\xi}_2, t\right) d\mathbf{x}_2 d\boldsymbol{\xi}_2. \end{aligned} \tag{10.2}$$

This result imposes a solution hierarchy since the marginal probability $f^{(1)}$ is dependent on $f^{(2)}$, $f^{(2)}$ is dependent on $f^{(3)}$, and so on. This hierarchy is called the BBGKY hierarchy after the works of Born, Bogoliubov, Green, Kirkwood and Yvon ([13, 17, 86, 87, 191]).

Considering $f = N f^{(1)}$ and $(N-1) \approx N$ and changing the notation for the target and the incident particles, the Liouville equation becomes, for large N,

$$\begin{aligned}&\partial_t f + \boldsymbol{\xi}.\partial_{\mathbf{x}} f + \mathbf{g}^{(e)}.\partial_{\xi} f \\ &= -\partial_{\xi} \int \cdots \int \mathbf{g}_{12} N^2 f^{(2)} \left(\mathbf{x}, \boldsymbol{\xi}, \mathbf{x}_1, \boldsymbol{\xi}_1, t\right) d\mathbf{x}_1 d\boldsymbol{\xi}_1 \\ &= \partial_{\xi} \int \cdots \int \frac{1}{m} \frac{\partial \Phi\left(|\mathbf{x}_1 - \mathbf{x}|\right)}{\partial \mathbf{x}} \times N^2 f^{(2)} \left(\mathbf{x}, \boldsymbol{\xi}, \mathbf{x}_1, \boldsymbol{\xi}_1, t\right) d\mathbf{x}_1 d\boldsymbol{\xi}_1, \end{aligned} \tag{10.3}$$

which is a Boltzmann equation for the distribution function f, with a collision term Ω.

This collision term will now be split in two collision terms[1] $\Omega = \Omega^{(sd)} + \Omega^{(ld)}$, where $\Omega^{(sd)}$ is referred to short distance interactions, $|\mathbf{x}_1 - \mathbf{x}| < \sigma$ and $\Omega^{(ld)}$ to long-range interactions $|\mathbf{x}_1 - \mathbf{x}| > \sigma$, σ being the distance related to the molecular diameter below which the attraction forces change to a strong repulsion among the molecules (Fig. 10.2).

Therefore

$$\partial_t f + \boldsymbol{\xi}.\partial_{\mathbf{x}} f + \mathbf{g}^{(e)}.\partial_{\xi} f = \Omega^{(sd)} + \Omega^{(ld)}, \tag{10.4}$$

where

$$\Omega^{(sd)} = \partial_{\xi} \int \int\limits_{|\mathbf{x}_1 - \mathbf{x}| < \sigma} \frac{1}{m} \frac{\partial \Phi\left(|\mathbf{x}_1 - \mathbf{x}|\right)}{\partial \mathbf{x}} N^2 f^{(2)} \left(\mathbf{x}, \boldsymbol{\xi}, \mathbf{x}_1, \boldsymbol{\xi}_1, t\right) d\mathbf{x}_1 d\boldsymbol{\xi}_1, \tag{10.5}$$

and

$$\Omega^{(ld)} = \partial_{\xi} \int \int\limits_{|\mathbf{x}_1 - \mathbf{x}| > \sigma} \frac{1}{m} \frac{\partial \Phi\left(|\mathbf{x}_1 - \mathbf{x}|\right)}{\partial \mathbf{x}} N^2 f^{(2)} \left(\mathbf{x}, \boldsymbol{\xi}, \mathbf{x}_1, \boldsymbol{\xi}_1, t\right) d\mathbf{x}_1 d\boldsymbol{\xi}_1. \tag{10.6}$$

[1] See X. He and G. D. Doolen, [66].

10.2.2 The Long-Range Term. Mean Field Theory

Consider first the long-range collision term, Eq. (10.6)

$$\Omega^{(ld)} = \frac{\partial}{\partial \boldsymbol{\xi}} \cdot \int \int_{|\mathbf{x}_1 - \mathbf{x}| > \sigma} \frac{1}{m} \frac{\partial \Phi\left(|\mathbf{x}_1 - \mathbf{x}|\right)}{\partial \mathbf{x}} N^2 f^{(2)}\left(\mathbf{x}, \boldsymbol{\xi}, \mathbf{x}_1, \boldsymbol{\xi}_1, t\right) d\mathbf{x}_1 d\boldsymbol{\xi}_1.$$

By making the assumption that, for $|\mathbf{x}_1 - \mathbf{x}| > \sigma$, the molecular chaos prevails

$$N^2 f^{(2)}\left(\mathbf{x}, \boldsymbol{\xi}, \mathbf{x}_1, \boldsymbol{\xi}_1\right) = f\left(\mathbf{x}, \boldsymbol{\xi}, t\right) f\left(\mathbf{x}_1, \boldsymbol{\xi}_1, t\right) \equiv f f_1,$$

one obtains

$$\begin{aligned} \Omega^{(ld)} &= \frac{\partial f\left(\mathbf{x}, \boldsymbol{\xi}, t\right)}{\partial \boldsymbol{\xi}} \cdot \int \int_{|\mathbf{x}_1 - \mathbf{x}| > \sigma} \frac{1}{m} \frac{\partial \Phi\left(|\mathbf{x}_1 - \mathbf{x}|\right)}{\partial \mathbf{x}} f\left(\mathbf{x}_1, \boldsymbol{\xi}_1, t\right) d\mathbf{x}_1 d\boldsymbol{\xi}_1 \\ &= \frac{1}{m} \frac{\partial f\left(\mathbf{x}, \boldsymbol{\xi}, t\right)}{\partial \boldsymbol{\xi}} \cdot \frac{\partial}{\partial \mathbf{x}} \int_{|\mathbf{x}_1 - \mathbf{x}| > \sigma} \Phi\left(|\mathbf{x}_1 - \mathbf{x}|\right) n\left(\mathbf{x}_1, t\right) d\mathbf{x}_1. \end{aligned} \tag{10.7}$$

The integrand in the above equation is the mean field, i.e., the potential energy related to the integrated action of each one of the n molecules placed in the neighborhood, on a single molecule at the position $\mathbf{x}$,

$$\Phi_m\left(\mathbf{x}\right) = \int_{|\mathbf{x}_1 - \mathbf{x}| > \sigma} \Phi\left(|\mathbf{x}_1 - \mathbf{x}|\right) n\left(\mathbf{x}_1, t\right) d\mathbf{x}_1. \tag{10.8}$$

Since the acceleration $\mathbf{g}^{(\ell d)}$ is the gradient of the potential energy divided by its mass,

$$\mathbf{g}^{(\ell d)} = -\frac{1}{m} \frac{\partial \Phi_m}{\partial \mathbf{x}}, \tag{10.9}$$

we get

$$\Omega^{(ld)} = -\mathbf{g}^{(\ell d)} \cdot \frac{\partial f\left(\mathbf{x}, \boldsymbol{\xi}, t\right)}{\partial \boldsymbol{\xi}}. \tag{10.10}$$

Therefore the Boltzmann equation can be written as

$$\partial_t f + \boldsymbol{\xi} . \partial_{\mathbf{x}} f + \left(\mathbf{g}^{(e)} + \mathbf{g}^{(\ell d)}\right) \cdot \partial_{\boldsymbol{\xi}} f = \Omega^{(sd)}$$

10.2.3 The Short-Range Collision Term

Starting from Eqs. (10.2) and (10.10), the integration of the Liouville equation gives, after dropping out the subscript index for the target particles and using the subscript 1 for the integration variables,

$$\begin{aligned}
&\partial_t f^{(1)} + \boldsymbol{\xi} \cdot \partial_{\mathbf{x}} f^{(1)} + \left(\mathbf{g}^{(e)} + \mathbf{g}^{(\ell d)}\right) \cdot \partial_{\xi} f^{(1)} \\
&= -(N-1) \\
&\times \partial_{\xi} \cdot \int \int_{|\mathbf{x}_1 - \mathbf{x}| < \sigma} \mathbf{g}_{\mathbf{x}\mathbf{x}_1} \cdot f^{(2)}\left(\mathbf{x}, \boldsymbol{\xi}, \mathbf{x}_1, \boldsymbol{\xi}_1, t\right) d\mathbf{x}_1 d\boldsymbol{\xi}_1.
\end{aligned} \tag{10.11}$$

Similarly

$$\begin{aligned}
&\partial_t f^{(2)} + \boldsymbol{\xi} \cdot \partial_{\mathbf{x}} f^{(2)} + \boldsymbol{\xi}_1 \cdot \partial_{\mathbf{x}} f^{(2)} \\
&\quad + \left(\mathbf{g}^{(e)} + \mathbf{g}^{(\ell d)} + \mathbf{g}_{\mathbf{x}\,\mathbf{x}_1}\right) \cdot \partial_{\xi} f^{(2)} + \left(\mathbf{g}_1^{e} + \mathbf{g}_{\mathbf{x}_1\mathbf{x}}\right) \cdot \partial_{\xi_1} f^{(2)} \\
&= -(N-2) \times \partial_{\xi} \cdot \int \int \mathbf{g}_{\mathbf{x}\,\mathbf{x}_2} f^{(3)}\left(\mathbf{x}, \boldsymbol{\xi}, \mathbf{x}_1, \boldsymbol{\xi}_1, \mathbf{x}_2, \boldsymbol{\xi}_2, t\right) d\mathbf{x}_2 d\boldsymbol{\xi}_2 \\
&\quad - (N-2) \times \partial_{\xi_1} \cdot \int \int \mathbf{g}_{\mathbf{x}_1\,\mathbf{x}_2} f^{(3)}\left(\mathbf{x}, \boldsymbol{\xi}, \mathbf{x}_1, \boldsymbol{\xi}_1, \mathbf{x}_2, \boldsymbol{\xi}_2, t\right) d\mathbf{x}_2 d\boldsymbol{\xi}_2.
\end{aligned} \tag{10.12}$$

The above equations and the related equations for $f^{(3)}$, $f^{(4)}$,...form the above referred BBGKY hierarchy where the solution of each distribution $f^{(i)}$ depends on the knowledge of $f^{(i+1)}$.

For evaluating the short-distance collision term, the following assumptions are considered [92].

- **Assumption 1**: Only binary collisions are considered

With this assumption the r.h.s of Eq. (10.12) is negligible and Eq. (10.12) becomes:

$$\begin{aligned}
&\partial_t f^{(2)} + \boldsymbol{\xi} \cdot \partial_{\mathbf{x}} f^{(2)} + \boldsymbol{\xi}_1 \cdot \partial_{\mathbf{x}} f^{(2)} \\
&+ \left(\mathbf{g}^{(e)} + \mathbf{g}^{(\ell d)} + \mathbf{g}_{\mathbf{x}\,\mathbf{x}_1}\right) \cdot \partial_{\xi} f^{(2)} \\
&+ \left(\mathbf{g}_1^{e} + \mathbf{g}_{\mathbf{x}\,\mathbf{x}_1}\right) \cdot \partial_{\xi_1} f^{(2)} = 0.
\end{aligned}$$

- **Assumption 2**: The external and attractive forces are very small when compared with the intermolecular forces during a collision

$$\begin{aligned}
&\partial_t f^{(2)} + \boldsymbol{\xi} \cdot \partial_{\mathbf{x}} f^{(2)} + \boldsymbol{\xi}_1 \cdot \partial_{\mathbf{x}} f^{(2)} + \left(\mathbf{g}_{\mathbf{x}\,\mathbf{x}_1}\right) \cdot \partial_{\xi} f^{(2)} \\
&+ \left(\mathbf{g}_{\mathbf{x}_1\mathbf{x}}\right) \cdot \partial_{\xi_1} f^{(2)} = 0,
\end{aligned} \tag{10.13}$$

or, equivalently, considering the time t to be the instant immediately after the collision

$$\begin{aligned} & f^{(2)}\left(\mathbf{x}, \boldsymbol{\xi}, \mathbf{x}_1, \boldsymbol{\xi}_1, t\right) \\ & = f^{(2)}\left(\mathbf{x} - \boldsymbol{\xi}\Delta t, \xi - \mathbf{g}_{\mathbf{x}}\Delta t, \mathbf{x}_1 - \boldsymbol{\xi}_1\Delta t, \boldsymbol{\xi}_1 - \mathbf{g}_{\mathbf{x}_1}\Delta t, t - \Delta t\right), \end{aligned} \tag{10.14}$$

where Δt is a time interval that is supposed to be small with respect to the time interval between any two collisions, $\mathbf{g}_{\mathbf{x}}$ is the intermolecular force evaluated at point $\mathbf{x}$ and $\mathbf{g}_{\mathbf{x}_1}$ is the intermolecular force evaluated at point $\mathbf{x}_1$.

- **Assumption 3**: Molecular chaos

$$\begin{aligned} & f^{(2)}\left(\mathbf{x}, \xi, \mathbf{x}_1, \xi_1, t\right) \\ & = f^{(2)}\left(\mathbf{x} - \xi\Delta t, \xi - \mathbf{g}_{\mathbf{x}}\Delta t, \mathbf{x}_1 - \xi_1\Delta t, \xi_1 - \mathbf{g}_{\mathbf{x}_1}\Delta t, t - \Delta t\right) \\ & = f^{(1)}\left(\mathbf{x} - \xi\Delta t, \xi - \mathbf{g}_{\mathbf{x}}\Delta t, t - \Delta t\right) \\ & \quad \times f^{(1)}\left(\mathbf{x}_1 - \xi_1\Delta t, \xi_1 - \mathbf{g}_{\mathbf{x}_1}\Delta t, t + \Delta t\right). \end{aligned} \tag{10.15}$$

- **Assumption 4**: It is supposed that the distribution function does not vary between points $\mathbf{x}$ and $\mathbf{x}_1$ at the beginning of the collision process.

Considering now that the time average of $f^{(1)}$ in Eq. (10.11) is given by

$$\overline{f}^{(1)} = \frac{1}{\Delta t}\int_0^{\Delta t} f^{(1)} ds \tag{10.16}$$

we get

$$\begin{aligned} & \partial_t \overline{f}^{(1)} + \boldsymbol{\xi}.\partial_{\mathbf{x}} \overline{f}^{(1)} + \left(\mathbf{g}^{(e)} + \mathbf{g}^{(\ell d)}\right) \cdot \partial_{\xi} \overline{f}^{(1)} \\ & = \frac{(N-1)}{\Delta t} \\ & \quad \times \int\int \left\{ \begin{bmatrix} f^{(1)}\left(\mathbf{x} - \boldsymbol{\xi}\Delta t, \boldsymbol{\xi} - \mathbf{g}\Delta t, t - \Delta t\right) \\ \times f^{(1)}\left(\mathbf{x} - \boldsymbol{\xi}_1\Delta t, \boldsymbol{\xi}_1 - \mathbf{g}_1\Delta t, t - \Delta t\right) \end{bmatrix} \\ - \left[f^{(1)}\left(\mathbf{x}, \boldsymbol{\xi}, t\right) \times f^{(1)}\left(\mathbf{x}, \boldsymbol{\xi}_1, t\right)\right] \right\} d\mathbf{x}_1 d\boldsymbol{\xi}_1. \end{aligned}$$

Since Δt is very small

$$\mathbf{x} - \boldsymbol{\xi}\Delta t \approx \mathbf{x} - \boldsymbol{\xi}_1\Delta t \approx \mathbf{x}, \tag{10.17}$$

$$t - \Delta t \approx t, \tag{10.18}$$

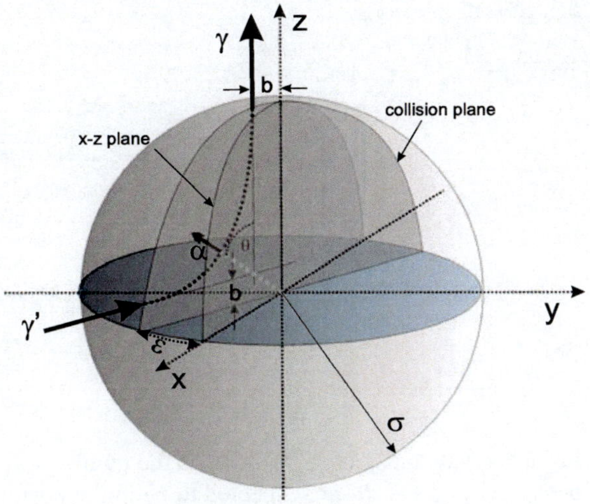

Fig. 10.3 Bullet particle trajectory inside the σ-sphere of the target particle

the collision process being considered to only modify the velocities of the target and bullet particles, respectively, $\xi - \mathbf{g}\Delta t = \xi'$,before the collision to ξ , after the collision and $\boldsymbol{\xi}_1 - \mathbf{g}_1\Delta t = \xi_1'$ before the collision to $\boldsymbol{\xi}_1$, after the collision.

Therefore, we can rewrite the Boltzmann equation as

$$\partial_t \overline{f}^{(1)} + \boldsymbol{\xi}.\partial_{\mathbf{x}} \overline{f}^{(1)} + \left(\mathbf{g}^{(e)} + \mathbf{g}^{(\ell d)}\right).\partial_{\boldsymbol{\xi}} \overline{f}^{(1)}$$
$$= \frac{(N-1)}{\Delta t} \int\int \begin{bmatrix} f^{(1)}\left(\mathbf{x}, \boldsymbol{\xi}', t\right) f^{(1)}\left(\mathbf{x}, \boldsymbol{\xi}_1', t\right) \\ -f^{(1)}\left(\mathbf{x}, \boldsymbol{\xi}, t\right) f^{(1)}\left(\mathbf{x}, \boldsymbol{\xi}_1, t\right) \end{bmatrix} d\mathbf{x}_1 d\boldsymbol{\xi}_1, \tag{10.19}$$

where $\boldsymbol{\xi}'$, $\boldsymbol{\xi}_1'$ are the velocities of the target and bullet particles before the collision (at time $t - \Delta t$) that will give the velocities $\boldsymbol{\xi}$, $\boldsymbol{\xi}_1$ after the collision (Fig. 10.3).

When evaluating the right-hand side of the above equation, the position of the target particle $\mathbf{x}$ is constant. Therefore when evaluating the integral we can replace $\mathbf{x}_1$ by $\mathbf{x}_1 - \mathbf{x}$, by placing the system of coordinates on the center of the target particle.

Writing the element of volume $d\boldsymbol{\eta} = d\left(\mathbf{x}_1 - \mathbf{x}\right)$ in cylindrical coordinates (Fig. 10.4),

$$\frac{1}{\Delta t} d\boldsymbol{\eta} = \gamma b db d\varepsilon, \tag{10.20}$$

where $\boldsymbol{\gamma}$ is the velocity of the bullet particle with respect to the target particle and $\gamma = \|\boldsymbol{\gamma}\|$.

When N is very large $N - 1 \approx N$ and, considering

$$N\overline{f}^{(1)}\left(\mathbf{x}, \boldsymbol{\xi}, t\right) = f\left(\mathbf{x}, \boldsymbol{\xi}, t\right), \tag{10.21}$$

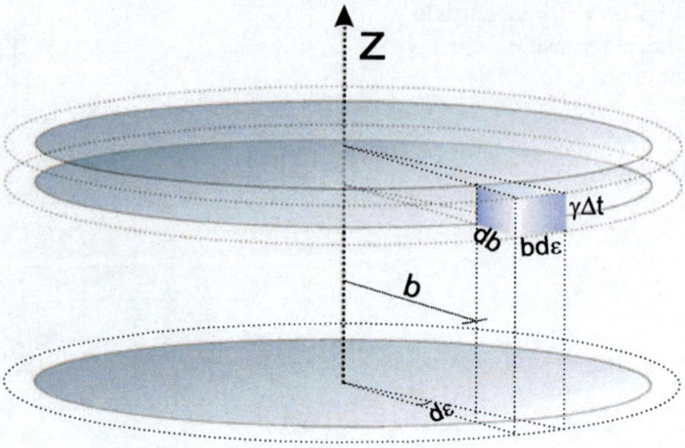

Fig. 10.4 Elementary volume in cylindrical coordinates

in such a way that $f(\mathbf{x}, \boldsymbol{\xi}, t)\, d\mathbf{x} d\boldsymbol{\xi}$ as the number of particles that, during the times between t and $t + dt$, are expected to be in the position between $\mathbf{x}$ and $\mathbf{x}+ d\mathbf{x}$ with velocity between $\boldsymbol{\xi}$ and $\boldsymbol{\xi}+ d\boldsymbol{\xi}$,

$$\partial_t f + \boldsymbol{\xi}.\partial_{\mathbf{x}} f + \left(\mathbf{g}^{(e)} + \mathbf{g}^{(\ell d)}\right).\partial_{\xi} f = \Omega, \tag{10.22}$$

where

$$\Omega = \Omega^{(sd)} = \int\int\int \begin{bmatrix} f\left(\mathbf{x}, \xi', t\right) f\left(\mathbf{x}, \xi_1', t\right) \\ -f\left(\mathbf{x}, \xi, t\right) f\left(\mathbf{x}, \xi_1, t\right) \end{bmatrix} \gamma b db d\varepsilon d\boldsymbol{\xi}_1$$

The above equation is the Boltzmann equation as originally derived.

10.2.3.1 The Equilibrium Solution

The Boltzmann equation is an integro-differential equation and the collision term gives the net balance between the molecules that acquire the velocity $\boldsymbol{\xi}$ and the ones that lose this velocity. In equilibrium conditions, this balance must be null,

$$\Omega = 0,$$

otherwise, the distribution f would vary in the course of the time and in the space. This equilibrium solution was found by Boltzmann and Maxwell in terms of the equilibrium moments as [115], [15, 16]

$$f^{eq} = n\left(\frac{1}{2\pi}\right)^{D/2} \frac{1}{\bar{\xi}^D} e^{-\frac{c_0^2}{2}}, \tag{10.23}$$

where $\bar{\xi} = \sqrt{kT/m}$ is related to a mean molecular speed, k is the Boltzmann constant, $\mathbf{c}_0 = \boldsymbol{\xi}_0 - \mathbf{v}_0$ is the dimensionless particle velocity with respect to an observer moving with $\mathbf{v}_0$, or the peculiar velocity, $\mathbf{v}_0 = \mathbf{v}/\bar{\xi}$ and $\boldsymbol{\xi}_0 = \boldsymbol{\xi}/\bar{\xi}$ are, respectively the dimensionless local and molecular velocity, or the peculiar velocity. Symbol n is used for the number density of the particles with mass m, symbol $\mathbf{v}$ is used for the local velocity, T for the thermodynamic temperature, and D for the dimension of the physical space, $D = 1, 2, 3$ for, respectively, one, two, and three-dimensional physical spaces.

10.2.4 Elementary Properties of the Collision Term

When integrated over all the velocity space domain, the Boltzmann collision term multiplied by any function of ξ gives, for each point $\mathbf{x}$ and time t,

$$I = \int \underbrace{\left(\int \left[f' f_1' - f f_1 \right] \gamma b db d\varepsilon d\boldsymbol{\xi}_1 \right) \varphi\left(\boldsymbol{\xi}\right)}_{F(\boldsymbol{\xi})} d\boldsymbol{\xi}. \tag{10.24}$$

By interchanging $\boldsymbol{\xi} \leftrightarrow \boldsymbol{\xi}_1$, the velocities $\boldsymbol{\xi}'$ and $\boldsymbol{\xi}_1'$ will be also interchanged because

$$\boldsymbol{\xi}' = \boldsymbol{\xi} + \cos\theta\boldsymbol{\alpha}, \tag{10.25}$$

becomes (see Fig. 10.5)

$$\boldsymbol{\xi}_1' = \boldsymbol{\xi}_1 - \cos\theta\boldsymbol{\alpha}, \tag{10.26}$$

and

$$\boldsymbol{\xi}_1' = \boldsymbol{\xi}_1 - \cos\theta\boldsymbol{\alpha}, \tag{10.27}$$

becomes

$$\boldsymbol{\xi} + \cos\theta\boldsymbol{\alpha} = \boldsymbol{\xi}'. \tag{10.28}$$

The only effect of this change in Eq. (10.24) will be

$$I_1 = \int \underbrace{\left(\int \left[f' f_1' - f f_1 \right] \gamma b db d\varepsilon d\boldsymbol{\xi} \right) \varphi\left(\boldsymbol{\xi}_1\right)}_{F(\boldsymbol{\xi}_1)} d\boldsymbol{\xi}_1 \tag{10.29}$$

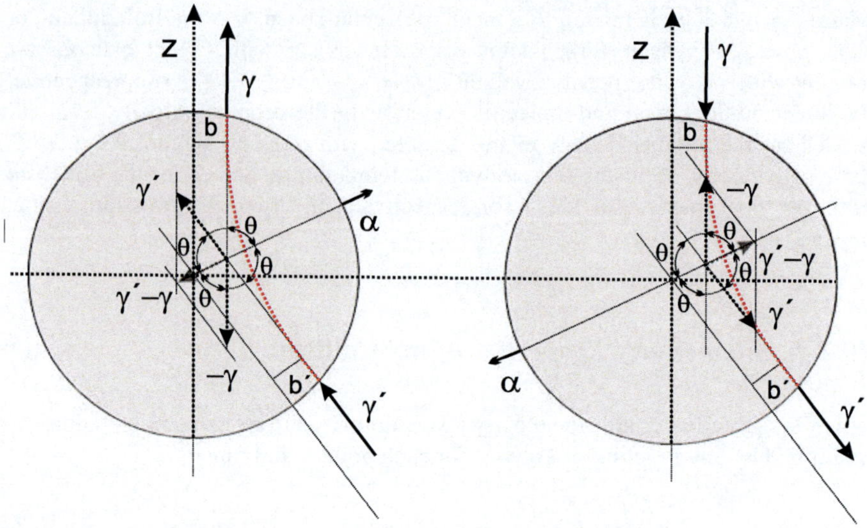

Fig. 10.5 Interchanging $\boldsymbol{\xi}$ and $\boldsymbol{\xi}_1$ invert $\boldsymbol{\gamma}$ and changes the sign of $\boldsymbol{\alpha}$ but θ remains unchanged

since the functional relation F will be the same in both cases, although $F\left(\boldsymbol{\xi}_1\right) \neq F\left(\boldsymbol{\xi}\right)$, unless $\boldsymbol{\xi}_1 = \boldsymbol{\xi}$.

Then, since both I and I_1 are integrals over the whole range of the velocities $\boldsymbol{\xi}_1$or $\boldsymbol{\xi}$,

$$I_1 = I. \tag{10.30}$$

Consider now that we interchange $\xi \leftrightarrow \xi'$ and $\xi_1 \leftrightarrow \xi_1'$. With this transformation

$$\boldsymbol{\xi}' = \boldsymbol{\xi} + \cos\theta\boldsymbol{\alpha}, \tag{10.31}$$

becomes

$$\underbrace{\boldsymbol{\xi}}_{before} = \underbrace{\boldsymbol{\xi}'}_{after} - \cos\theta\boldsymbol{\alpha}, \tag{10.32}$$

and

$$\underbrace{\boldsymbol{\xi}_1}_{before} = \underbrace{\boldsymbol{\xi}_1'}_{after} + \cos\theta\boldsymbol{\alpha}. \tag{10.33}$$

In addition

$$\gamma' = \underbrace{\boldsymbol{\xi}_1}_{before} - \underbrace{\boldsymbol{\xi}}_{before} = \underbrace{\underbrace{\boldsymbol{\xi}'_1}_{after} - \underbrace{\boldsymbol{\xi}'}_{after}}_{\gamma} + 2\cos\theta\boldsymbol{\alpha}, \tag{10.34}$$

and since the only effect of this transformation is that α changes its sign, the same functional relation F is kept as before but with a change in sign, because the amount $ff_1 - f'f'_1 = -\,(f'f'_1 - ff_1)$

$$I_2 = \underbrace{\int \left(\int \left[ff_1 - f'f'_1 \right] gbdbd\varepsilon d\boldsymbol{\xi}'_1 \right) \varphi\left(\boldsymbol{\xi}'\right)}_{-F(\boldsymbol{\xi}')} d\boldsymbol{\xi}' = -\,I. \tag{10.35}$$

For the same reasons above

$$I_3 = \underbrace{\int \left(\int \left[ff_1 - f'f'_1 \right] gbdbd\varepsilon d\boldsymbol{\xi}' \right) \varphi\left(\boldsymbol{\xi}'_1\right)}_{-F(\boldsymbol{\xi}'_1)} d\boldsymbol{\xi}'_1 = I_2. \tag{10.36}$$

Consequently,

$$\int \left(\int \left[f'f'_1 - ff_1 \right] gbdbd\varepsilon d\boldsymbol{\xi}_1 \right) \varphi\left(\boldsymbol{\xi}\right) d\boldsymbol{\xi}$$
$$= \frac{1}{4} \int \left[f'f'_1 - ff_1 \right] \left[\varphi + \varphi_1 - \varphi' - \varphi'_1 \right] gbdbd\varepsilon d\boldsymbol{\xi} d\boldsymbol{\xi}_1. \tag{10.37}$$

10.2.5 The H-Theorem

Consider

$$\varphi = \ln f,$$

in Eq. (10.37) . When $f'f'_1 \neq ff_1$, Eq. (10.37) results in an inequality,

$$I = \frac{1}{4} \int \left[f'f'_1 - ff_1 \right] \ln \left(\frac{ff_1}{f'f'_1} \right) gbdbd\varepsilon d\boldsymbol{\xi} d\boldsymbol{\xi}_1 \leq 0,$$

since for $x > 0$ and $y > 0$

$$(x - y) \ln \frac{y}{x} \leq 0. \tag{10.38}$$

We consider, now, the Boltzmann equation

$$\begin{aligned} &\partial_t f + \xi \cdot \partial_{\mathbf{x}} f + \mathbf{g} \cdot \partial_\xi f \\ &= \int\int\int \left[f' f_1' - f f_1 \right] \gamma b db d\varepsilon d\boldsymbol{\xi}_1. \end{aligned} \tag{10.39}$$

By multiplying the Boltzmann equation by $(1 + lnf)$ and integrating it in the velocity space we obtain:

(a) For the time derivative term

$$\begin{aligned} \int (1 + \ln f)\, \partial_t f d\boldsymbol{\xi} &= \partial_t \int f d\boldsymbol{\xi} + \int (\partial_t f) \ln f d\boldsymbol{\xi} \\ &= \partial_t \int f d\boldsymbol{\xi} + \int \partial_t (f \ln f)\, d\boldsymbol{\xi} - \underbrace{\int f \partial_t (\ln f)\, d\boldsymbol{\xi}}_{= \int f \frac{\partial_t (f)}{f} d\xi} \\ &= \partial_t \int (f \ln f)\, d\boldsymbol{\xi}. \end{aligned} \tag{10.40}$$

We define

$$\mathrm{H} = \int f \ln f d\boldsymbol{\xi}.$$

(b) For the spatial gradient term

$$\begin{aligned} \int (1 + \ln f)\, \xi_\alpha \partial_\alpha f d\boldsymbol{\xi} &= \partial_\alpha \int \xi_\alpha f d\boldsymbol{\xi} + \int \xi_\alpha (\partial_\alpha f) \ln f d\boldsymbol{\xi} \\ &= \partial_\alpha \int \xi_\alpha f d\boldsymbol{\xi} + \int \xi_\alpha \partial_\alpha (f \ln f)\, d\boldsymbol{\xi} - \underbrace{\int \xi_\alpha f \partial_\alpha (\ln f)\, d\boldsymbol{\xi}}_{= \int \xi_\alpha f \frac{\partial_\alpha (f)}{f} d\xi} \\ &= \partial_\alpha \int \xi_\alpha (f \ln f)\, d\boldsymbol{\xi}. \end{aligned}$$

Now we decompose ξ_α into an advection velocity v_α and a peculiar velocity $\xi_\alpha - v_\alpha$

$$\xi_\alpha = v_\alpha + (\xi_\alpha - v_\alpha).$$

So

$$\begin{aligned}&\partial_\alpha \int \xi_\alpha (f \ln f)\, d\boldsymbol{\xi}\\ &= \partial_\alpha \left(v_\alpha \underbrace{\int (f \ln f)\, d\boldsymbol{\xi}}_{=H} \right) + \partial_\alpha \left(\underbrace{\int (\xi_\alpha - v_\alpha)(f \ln f)\, d\boldsymbol{\xi}}_{=j_\alpha(H)} \right). \end{aligned} \tag{10.41}$$

(c) For the velocity gradient term, since the acceleration g_α is only dependent on the spatial coordinate

$$\int (1 + \ln f)\, g_\alpha \partial_{\xi_\alpha} f d\boldsymbol{\xi} = g_\alpha \int \partial_{\xi_\alpha} f d\boldsymbol{\xi} + g_\alpha \int \ln f \partial_{\xi_\alpha} f d\boldsymbol{\xi}.$$

Now, remark that

$$\partial_{\xi_\alpha} (f \ln f) = \ln f \partial_{\xi_\alpha} f + \partial_{\xi_\alpha} f,$$

and so

$$\begin{aligned}&\int (1 + \ln f)\, g_\alpha \partial_{\xi_\alpha} f d\boldsymbol{\xi}\\ &= g_\alpha \int \partial_{\xi_\alpha} (f \ln f)\, d\boldsymbol{\xi}\\ &= 0.\end{aligned} \tag{10.42}$$

Therefore, from Eqs. (10.40), (10.41) and (10.42)

$$\partial_t \mathrm{H} + \nabla \cdot (\mathbf{u}\mathrm{H}) = -\nabla \cdot (\mathbf{j}\,(\mathrm{H})) + S\,(\mathrm{H})\,, \tag{10.43}$$

where

$$S(H)=\frac{1}{4}\int \left[f'f_1' - ff_1\right] \ln \frac{ff_1}{f'f_1'} gbdbd\varepsilon d\boldsymbol{\xi}_1 d\boldsymbol{\xi} \le 0. \tag{10.44}$$

is a source of H.

Consider now that Eq. (10.43) is integrated inside a material volume $V(t)$. From the Reynolds transport equation

$$\int_{V(t)} \partial_t \mathrm{H} dV = \frac{d}{dt}\int_{V(t)} \mathrm{H} dV - \int_{A(t)} \mathrm{H}\mathbf{v} \cdot \mathbf{n} dA.$$

Using the divergence theorem

$$\int_{V(t)} \nabla \cdot (\mathbf{v}\mathrm{H})\, dV = \int_{A(t)} \mathrm{H}\mathbf{v} \cdot \mathbf{n} dA,$$

and

$$\int_{V(t)} \nabla \cdot (\mathbf{j}\,(\mathrm{H}))\, dV = \int_{A(t)} \mathbf{j}\,(H) \cdot \mathbf{n} dA,$$

where $A(t)$ is the surface that encloses the material volume $V(t)$.

So, when integrated Eq. (10.43) becomes

$$\frac{d}{dt}\int_{V(t)} \mathrm{H} dV = -\int_{A(t)} \mathbf{j}\,(\mathrm{H}) \cdot \mathbf{n} dA + \underbrace{\int_{V(t)} S\,(\mathrm{H})\, dV}_{\le 0}. \tag{10.45}$$

Equation (10.45) is the celebrated H-theorem of Boltzmann which states that for an isolated system, when there is no flux of H through $A(t)$, $\mathbf{j}\,(\mathrm{H}) = 0$ for $\mathbf{x} \in A$, the total

$$\mathcal{H} = \int_{V(t)} \mathrm{H} dV,$$

decreases with time in non-equilibrium states and has a minimum when $f'f_1' = ff_1$, in equilibrium states.

This finding gave a statistical meaning to the entropy of a mechanical system of particles. On a molecular basis, the local entropy can be written as

$$s = -k\bar{\mathrm{H}} = -k \int \bar{f} \ln \bar{f} d\boldsymbol{\xi}_0,$$

where $\boldsymbol{\xi}_0 = \boldsymbol{\xi}/\bar{\xi}$ and $\bar{f} = f\bar{\xi}^D/n$ are dimensionless variables.

10.2.6 Macroscopic Transport Equations

When the Boltzmann equation, Eq. (10.22), is multiplied by the mass m of each molecule and integrated in the velocity space we retrieve the mass conservation equation

$$\partial_t \rho + \nabla \cdot (\rho \mathbf{v}) = 0, \tag{10.46}$$

The momentum balance equations are retrieved as first-order moments of Eq. (10.22)

$$\partial_t (\rho \mathbf{v}) + \nabla \cdot (\rho \mathbf{v}\mathbf{v} + P\boldsymbol{\delta} + \boldsymbol{\tau}) = \rho \mathbf{g}^{(e)}, \tag{10.47}$$

where $P = nkT$ is the thermodynamic pressure and

$$\tau_{\alpha\beta} = \int m \left(\xi_\alpha - v_\alpha\right) \left(\xi_\beta - v_\beta\right) f^{neq} d\boldsymbol{\xi},$$

is the viscous stress tensor or the flux of the peculiar momentum $m\left(\xi_\alpha - v_\alpha\right)$ that is transported along the direction β by velocity fluctuations, and

$$f^{neq} = f - f^{eq},$$

is the non-equilibrium part of the distribution f.

The thermodynamic internal energy is the statistical average of the peculiar kinetic energy

$$\left\langle e_{c,f} \right\rangle = \frac{\int \frac{1}{2} m \left(\boldsymbol{\xi} - \mathbf{v}\right)^2 f d\boldsymbol{\xi}}{\int f d\boldsymbol{\xi}},$$

and the internal energy balance equation

$$\partial_t \left(\rho \left\langle e_{c,f} \right\rangle\right) + \nabla \cdot \left(\rho \left\langle e_{c,f} \right\rangle \mathbf{v} + \mathbf{q}\right) = -\boldsymbol{\tau} : \nabla \mathbf{v} - P \nabla \cdot \mathbf{v}, \tag{10.48}$$

may be retrieved by subtracting the balance equation for the advection energy

$$\partial_t \left(\frac{1}{2}\rho v^2\right) + \nabla \cdot \left(\frac{1}{2}\rho v^2 \mathbf{v} + P\mathbf{v} + \boldsymbol{\tau} \cdot \mathbf{v}\right) = \rho \mathbf{g}^{(e)} \cdot \mathbf{v} + \boldsymbol{\tau} : \nabla \mathbf{v} + P \nabla \cdot \mathbf{v}, \tag{10.49}$$

from the kinetic energy balance equation

$$\partial_t \left(\rho \left\langle e_c \right\rangle\right) + \nabla \cdot \left(\rho \left\langle e_c \right\rangle \mathbf{v} + P\mathbf{v} + \boldsymbol{\tau} \cdot \mathbf{v} + \mathbf{q}\right) = \rho \mathbf{g}^{(e)} \cdot \mathbf{v}. \tag{10.50}$$

In the above equations,

$$\left\langle e_c \right\rangle = \frac{\int \frac{1}{2} m \xi^2 f d\boldsymbol{\xi}}{\int f d\boldsymbol{\xi}},$$

is the statistical average of the kinetic energy and

$$\mathbf{q} = \int \frac{1}{2} m \left(\boldsymbol{\xi} - \mathbf{v}\right)^2 \left(\boldsymbol{\xi} - \mathbf{v}\right) f d\boldsymbol{\xi},$$

is the flux of the peculiar kinetic energy transported by velocity fluctuations, corresponding to the Fourier heat flow vector.

10.2.7 Exercises

1. Starting from the Liouville Eq. (10.1), try to derive Eq. (10.2), considering the particles to be indistinguishable.
2. Maxwell [115] made a fully statistical model of gases, in which, in equilibrium, the velocities were "*distributed according to the same formula as the errors are distributed* in observations (the normal distribution)"

$$f^{eq} = A e^{-\frac{\xi_f^2}{B}},$$

where $\xi_f = \xi - \mathbf{v}$ is the peculiar molecular velocity and A and B can be determined by considering that the number density of molecules is given by

$$n = \int f^{eq} d\xi = A \int_{-\infty}^{\infty} \int_{-\infty}^{\infty} \int_{-\infty}^{\infty} e^{-\frac{\xi_{f,x}^2 + \xi_{f,y}^2 + \xi_{f,z}^2}{B}} d\xi_x d\xi_y d\xi_z,$$

and that the kinetic energy related to the '*molecular agitation*' is given by

$$\mathbf{e}_{c,f} = \int f^{eq} \frac{1}{2} m \xi_f^2 d\xi$$
$$= \int_{-\infty}^{\infty} \int_{-\infty}^{\infty} \int_{-\infty}^{\infty} A e^{-\frac{\xi_{f,x}^2 + \xi_{f,y}^2 + \xi_{f,z}^2}{B}} \frac{1}{2} m \left(\xi_{f,x}^2 + \xi_{f,y}^2 + \xi_{f,z}^2\right) d\xi_x d\xi_y d\xi_z.$$

Find A and B in terms of the number density n and the mean kinetic energy of fluctuations $\mathbf{e}_{c,f}$. What are the independent variables of n and $\mathbf{e}_{c,f}$?
Hint. Define new integration variables such as $\frac{\xi_{f,x}}{\sqrt{B}} = x$, $\frac{\xi_{f,y}}{\sqrt{B}} = y$, $\frac{\xi_{f,z}}{\sqrt{B}} = z$ and remember that

$$\int_{-\infty}^{\infty} e^{-x^2} dx = \sqrt{\pi}, \quad \int_{-\infty}^{\infty} e^{-x^2} x^2 dx = \frac{1}{2}\sqrt{\pi}.$$

3. Boltzmann derived the equilibrium distribution in the form

$$f^{(eq)} = e^a e^{-b(\xi - \mathbf{c})^2} = d e^{-b(\xi - \mathbf{c})^2}.$$

Find d, b and $\mathbf{c}$ using

$$n = \int f^{(eq)} d\xi = \text{number density of molecules},$$
$$\mathbf{v} = \frac{1}{n} \int f^{(eq)} \xi d\xi = \text{mean molecular velocity},$$
$$e_{c,f} = \frac{1}{n} \int f^{(eq)} \frac{1}{2} m \left(\xi - \mathbf{v}\right)^2 d\xi = \text{mean peculiar kinetic energy}.$$

4. Consider a material volume $\mathrm{V}(t)$, i.e., a volume that is composed of the same molecules. The total number of molecules in this volume is

$$N = \int_{V(t)} n\left(\mathbf{x}, t\right) dV,$$

and the time derivative of N is null

$$\frac{dN}{dt} = \frac{d}{dt} \int_{V(t)} n dV = 0,$$

Use the Reynolds transport theorem

$$\frac{d}{dt}\int_{V(t)} n dV = \int_{V(t)} \frac{\partial n}{\partial t} dV + \int_{A(t)} n\mathbf{v}\cdot\mathbf{e}_n dA$$

where $A\ (t)$ is the area of the surface around $\mathrm{V}(t)$ and $\mathbf{e}_n$ is the unitary vector pointing outside $A\ (t)$, to show that

$$\frac{\partial \rho}{\partial t} + \nabla\cdot(\rho\mathbf{v}) = 0 \tag{10.51}$$

where $\rho = nm$ and m is the mass of each molecule.

5. Equation (10.51) can be also written as

$$\frac{\partial \rho}{\partial t} + \mathbf{v}\cdot\nabla\rho = -\rho\nabla\cdot\mathbf{v}$$

Consider now a Taylor expansion of $\rho\ (\mathbf{x}+\Delta\mathbf{x}, t+\Delta t)$ around $\rho\ (\mathbf{x}, t)$

$$\rho\ (\mathbf{x}+\Delta\mathbf{x}, t+\Delta t) = \rho\ (\mathbf{x}, t) + \frac{\partial \rho}{\partial t}\Delta t + \nabla\rho\cdot\Delta\mathbf{x} + \cdots$$

In the limit $\Delta t \to 0$ we get

$$\frac{\rho\ (\mathbf{x}+\Delta\mathbf{x}, t+\Delta t) - \rho\ (\mathbf{x}, t)}{\Delta t} = \frac{\partial \rho}{\partial t} + \mathbf{v}\cdot\nabla\rho$$

Show that when $\nabla\cdot\mathbf{v} = 0$, the density does not vary along a fluid trajectory in the *physical space* (we say that the flow is incompressible).

6. A system of particles is described by giving the expected value $f\ (\mathbf{x}, \xi, t) = f\left(x, y, z, \xi_x, \xi_y, \xi_z, t\right)$ of the number of molecules inside an elementary volume $d\mathbf{x} = dxdydz$ that have velocities between ξ_x and $\xi_x + d\xi_x$, ξ_y and $\xi_y + d\xi_y$, ξ_z and $\xi_z + d\xi_z$. Based on Exercise 4 and using words, try to give a physical meaning for

$$\lim_{\Delta t\to 0}\frac{f(\mathbf{x}+\Delta\mathbf{x}, \xi+\Delta\xi, t+\Delta t) - f(\mathbf{x}, \xi, t)}{\Delta t},$$

when this system is in equilibrium.

7. The pressure tensor P can be written as the flux of the momentum $m\ (\xi_\alpha - v_\alpha)$ that is transported along a direction β with the velocity fluctuation $\left(\xi_\beta - v_\beta\right)$

$$P_{\alpha\beta} = \int fm\left(\xi_\alpha - v_\alpha\right)\left(\xi_\beta - v_\beta\right)d\xi. \quad (10.52)$$

(a) Show that *in equilibrium*, this tensor is isotropic

$$P_{\alpha\beta} = P\delta_{\alpha\beta}$$

and reduces to the thermodynamic pressure $P = nkT$ (*the Clapeyron equation*).

(b) Under the light of Eq. (10.52) try to describe the meaning of an isotropic tensor, using your own words.

(c) Consider a 2D case and the integral

$$\int_{\xi_y=0}^{\infty}\int_{\xi f_x=-\infty}^{\infty} f^{eq} m\xi_{fx}\xi_y d\xi_x d\xi_y$$

giving the momentum $m\xi_{fx}$ that is transported into the positive direction ξ_y and the integral

$$\int_{\xi_y=-\infty}^{0}\int_{\xi f_x=-\infty}^{\infty} f^{eq} m\xi_{fx}\xi_y d\xi_x d\xi_y$$

giving the momentum $m\xi_{fx}$ that is transported into the negative direction ξ_y and answer the question:
In equilibrium, the momentum $m\xi_{fx}$ is transferred along direction y?...Justify your answer.

10.3 Kinetic Models for the Collision Term

As originally derived, the Boltzmann equation is restricted to material points considered sources of repulsion fields and collisions are binary involving, solely, a pair of particles. It is thus limited to rarefied gases.

Molecules are not material points and, in a liquid, the mean free path has the same order of magnitude as the molecular diameter, multiple collisions are frequent, and long-range interactions are important.

Considering their molecular nature and the great difficulty of the treatment of multiple collisions, the numerical simulation of two-phase systems and solutions is still a scientific challenge.

A kinetic model may be understood as a model that, in the reverse sense uses the information of the macroscopic scale to improve the simplifying assumptions that were required when deriving a model considering solely the information of the molecular scale.

10.3.1 The BGK Collision Model

The Boltzmann equation, Eq. (10.22) is an integrodifferential equation and the evaluation of the collision term Ω for each point and time is a very difficult task and never solved except for very simple problems. This gave rise to kinetic models that replace the integral form for Ω in Eq. (10.22), by a simpler model. The more widely used model for Ω is the *Bhatnagar, Gross and Krook*, BGK model [10]

$$\Omega = \frac{f^{eq} - f}{\tau}, \tag{10.53}$$

meaning that the Boltzmann collision term is replaced by a relaxation term with a single relaxation time τ. Although much simpler, this model preserves the main effect of collisions leading the distribution of particles towards equilibrium, when a fluid is in a non-equilibrium state.

Kinetic models with the BGK relaxation term have a single independent parameter, the relaxation time τ. In the numerical simulation of isothermal, incompressible flow, this parameter can be related to the kinematic viscosity υ of the fluid, but the BGK model is unable to be used in the simulation of non-isothermal flows, except in the case of a unitary Prandtl number.

If we consider numerical simulations performed with discrete forms of the Boltzmann equation (the lattice-Boltzmann method-LBM), the strongest weakness of the BGK model is, nevertheless, that, in the course of the simulation, it very soon leads to instability issues.

In this field, several alternatives have been proposed in the last two decades for avoiding or reducing these instability issues.

As a first alternative, we can replace the BGK single relaxation-time collision term with a sum of relaxation terms for the non-equilibrium moments. This method is known as the multiple relaxation times method (MRT). This method includes physical relaxation times related to the relaxation of the viscous stress moment and supplementary high-order relaxation times without any physical meaning but introduced into the model for increasing stability, the corresponding relaxation times being tuned with the help of a linear stability analysis [40–42, 96, 188].

The entropic LB scheme [84] appears as a second solution and was conceived based on the maximization of entropy by locally tuning the single relaxation time of the Bhatnagar–Gross– Krook (BGK) collision model. Recently, a new extension of LB schemes was proposed, namely, the entropic stabilizer [83]. Unlike the entropic LB scheme, the entropic stabilizer does not locally alter the viscosity but rather relies on modifying the relaxation time for the higher-order moments (i.e., the moments beyond the stress tensor) which do not contribute to the viscosity. In this respect, this extension is akin to the already mentioned relaxation parameter tuning for MRT schemes.

The stability of lattice-Boltzmann schemes can also be improved by adopting larger discrete velocity sets. This option was investigated by Siebert et al. [158]. Another alternative is to add high-order Hermite polynomial tensors to the equilibrium distribution, trying to reduce the effect of their related moments on stability. This approach is commonly used in MRT and entropic models and was utilized by Siebert et al. [158] for improving the stability of second-order lattice-Boltzmann equations (LBE). One can also resort to regularized LBEs, i.e., rewrite a LBE in such a manner as to filter the undesirable ghost moments from the numerical scheme (see, e.g., [3] for a further discussion on the ghost moments). This choice was investigated by Ladd [95], Chen et al. [31], Zhang et al. [192] and by Latt and Chopard [100] for second-order models; very recently this method received contributions from Mattila et al. [113], Malaspinas [109], Mattila et al. [114], Hegele et al. [81] and Bazarin et al. [9].

In the next section, we present a method for building collision models with two relaxation times. The main idea is the same as the MRT method, i.e., to relax each non-equilibrium moment using independent relaxation parameters, but the purpose is here to use this model for non-isothermal problems in such a way as to work with independent parameters for describing the viscous and thermal diffusivities.

10.3.2 Beyond BGK

In its dimensionless form, the equilibrium distribution can be written as

$$\bar{f}^{eq} = \frac{f^{eq}\bar{\xi}^D}{n_0},$$

where n_0 is a number density of reference.

Writing the distribution f in, also, a dimensionless form

$$\bar{f} = \frac{f\bar{\xi}^D}{n_0},$$

the Boltzmann equation, Eq. (10.22). becomes,

$$\partial_t \bar{f} + \boldsymbol{\xi} \cdot \nabla \bar{f} + \left(\mathbf{g}^{(e)} + \mathbf{g}^{(\ell d)}\right) . \partial_{\xi} \bar{f} = \bar{\Omega}. \tag{10.54}$$

When $\bar{f}$ is near $\bar{f}^{eq}$, we write the collision term as

$$\bar{\Omega} = \mathscr{L}\left(\bar{f}\right),$$

where $\mathscr{L}$ is a linear operator.

We now write the distribution

$$\bar{f} = \bar{f}^{eq} + \bar{f}^{neq},$$

as the sum of an equilibrium and a non-equilibrium distribution.

Since $\mathscr{L}\left(\bar{f}^{eq}\right) = 0$, the collision operator becomes

$$\bar{\Omega} = \mathscr{L}\left(\bar{f}^{neq}\right), \tag{10.55}$$

as would be expected, since the equilibrium distribution is the solution of $\bar{\Omega} = 0$.

The distribution $\bar{f}^{neq}$ can be written in terms of an orthogonal basis of the Hilbert space $\mathscr{H}$ that maps the velocity space into the positive real numbers set $\mathscr{R}^{+}$. Due to its cartesian symmetry, we choose the Hermite polynomials set

$$\bar{f}^{neq} = \omega_0 \sum_{\theta} \frac{a_{\theta,(r_\theta)}}{\left(H_{\theta,(r_\theta)}\right)^2} H_{\theta,(r_\theta)}\left(\boldsymbol{c}_0\right),$$

where $\omega_0 = \frac{e^{-\frac{c_0^2}{2}}}{(2\pi)^{\frac{D}{2}}}$ is a weight function, and the Hermite polynomial tensor $H_{\theta,(r_\theta)}\left(\boldsymbol{c}_0\right)$ is written in terms of the peculiar velocity. The symbol (r_θ) designate a set of indices $r_1, r_2, \ldots, r_\theta$. For instance, $H_{2,xx}$, $H_{2,xy}$,$H_{2,xz}$,$H_{2,yy}$, $H_{2,yz}$, $H_{2,zz}$ are all designated by $H_{2,(r_2)}$ and, for each θ, we use Einstein notation for repeated symbols (r_θ). Therefore, for $\theta = 2$

$$\frac{a_{\theta,(r_\theta)}}{\left(H_{\theta,(r_\theta)}\right)^2} H_{\theta,(r_\theta)} = \frac{a_{2,\alpha\beta}}{\left(H_{2,\alpha\beta}\right)^2} H_{2,\alpha\beta}.$$

In two-dimensions

$$\frac{a_{2,\alpha\beta}}{\left(H_{2,\alpha\beta}\right)^2} H_{2,\alpha\beta} = \frac{a_{2,xx}}{2} H_{2,xx} + \frac{a_{2,yy}}{2} H_{2,yy} + \frac{a_{2,xy}}{1} H_{2,xy}$$

The norm of $H_{\theta,(r_\theta)}$is given by

$$\| H_{\theta,(r_\theta)} \|^2 = \int \omega_0 H^2_{\theta,(r_\theta)} d\mathbf{c}_0.$$

So, the Hermitian central moments $a_{\theta,(r_\theta)}$ can be found as

$$a_{\theta,(r_\theta)} = \int \bar{f}^{neq} H_{\theta,(r_\theta)} d\mathbf{c}_0. \tag{10.56}$$

On the other hand, following Eq. (10.55) the collision operator $\bar{\Omega}$ can be written as

$$\bar{\Omega} = \sum_\theta \frac{a_{\theta,(r_\theta)}}{\left(H_{\theta,(r_\theta)}\right)^2} \mathscr{L}\left(\omega_0 H_{\theta,(r_\theta)}\right). \tag{10.57}$$

Since $\mathscr{L}$ is a linear operator, it maps the Hilbert space $\mathscr{H}$ onto the same Hilbert space $\mathscr{H}$ and $\mathscr{L}\left(\omega_0 H^{(\theta)}_{n_x,n_y,n_z}\right)$can be written in terms of the Hermitian basis of $\mathscr{H}$

$$\mathscr{L}\left(\omega_0 H_{\theta,(r_\theta)}\right) = \omega_0 \gamma_{(r_\theta),(s_\theta)} H_{\theta,(s_\theta)}, \tag{10.58}$$

where $\gamma_{(r_\theta),(s_\theta)}$designates the (r_θ) , (s_θ) components of a 2θ-rank relaxation tensor. For instance, $\gamma_{xxxx}, \gamma_{xxxy}, \gamma_{xyxx}, \gamma_{xxyy}$, and γ_{yyyy} for $\theta = 2$ in two-dimensional spaces.
These tensors can be found as

$$\gamma_{(r_\theta),(s_\theta)} = \frac{\int H_{\theta,(s_\theta)} \mathscr{L}\left(\omega_0 H_{\theta,(r_\theta)}\right) d\mathbf{c}_0}{\| H_{\theta,(s_\theta)} \|^2},$$

due to the orthogonality of the Hermitian polynomials .
Replacing Eq. (10.58) into Eq. (10.57)

$$\bar{\Omega} = \omega_0 \sum_\theta \gamma_{(r_\theta),(s_\theta)} \frac{a_{\theta,(r_\theta)}}{\| H_{\theta,(r_\theta)} \|^2} H_{\theta,(s_\theta)}. \tag{10.59}$$

This sum can be rewritten as

$$\bar{\Omega} = \omega_0 \left(\sum_{\theta=0}^{N} \gamma_{(r_\theta),(s_\theta)} \frac{a_{\theta,(r_\theta)}}{\| H_{\theta,(r_\theta)} \|^2} H_{\theta,(s_\theta)} + \sum_{\theta=N+1}^{\infty} \gamma_{(r_\theta),(s_\theta)} \frac{a_{\theta,(r_\theta)}}{\| H_{\theta,(r_\theta)} \|^2} H_{\theta,(s_\theta)} \right). \tag{10.60}$$

Now, following Gross-Jackson [62], we diagonalize this expansion after a certain order N by replacing the tensor $\gamma_{(r_\theta),(s_\theta)}$ by $\gamma_N \delta_{(r_\theta),(s_\theta)}$

$$\bar{\Omega} = \omega_0 \left(\sum_{\theta=0}^{N} \left(\gamma_{(r_\theta),(s_\theta)} - \gamma_N \delta_{(r_\theta),(s_\theta)} \right) \frac{a_{\theta,(r_\theta)}}{\| H_{\theta,(r_\theta)} \|^2} H_{\theta,(s_\theta)} \right)$$

$$+ \gamma_N \omega_0 \left(\sum_{\theta=0}^{N} \frac{a_{\theta,(s_\theta)}}{\| H_{\theta,(r_\theta)} \|^2} H_{\theta,(s_\theta)} + \sum_{\theta=N+1,s_\theta}^{\infty} \frac{a_{\theta,(s)}}{\| H_{\theta,(r_\theta)} \|^2} H_{\theta,(s_\theta)} \right)$$

or

$$\bar{\Omega} = -\omega_0 \left(\sum_{\theta=0}^{N} \lambda_{(r_\theta),(s_\theta)} \frac{a_{\theta,(r_\theta)}}{\| H_{\theta,(r_\theta)} \|^2} H_{\theta,(s_\theta)} \right) - \lambda_N \bar{f}^{neq} \tag{10.61}$$

where

$$\lambda_{(r_\theta),(s_\theta)} = -\left(\gamma_{(r_\theta),(s_\theta)} - \gamma_N \delta_{(r_\theta),(s_\theta)} \right), \quad \lambda_N = -\gamma_N.$$

The eigenvalues of $\mathscr{L}$ are null for H_0, $H_{1,\alpha}$ and non-positive for $\theta \geq 2$ as a consequence of the Boltzmann H-theorem [62]. In addition, $\lambda_{(r_\theta),(s_\theta)} = -\gamma_{(r_\theta),(s_\theta)}$ for all off-diagonal components and $\gamma_{(r_\theta),(r_\theta)}$ is greater in absolute value than γ_N for the diagonal components when $\theta \leq N$. Therefore,

$$\lambda_{(r_\theta),(s_\theta)} > 0, \quad \lambda_N > 0,$$

for all $\theta \leq N$.

Equation (10.61) can be considered as an Nth-order kinetic model for the collision term, with an absorption term $\lambda_N \bar{f}^{neq}$ resulting from the diagonalization of the relaxation tensors after a given order N, all the moments of order higher than N being collapsed into a single non-equilibrium term.

When N = 0 or N = 1, Eq. (10.61) gives the well-known BGK model, when all the collision operator spectra are replaced by a single relaxation term.

Each term in the sum, Eq. (10.61), gives the relaxation to the equilibrium of second or higher-order kinetic moments that are not preserved in collisions.

10.3.2.1 Hermitian Non-equilibrium Moments

The tensors $a_{\theta,(r_\theta)} = \{a_0, a_{1,\alpha}, a_{2,\alpha\beta}, a_{3,\alpha\beta\gamma}, \ldots\}$given by Eq. (10.56) are the Hermitian non-equilibrium moments. Both a_0 and $a_{1,\alpha}$ are null because ρ^* and $\mathbf{v}_0$ are equilibrium moments,

On the other hand, the viscous stress tensor is given by

$$\tau_{\alpha\beta} = \int f^{neq} m c_\alpha c_\beta d\mathbf{c} = \rho_0 \bar{\xi}^2 \int \bar{f}^{neq} c_{0,\alpha} c_{0,\beta} d\mathbf{c}_0,$$

and so, the dimensionless viscous stress tensor can be written as

$$\tau_{0,\alpha\beta} = \frac{\tau_{\alpha\beta}}{\rho_0 \bar{\xi}^2} = \int \bar{f}^{neq} c_{0,\alpha} c_{0,\beta} d\mathbf{c}_0.$$

Now, from Eq. (10.56)

$$a_{2,\alpha\beta} = \int \bar{f}^{neq} H_{2,\alpha\beta} d\mathbf{c}_0 = \tau_{0,\alpha\beta}. \tag{10.62}$$

The heat flow vector gives the flux with which the peculiar energy is transported by diffusion along a given direction. Its dimensionless form is given by

$$q_{0,\alpha} = \frac{1}{2} \int \overline{f}^{neq} \mathbf{c}_0^2 c_{0,\alpha} d\mathbf{c}_0 = \frac{1}{2} \int \overline{f}^{neq} H_{3,\alpha\beta\beta} (\mathbf{c}_0) \, d\mathbf{c}_0 = \frac{1}{2} a_{3,\alpha\beta\beta},$$

as a sum of non-equilibrium third-order Hermitian moments.

10.3.2.2 Collision Models with Two Relaxation Times

We focus our analysis on second-order collision models (N = 2). In Eq. (10.61)

$$\lambda_{(r_2),(s_2)} a_{\theta,(r_2)} H_{\theta,(s_2)} = \lambda_{\alpha\beta\gamma\delta} a_{2,\alpha\beta} H_{2,\gamma\delta}.$$

For Newtonian fluids, we can require isotropy of the 4th-rank tensor $\lambda_{\alpha\beta\gamma\delta}$ writing these tensors as

$$\lambda_{\alpha\beta\gamma\delta} = \lambda_1 \delta_{\alpha\beta} \delta_{\gamma\delta} + \lambda_2 \delta_{\alpha\gamma} \delta_{\beta\delta} + \lambda_3 \delta_{\alpha\delta} \delta_{\beta\gamma}.$$

Therefore,

$$\lambda_{\alpha\beta\gamma\delta} a_{2,\alpha\beta} H_{2,\gamma\delta} = \lambda_1 a_{2,\alpha\alpha} H_{2,\gamma\gamma} + (\lambda_2 + \lambda_3) \, a_{2,\alpha\beta} H_{2,\alpha\beta}.$$

From Eq. (10.62), it is seen that the term $a_{2,\alpha\alpha}$ is related to the trace of the viscous stress tensor and so

$$a_{2,\alpha\alpha} = 0,$$

because collisions must preserve the kinetic energy.

Therefore,

$$\bar{\Omega} = -\frac{1}{2}\omega_0\lambda_\mu \tau_{0,\alpha\beta}\left(c_{0,\alpha}c_{0,\beta} - \delta_{\alpha\beta}\right) - \lambda_N \bar{f}^{neq}, \tag{10.63}$$

where $\lambda_\mu = \lambda_2 + \lambda_3$.

Equation (10.63) can also be written as

$$\bar{\Omega} = -\frac{1}{2}\omega_0\lambda_\mu \tau_{0,\alpha\beta}c_{0,\alpha}c_{0,\beta} - \lambda_N \bar{f}^{neq}, \tag{10.64}$$

because $\tau_{0,\alpha\beta}\delta_{\alpha\beta} = \tau_{0,\alpha\alpha}$ or the trace of $\tau_{0,\alpha\beta}$, which is null for models that are required to preserve energy.

In two dimensions the collision model would be written as

$$\bar{\Omega} = -\omega_0\lambda_\mu\left[\frac{\tau_{0,xx}}{2}c_{0,x}^2 + \frac{\tau_{0,yy}}{2}c_{0,y}^2 + \tau_{0,xy}c_{0,x}c_{0,y}\right] - \lambda_N \bar{f}^{neq}. \tag{10.65}$$

The present kinetic model with two independent relaxation frequencies is thermodynamically consistent and a Chapman-Enskog analysis showed that it is able to analyze non-isothermal and fully compressible flows [130].

10.3.3 The Stokes Hypothesis

For Newtonian fluids satisfying the conditions of isotropy and symmetry, the relationship between the viscous stress tensor $\tau_{\alpha\beta}$ and the strain rate tensor is given by

$$\tau_{\alpha\beta} = -\mu_1\left(\partial_\beta v_\alpha + \partial_\alpha v_\beta\right) + \mu_2 \nabla\cdot\mathbf{v}\delta_{\alpha\beta}$$

whose trace results in

$$\tau_{\alpha\alpha} = -\left(2\mu_1 - \mu_2 D\right)\left(\partial_\eta v_\eta\right) = -\chi D(\partial_\eta v_\eta),$$

where χ is the bulk viscosity.

For systems of particles with solely translational degrees of freedom, the trace of the viscous stress tensor is related to the kinetic energy of fluctuations, which is a conserved quantity. Therefore, the viscous stress tensor is traceless. This is the Stokes hypothesis meaning that the bulk viscosity is null. In other words,

$$\mu_2 = \frac{2}{D}\mu_1.$$

So the viscosity coefficients μ_1 and μ_2 cannot be considered as independent properties unless the kinetic model takes the internal degrees of freedom of the molecules into account.

Stokes's assumption is reasonably accurate for Newtonian fluids. So, Stokes's assumption is commonly taken as just another characteristic of Newtonian fluids.

If compression or expansion of a fluid is very rapid, such as a shock or sound wave, and the molecules have internal degrees of freedom, such as vibration or rotation, the kinetic model should take account of the several additional relaxation processes related to the internal modes.

Hanson and Morse [63] built a kinetic model for polyatomic gases by the introduction of the Wang-Chang and Uhlenbeck [185] polynomials in the Gross-Jackson expansion. These kinetic models were extended for polyatomic mixtures by Philippi and Brun [129].

At equilibrium, the internal energy should be distributed equally to all internal modes of motion. However, after a rapid change of state, the energy appears first in the translation mode, and only after several additional molecular collisions is energy distributed to the rotation and later on, after more collisions, to the vibration modes.

In certain situations (when the relaxation time is long compared to the flow time) the bulk viscosity coefficient can be used to model these nonequilibrium effects. The absorption of sound waves is such a process. Sound absorption in noble gases follows Stokes's assumption in agreement with kinetic theory. Sound absorption in the air has a nonzero bulk viscosity, but its value depends strongly on the water vapor content (which greatly modifies the relaxation times).

10.3.4 Exercises

(1) Starting with the Boltzmann equation with the BGK collision term, find the mass conservation equation and the balance equations for the momentum and internal energy for a thermodynamic system with a single component in non-equilibrium conditions.

(2) Repeat Exercise 1 when the collision term is replaced by the kinetic model given by Eq. (10.65).

10.4 Non-ideal Fluids

When molecules are close, the intermolecular attraction forces $\mathbf{g}^{(\ell d)}$ are important and deviate the fluid from the ideality. In this section, we derive a kinetic model for non-ideal fluids taking into account two other additional parameters when compared to single-component thermodynamic systems: the van der Waals bulk-phase force parameter a and the surface parameter κ in the liquid-vapor transition layer. While the first is related to the deviations of the equation of state from the ideal gas behavior, the second is the molecular parameter for the surface tension.

Since the distribution f is close to the equilibrium distribution f^{eq}, we can write the force term in the Boltzmann equation as

$$\left(\mathbf{g}^{(e)}+\mathbf{g}^{(\ell d)}\right)\cdot\partial_{\xi} f=\left(\mathbf{g}^{(e)}+\mathbf{g}^{(\ell d)}\right)\cdot\partial_{\xi} f^{eq},$$

without any effects on the macroscopic equations.

Therefore, we write the kinetic model in the form

$$\partial_t f+\boldsymbol{\xi}\cdot\partial_x f=\Omega+\mathbf{g}^{(\ell d)}\cdot\frac{(\boldsymbol{\xi}-\mathbf{u})}{\frac{kT}{m}} f^{eq}+\mathbf{g}^{(e)}\cdot\frac{(\boldsymbol{\xi}-\mathbf{u})}{\frac{kT}{m}} f^{eq}, \tag{10.66}$$

where $\mathbf{g}^{(e)}$ corresponds to the force per unit mass due to external body forces, $\mathbf{g}^{(\ell d)}$ is related to the intermolecular attraction between the molecules—which we will suppose to be only dependent on the distance between the centers of the molecules—and, Ω, to the collision term related to the short-range repulsion forces.

The term

$$\mathbf{g}^{(\ell d)}\cdot\frac{(\xi-\mathbf{v})}{\frac{kT}{m}} f^{eq}, \tag{10.67}$$

related to the long-range interaction, deserves some discussion.

Consider a molecule labelled as 1 located in point $\mathbf{x}_1$ and all the molecules labeled as 2 around it (Fig. 10.6). The attractive potential energy related to the interaction of all molecules 2 with the single molecule 1 is given by

$$\Phi_1=\int\limits_{|\mathbf{x}_2-\mathbf{x}_1|>\sigma}\Phi^{(12)} n\left(\mathbf{x}_2\right) d\mathbf{x}_2, \tag{10.68}$$

where $\Phi^{(12)}$ is the potential energy related to the electrostatic interaction between particle 1 and a single particle 2.

Defining $\eta=\mathbf{x}_2-\mathbf{x}_1$, considering that the interaction length has molecular dimensions and the number of particles per unit volume n does not have strong variations, it is possible to consider $n\left(\mathbf{x}_2\right)=n\left(\mathbf{x}_1\right)$ inside the whole integration domain and we obtain,

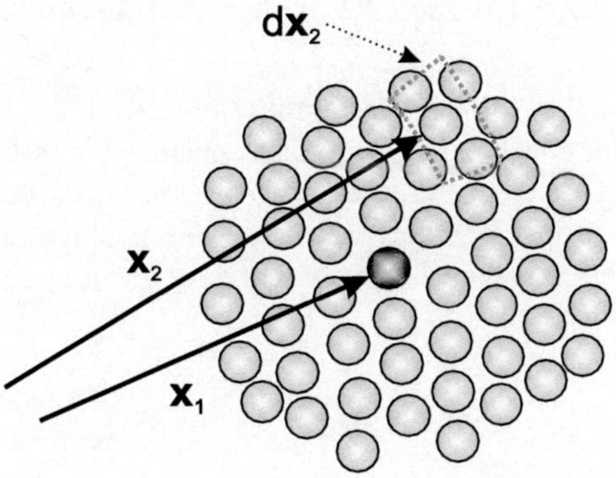

Fig. 10.6 Attractive potential related to the interaction of all molecules 2 with the single molecule 1

$$\Phi_1 = -2an(\mathbf{x}_1), \tag{10.69}$$

where

$$a = -\frac{1}{2}\int_{|\eta|>\sigma} \Phi^{(12)} d\eta, \tag{10.70}$$

is a positive constant, because $\Phi^{(12)} < 0$. This constant depends solely on the electrostatic field generated by the molecules being a molecular property of the substance that is being analyzed.

The force $\mathbf{g}^{(\ell d)}$ on molecule 1 will be thus

$$\mathbf{g}^{(\ell d)} = -\frac{\nabla\Phi_1}{m}. \tag{10.71}$$

Therefore, the Boltzmann equation takes the form

$$\partial_t f + \xi \cdot \partial_x f = \underbrace{\Omega}_{\text{repulsion}} + \underbrace{\Omega^{(\ell d)}}_{\text{intermolecular attraction}} \underbrace{+\, \mathbf{g}^{(e)} \cdot \frac{(\xi - \mathbf{v})}{\frac{kT}{m}} f^{eq}}_{\text{external body forces}}, \tag{10.72}$$

where

$$\Omega^{(\ell d)} = \frac{2a}{m}\nabla n \cdot \frac{(\xi - \mathbf{v})}{\frac{kT}{m}} f^{eq}, \tag{10.73}$$

represents the net gain of particles that acquire the velocity ξ due to the intermolecular forces from the other particles around it.

10.4.1 Two Phase Liquid-Vapor Systems

When a liquid has a phase transition, the interface liquid-vapor is too much thin to enable the density n to be considered constant throughout it, as we have done above. Due to to the interaction between a molecule in this point and all the molecules around it, when calculating the potential energy at a point $\mathbf{x}_1$ the expression to be used is the same as before, still given by Eq. (10.68)

$$\Phi_1 = \int\limits_{|\mathbf{x}_2 - \mathbf{x}_1| > \sigma} \Phi^{(12)} n\left(\mathbf{x}_2\right) d\mathbf{x}_2.$$

but $n\left(\mathbf{x}_2\right)$ is, now, dependent on $|\mathbf{x}_2 - \mathbf{x}_1|$.

Nevertheless, using $\boldsymbol{\eta} = \mathbf{x}_2 - \mathbf{x}_1$ we can consider a Taylor series for n,

$$n\left(\mathbf{x}_2\right) = n\left(\mathbf{x}_1\right) + \frac{\partial n}{\partial x_\alpha}\underbrace{\left(x_{2\alpha} - x_{1\alpha}\right)}_{=\eta_\alpha} + \frac{1}{2}\frac{\partial n}{\partial x_\alpha}\frac{\partial n}{\partial x_\beta}\underbrace{\left(x_{2\alpha} - x_{1\alpha}\right)}_{=\eta_\alpha}\underbrace{\left(x_{2\beta} - x_{1\beta}\right)}_{=\eta_\beta} + \cdots,$$

giving,

$$\begin{aligned}\Phi_1 = 2n\left(\mathbf{x}_1\right)\underbrace{\frac{1}{2}\int\limits_{|\eta|>\sigma} \Phi^{(12)} d\eta}_{=-a} \\ + \frac{\partial n}{\partial x_\alpha}\underbrace{\int\limits_{|\eta|>\sigma} \Phi^{(12)} \eta_\alpha d\eta}_{=0} \\ + \frac{1}{2}\frac{\partial n}{\partial x_\alpha}\frac{\partial n}{\partial x_\beta}\int\limits_{|\eta|>\sigma} \Phi^{(12)} \eta_\alpha \eta_\beta d\eta.\end{aligned}$$

The potential $\Phi^{(12)}$ depends solely on the norm of η being an even function of η_x, η_y and η_z. Therefore

$$\int\limits_{|\eta|>\sigma} \Phi^{(12)} \eta_\alpha \eta_\beta d\eta = \delta_{\alpha\beta} \int\limits_{|\eta|>\sigma} \Phi^{(12)} \frac{\eta^2}{3} d\eta.$$

On the other hand, in the same way as for the force parameter a, the term $\int_{|\eta|>\sigma} \Phi^{(12)} \frac{\eta^2}{3} d\eta$ depends only on the electrical properties of the molecules, being a constant

$$\int_{|\eta|>\sigma} \Phi^{(12)} \frac{\eta^2}{3} d\eta = 2 \int_{|\eta|>\sigma} \Phi^{(12)} \frac{\eta^2}{6} d\eta = -2\kappa. \tag{10.74}$$

So

$$\Phi_1 = -2an(\mathbf{x}_1) - \kappa \nabla^2 n(\mathbf{x}_1) + \cdots ,$$

and the intermolecular attraction force will be given by

$$\mathbf{g}^{(\ell d)} = -\frac{\nabla \Phi_1}{m},$$

or

$$\rho \mathbf{g}^{(\ell d)} = -n \nabla \Phi_1.$$

So

$$\rho \mathbf{g}^{(\ell d)} = 2an\nabla n + n\kappa \nabla \left(\nabla^2 n\right). \tag{10.75}$$

In this case, the kinetic equation will be written as

$$\partial_t f + \xi \cdot \partial_x f = \underbrace{\Omega}_{\text{repulsion}} + \underbrace{\Omega^{(\ell d)}}_{\text{intermolecular attraction}} \underbrace{+ \mathbf{g}^{(e)} \cdot \frac{(\xi - \mathbf{v})}{\frac{kT}{m}} f^{eq}}_{\text{external body forces}}, \tag{10.76}$$

with

$$\begin{aligned} \Omega^{(\ell d)} &= \mathbf{g}^{(\ell d)} \cdot \frac{(\xi - \mathbf{v})}{\frac{kT}{m}} f^{(eq)} \\ &= \left(\frac{2a}{m}\nabla n + \frac{\kappa}{m}\nabla\left(\nabla^2 n\right)\right) \cdot \frac{(\xi - \mathbf{v})}{\frac{kT}{m}} f^{(eq)}. \end{aligned} \tag{10.77}$$

The momentum balance equation will be also modified. Considering that,

$$\begin{aligned} n\nabla\left(\nabla^2 n\right) &= n\partial_\alpha\left(\partial_\beta\left(\partial_\beta n\right)\right) = \partial_\alpha\left[n\left(\partial_\beta\left(\partial_\beta n\right)\right)\right] - \partial_\beta\left(\partial_\beta n\right)\partial_\alpha n \\ &= \partial_\alpha\left[n\left(\partial_\beta\left(\partial_\beta n\right)\right)\right] - \partial_\beta\left[\partial_\beta n\partial_\alpha n\right] + \left(\partial_\beta n\right)\partial_\beta\partial_\alpha n, \end{aligned} \tag{10.78}$$

and

$$\left(\partial_\beta n\right)\partial_\alpha\partial_\beta n = \frac{1}{2}\partial_\alpha\left[\partial_\beta n\partial_\beta n\right], \tag{10.79}$$

we obtain

$$\begin{aligned} \rho\mathbf{g}^{(\ell d)} &= 2an\nabla n + \kappa\nabla\left(n\nabla^2 n\right) + \frac{1}{2}\kappa\nabla\left(\nabla n\cdot\nabla n\right) - \kappa\nabla\cdot\left(\nabla n\nabla n\right) \\ &= \nabla\cdot\left[\left(an^2 + \kappa n\nabla^2 n + \frac{1}{2}\kappa\left(\nabla n\cdot\nabla n\right)\right)\boldsymbol{\delta} - \kappa\left(\nabla n\nabla n\right)\right]. \end{aligned} \tag{10.80}$$

The momentum balance equation will thus have the form

$$\begin{aligned} &\partial_t\left(\rho\mathbf{v}\right) + \nabla.\left(\rho\mathbf{v}\mathbf{v}\right) \\ &= \underbrace{\rho\mathbf{g}^{(e)}}_{\text{external forces}} - \underbrace{\nabla P_s}_{\text{scalar pressure}} - \underbrace{\nabla\cdot\tau}_{\text{viscous forces}} - \underbrace{\nabla\cdot\mathbf{S}}_{\text{surface forces}}, \end{aligned} \tag{10.81}$$

where the scalar pressure is given by

$$P_s = \underbrace{\underbrace{nkT}_{\text{repulsion}} - \underbrace{an^2}_{\text{attraction}}}_{P=\text{thermodynamic pressure}} \underbrace{- \kappa n\nabla^2 n - \frac{1}{2}\kappa\nabla n\cdot\nabla n}_{\text{surface pressure}}, \tag{10.82}$$

and the surface tensor by

$$\mathbf{S} = \kappa\nabla n\nabla n. \tag{10.83}$$

The surface forces have a role solely in the interface liquid-vapor where the gradients are strong.

The kinetic equation

$$\partial_t f + \xi\cdot\partial_x f = \underbrace{\Omega}_{\text{repulsion}} + \underbrace{\Omega^{(\ell d)}}_{\text{intermolecular attraction}} \underbrace{+ \mathbf{g}^{(e)}\cdot\frac{(\xi - \mathbf{v})}{\frac{kT}{m}}f^{eq}}_{\text{external body forces}}, \tag{10.84}$$

does not consider that the molecules have a volume b. In consequence, the equation of state is given by

$$P = \underbrace{nkT}_{\text{repulsion part}} - \underbrace{an^2}_{\text{attraction part}}, \tag{10.85}$$

without any volume contribution.

In numerical simulation of kinetic models, it is important to consider this contribution to avoid the *colapse of mass*[2] at certain points in the course of the simulation.

We can, nevertheless, correct the kinetic equation in such a way as to retrieve the equation of state we want. David Enskog[3] was the first who considered a correction in the Boltzmann equation by taking the molecular volume b into account. His treatment is given in the Chapman and Cowling book [30]. For isothermal problems, Enskog's correction can be written as

$$\Omega = \underbrace{\Omega_{b=0}}_{\text{repulsion field contribution}} - \underbrace{\frac{1}{nkT}\nabla\left(nkT\left(bn\chi\right)\right)\cdot\left(\xi - \mathbf{v}\right) f^{eq}}_{\text{volume contribution}}. \tag{10.86}$$

When volume is attributed to the molecules, the momentum balance equation will thus have the form

$$\partial_t\left(\rho\mathbf{v}\right) + \nabla.\left(\rho\mathbf{v}\mathbf{v}\right) = \rho\mathbf{g}^{(e)} - \nabla P_s - \nabla\cdot\tau - \nabla\cdot\mathbf{S}, \tag{10.87}$$

where the scalar pressure is given by

$$P_s = P - \kappa n\nabla^2 n - \frac{1}{2}\kappa\nabla n\cdot\nabla n, \tag{10.88}$$

and the thermodynamic pressure P is here written in terms of the temperature T and number density n, in accordance with a van der Waals equation of state

$$P = \frac{nkT}{1 - bn} - an^2.$$

This suggests to write the kinetic model in a generic form as

$$\begin{aligned}&\partial_t f + \boldsymbol{\xi}\cdot\nabla f\\ &= \Omega_{b=0} + \left(\rho\mathbf{g}^{(e)} - \nabla\left(P - nkT\right) + n\kappa\nabla\nabla^2 n\right)\cdot\frac{(\boldsymbol{\xi} - \mathbf{v})}{nkT} f^{eq}.\end{aligned} \tag{10.89}$$

[2] The collapse of mass happens in a given point when the repulsion forces are weak and do not equilibrate the attraction forces. When this happens, the numerical scheme loses stability.

[3] David Enskog (22 April 1884, Västra Ämtervik, Sunne—1 June 1947, Stockholm) was a Swedish mathematical physicist. Enskog helped to develop the kinetic theory of gases by extending the Maxwell–Boltzmann equations.

where the relationship $P(n, T)$ can be considered as related to any arbitrary equation of state judged to be suitable for describing liquid-vapor two-phase systems.

10.4.2 Exercises

1. Using the long-range force term written as in Eq. (10.73), derive the momentum balance equation as a first-order moment of the kinetic model given by Eq. (10.72). What is the equation of state for this model?
2. Repeat Exercise 1 using the model given by Eq. (10.77) instead of the one given by Eq. (10.73). Describe what are the main modifications when the momentum balance equations are compared.
3. The momentum balance equation can be written in a generic form as

$$\partial_t (\rho \mathbf{v}) + \nabla . (\rho \mathbf{v}\mathbf{v} + \mathbf{P}) = \rho \mathbf{g}^{(e)},$$

where the pressure tensor is given by

$$P_{\alpha\beta} = \int f m \xi_\alpha \xi_\beta d\boldsymbol{\xi}.$$

Starting from Eq. (10.89) find the momentum balance equation as the first order moment of the kinetic model and the pressure tensor $\mathbf{P}$ for a two-phase liquid-vapor system.

10.5 Multicomponent Systems

The Boltzmann equation for a non-ideal r-component system can be written in continuous variables as

$$\partial_t f^{(p)} + \xi \cdot \partial_{\mathbf{x}} f^{(p)} + \left(\mathbf{g}^{(e)} + \sum_s \mathbf{g}^{(ps)} \right) \cdot \partial_\xi f^{(p)} = \Omega^{(p)}, \quad p = 1, \ldots, r, \tag{10.90}$$

where $f^{(p)}(\mathbf{x}, \xi, t)$ is the probable amount of p-particles, which at time t, dt are found inside the elementary volume between $\mathbf{x}$ and $\mathbf{x} + d\mathbf{x}$ with velocities between ξ and $\xi + d\xi$ and $\mathbf{g}^{(e)}$ is an external body force, usually the gravity force. The short-range collision term $\Omega^{(p)}$ means the net amount of p-particles that acquire velocities between ξ and $\xi + d\xi$ in the time interval dt, due to the collisions they have with the other s-particles, $s = 1, \ldots, r$ that are found inside $\mathbf{x}$ and $\mathbf{x} + d\mathbf{x}$, during this time interval. The term $\Omega^{(p)}$ can be written as a sum of $p - s$ collision terms

$$\Omega^{(p)} = \sum_{s=1}^{r} \Omega^{(ps)}, \tag{10.91}$$

In this section, a heuristic approach is adopted by writing kinetic models for $\Omega^{(ps)}$ in such way as to retrieve the macroscopic equations and the equations of state for the mixture and pure compounds. In fact, the kinetic level of description, leading to discrete LB equations, may be thought of as a two-way bridge linking the molecular and the macroscopic levels of description. Therefore, although derived from the molecular domain and, in consequence, featuring some properties of fluid interfaces that are inaccessible to the macroscopic level of description, the kinetic model may be written in such a manner as to recover the correct macroscopic equations. Along this line of approach, the repulsion term is written considering the Enskog's volume correction [30].

$$\Omega^{(ps)} = \Omega^{(ps)}_{b=0} - x^{(s)} \frac{\nabla \left[\frac{bn^2kT}{1-bn} \right]}{nkT} \cdot (\xi - \mathbf{v})\, f^{(p)}_{eq}(\mathbf{v}), \tag{10.92}$$

where

$$b = \sum_{s} x^{(s)} b^{(s)}, \tag{10.93}$$

the parameter $b^{(s)}$ is related to the volume of s-particles, $x^{(s)}$ is the number fraction of s-particles, $x^{(s)} = \frac{n^{(s)}}{n}$, $f^{(p)}_{eq}(\mathbf{v})$ is the Maxwell-Boltzmann equilibrium distribution, written around the center of mass velocity,

$$\mathbf{v} = \sum_{s} \omega^{(s)} \mathbf{v}^{(s)}, \tag{10.94}$$

$\omega^{(s)}$ is the mass fraction of species s, $\omega^{(s)} = \frac{\rho^{(s)}}{\rho}$. The term $\Omega^{(ps)}_{b=0}$ represents the short-range repulsion term when the particles are considered without volume, $b = 0$, and must satisfy the following conservation restrictions,

$$\int m^{(p)} \Omega^{(ps)}_{b=0} d\xi = 0, \tag{10.95}$$

meaning that the mass of $p-$species is not affected in their interaction with the other species,

$$\sum_{s,p} \int m^{(p)} \xi \Omega^{(ps)}_{b=0} d\xi = 0, \tag{10.96}$$

and

$$\sum_{s,p} \int \frac{1}{2} m^{(p)} \xi^2 \Omega_{b=0}^{(ps)} d\xi = 0, \tag{10.97}$$

related, respectively, to the conservation of momentum and kinetic energy in short-range interactions of material points.

The force term

$$\mathbf{g}^{(ps)} = -\frac{\nabla \Phi_m^{(ps)}}{m^{(p)}} \tag{10.98}$$

is the force per unit mass on the p-particles resulting from their intermolecular interaction with the s-particles. This force is related to the mean potential energy $\Phi_m^{(ps)}$,

$$\begin{aligned} \Phi_m^{(ps)} &= \int\limits_{|\eta|>\sigma_{ps}} n^{(s)} (\mathbf{x} + \eta)\, \Phi^{(ps)} (|\eta|)\, d\eta \\ &= -2a^{(ps)} n^{(s)} (\mathbf{x}) - \kappa^{(ps)} \nabla^2 \left[n^{(s)} (\mathbf{x}) \right] - \cdots, \end{aligned} \tag{10.99}$$

with σ_{ps} being the distance where the two-particles p-s interaction potential $\Phi^{(ps)}$ changes from repulsion to attraction,

$$a^{(ps)} = -\frac{1}{2} \int\limits_{|\eta|>\sigma_{ps}} \Phi^{(ps)} (|\eta|)\, d\eta. \tag{10.100}$$

and

$$\kappa^{(ps)} = -\frac{1}{6} \int\limits_{|\eta|>\sigma_{ps}} \eta^2 \Phi^{(ps)} (|\eta|)\, d\eta. \tag{10.101}$$

Therefore

$$\rho^{(p)} \mathbf{g}^{(ps)} = 2a^{(ps)} n^{(p)} \nabla n^{(s)} (\mathbf{x}) + \kappa^{(ps)} n^{(p)} \nabla \nabla^2 \left[n^{(s)} (\mathbf{x}) \right] \tag{10.102}$$

Finally, using $f^{(p)} = f_{eq}^{(p)} (\mathbf{v})$ for the external force term in Eq. (10.90), and Eq. (10.92), the following kinetic model is obtained

$$\partial_t f^{(p)} + \xi . \partial_{\mathbf{x}} f^{(p)}$$
$$= \sum_s \Omega_{b=0}^{(ps)}$$
$$+ \left(\rho^{(p)} \mathbf{g}^{(e)} - x^{(p)} \nabla \left(\frac{bn^2 kT}{1 - bn} \right) + \sum_s \rho^{(p)} \mathbf{g}^{(ps)} \right) \cdot \frac{(\xi - \mathbf{v})}{n^{(p)} kT} f_{eq}^{(p)} (\mathbf{v}) . \quad (10.103)$$

It is important to stress that the purpose of a kinetic model is not to solve the full Boltzmann equation itself, which is, in fact, unknown, but to consistently retrieve the macroscopic equations describing, in the present case, the non-ideal behavior of a mixture. Therefore, numerically solving a kinetic equation, must, firstly, be thought of as a method for solving a physical problem. In the reverse sense, kinetic models enable to reveal, or put in evidence, the influence of a number of molecular processes on the macroscopic behavior of a physical system.

10.5.1 Macroscopic Equations

The first moment of Eq. (10.103) can be found as

$$\partial_t \rho^{(p)} + \nabla \cdot \left(\rho^{(p)} \mathbf{v}^{(p)} \right) = 0, \quad p = 1, \ldots, r, \quad (10.104)$$

which is the advection-diffusion equation for species p.

After summing over all the components, the second moment of Eq. (10.103) is found to be

$$\partial_t (\rho \mathbf{v}) + \nabla \cdot (\rho \mathbf{v} \mathbf{v}) = \rho \mathbf{g}^{(e)} - \nabla \cdot \mathbf{P}, \quad (10.105)$$

the pressure tensor $\mathbf{P}$ satisfies

$$\nabla \cdot \mathbf{P} = \nabla \cdot \tau + \nabla \left[\frac{nkT}{1 - bn} \right] - \sum_{s,p} \rho^{(p)} \mathbf{g}^{(ps)}, \quad (10.106)$$

and the viscous stress tensor is given by,

$$\tau = \sum_p \int m^{(p)} (\xi - \mathbf{v}) (\xi - \mathbf{v}) f_{neq}^{(p)} d\xi, \quad (10.107)$$

with the non-equilibrium distribution $f_{neq}^{(p)}$ given by

$$f_{neq}^{(p)} = f^{(p)} - f_{eq}^{(p)} (\mathbf{v}) . \quad (10.108)$$

The pressure tensor is found as

$$\mathbf{P} = \boldsymbol{\tau} + P_s \boldsymbol{\delta} + \mathbf{S}, \tag{10.109}$$

where P_s is a scalar pressure, which, in the present multicomponent case, is to be written as

$$P_s = P - \frac{1}{2}\left[\nabla^2\left(\kappa n^2\right) - \sum_{p,s} \kappa^{(ps)} \nabla n^{(p)} \cdot \nabla n^{(s)}\right], \tag{10.110}$$

and $\mathbf{S}$ is the Korteweg stress tensor [91] that can be written as

$$\mathbf{S} = \sum_{p,s} \kappa^{(ps)} \nabla n^{(p)} \nabla n^{(s)}, \tag{10.111}$$

with

$$\kappa = \sum_{p,s} x^{(p)} x^{(s)} \kappa^{(ps)}. \tag{10.112}$$

The scalar pressure P_s differs from the thermodynamic pressure P, by terms whose action is restricted to physical regions where the number density $n^{(s)}$, $s = 1, \ldots, r$ have strong gradients, i.e., in the interfaces. In the bulk phases, without gradients, $P_s = P$ and $\mathbf{S} = 0$.

Equation (10.105) could be also written as

$$\partial_t\left(\rho \mathbf{v}\right) + \nabla \cdot \left[\rho \mathbf{v}\mathbf{v} + \boldsymbol{\tau} + P_s \boldsymbol{\delta}\right] = \rho \mathbf{g}^{(e)} - \nabla \cdot \mathbf{S}, \tag{10.113}$$

putting in evidence the role of the non-spherical interface tension tensor $\mathbf{S}$ in the mechanical properties of interfaces.

The first term on the right-hand side of Eq. (10.109) is a scalar pressure. In the macroscopic approach, the scalar pressure is made equal to the thermodynamic pressure P and the force resulting from the Korteweg stress tensor is considered a source term in the momentum equation with two components: a radial component, proportional to the interface tension and to the interface curvature and a tangential component, the Marangoni force. In the present kinetic approach, these forces are given in terms of the molecular properties of the involved fluids.

10.5.2 Using a Reverse Standpoint to Find the Microscopic Parameters and the Intermolecular Force

The equation of state is a van der Waals equation of state

$$P = \frac{nkT}{1 - bn} - an^2, \tag{10.114}$$

with the force parameter a given by

$$a = \sum_{p,s} x^{(p)} x^{(s)} a^{(ps)}, \tag{10.115}$$

and the volume parameter b by Eq. (10.93).

Adopting a reverse standpoint, Eq. (10.85) can be used to find the intermolecular force $\mathbf{g}^{(ps)}$ for any given cubic equation of state $P = P(n, T)$. Indeed, the kinetic model, Eq. (10.103), was derived considering a number of assumptions: molecules were first considered as material points and although the Enskog's volume exclusion theory leads to the correct repulsion part of the thermodynamic pressure, the molecular shapes were not taken into account.

In this case, the equation of state suitable for the problem is supposed to be known and the inverse problem consists in finding the intermolecular force $\mathbf{g}^{(ps)}$ that retrieves this equation of state.

Therefore, it is considered that, in the interface, the pressure tensor $\mathbf{P}$ is still given by Eq. (10.109) as a sum of a scalar pressure $P\delta$ and the Korteweg stress tensor $\mathbf{S}$, but it is supposed that the thermodynamic pressure P in Eq. (10.110) follows a given and known equation of state. Therefore, from Eq. (10.85) it results

$$\sum_{s,p} \rho^{(p)} \mathbf{g}^{(ps)} = -\nabla \begin{pmatrix} \left(P - \frac{nkT}{1-bn}\right) \\ -\frac{1}{2}\left[\nabla^2\left(\kappa n^2\right) - \sum_{p,s} \kappa^{(ps)} \nabla n^{(p)} \cdot \nabla n^{(s)}\right] \end{pmatrix} - \nabla \cdot \mathbf{S}. \tag{10.116}$$

Consider, for instance, that the mixture is well-described by a cubic equation of state,

$$P = \frac{nkT}{1 - bn} - \frac{an^2}{f\,(bn)}, \tag{10.117}$$

where $f(bn)$ is a dimensionless function of the volume term bn, with a quadratic mixing rule for the force parameter a, as given by Eq. (10.115).

The force parameter $a^{(ss)}$ for each s-component is dependent on the temperature and on an acentric factor $\omega^{(s)}$ which is a measure of the non-sphericity of the s-molecules. The cross parameters $a^{(ps)}$, with $p \neq s$, is given as a geometric average of the intrinsic force parameters, weighted by a cross interaction-term $\varsigma^{(ps)}$,

$$a^{(ps)} = \varsigma^{(ps)} \sqrt{a^{(pp)} a^{(ss)}}, \tag{10.118}$$

in such a manner that when $p = s$, $\varsigma^{(ps)} = 1$.

It can be easily found the intermolecular force $\mathbf{g}^{(ps)}$ in Eq. (10.103) must satisfy

$$\rho^{(p)} \mathbf{g}^{(ps)} = a^{(ps)} \nabla \left(\frac{n^{(p)} n^{(s)}}{f\,(bn)} \right) + \frac{\kappa^{(ps)}}{2} \left[n^{(p)} \nabla \nabla^2 n^{(s)} + n^{(s)} \nabla \nabla^2 n^{(p)} \right], \tag{10.119}$$

where the following relationship was used

$$\begin{aligned} & n^{(p)} \nabla \nabla^2 n^{(s)} + n^{(s)} \nabla \nabla^2 n^{(p)} \\ & = \nabla \cdot \left(\left[\nabla^2 \left(n^{(p)} n^{(s)} \right) - \nabla n^{(p)} \cdot \nabla n^{(s)} \right] \widehat{\delta} - 2 \nabla n^{(p)} \otimes \nabla n^{(s)} \right). \end{aligned} \tag{10.120}$$

10.5.3 Exercises

1. Consider a system of two components p and q. Using Eq. (10.102) and the relationship given by Eq. (10.120) show that

$$\begin{aligned} & \sum_{p,s} \rho^{(p)} \mathbf{g}^{(ps)} \\ & = \nabla \left(\begin{array}{c} a n^2 + \frac{1}{2} \nabla^2 \left(\kappa n^2 \right) \\ -\frac{1}{2} \nabla \left(\begin{array}{c} \kappa^{(pp)} \nabla n^{(p)} \cdot \nabla n^{(p)} + 2 \kappa^{(pq)} \nabla n^{(p)} \cdot \nabla n^{(q)} \\ + \kappa^{(qq)} \nabla n^{(q)} \cdot \nabla n^{(q)} \end{array} \right) \end{array} \right) \\ & - \nabla \cdot \mathbf{S} \end{aligned}$$

where

$$a = a^{(pp)} \left(x^{(p)} \right)^2 + 2 a^{(pq)} \left(x^{(p)} x^{(q)} \right) + a^{(qq)} \left(x^{(q)} \right)^2,$$

is the force term in the equation of state,

$$\kappa = \kappa^{(pp)} \left(x^{(p)} \right)^2 + 2 \kappa^{(pq)} \left(x^{(p)} x^{(q)} \right) + \kappa^{(qq)} \left(x^{(q)} \right)^2,$$

is the surface force term, and

$$\mathbf{S} = \kappa^{(pp)} \left(\nabla n^{(p)} \nabla n^{(p)} \right) + 2 \kappa^{(pq)} \nabla n^{(p)} \nabla n^{(q)} + \kappa^{(qq)} \left(\nabla n^{(q)} \nabla n^{(q)} \right),$$

is the Korteweg surface tensor.

2. By replacing the above expression for

$$\sum_{p,s} \rho^{(p)} \mathbf{g}^{(ps)},$$

into Eq. (10.85) show that the scalar pressure is given by Eq. (10.110) and that the equation of state is the one of van der Waals.

Appendix
The Euler–Lagrange Equations

The problem we wanted to solve in Sect. 6.5.4 was to find $\psi_{exc}(n, T)$ in such a way as to minimize the integral

$$\Psi_{exc} = \int_{-\infty}^{\infty} \left(\psi_{exc}(n, T) + \frac{\kappa}{2} \left(\frac{dn}{dx} \right)^2 \right) dx, \tag{A.1}$$

knowing that ψ_{exc} is null outside the liquid–vapor transition layer or that $\psi_{exc} \to 0$ when $x \to \pm\infty$ and the number density of molecules $n \to n_{vs}$ when $x \to -\infty$ and $n \to n_{\ell s}$ when $x \to +\infty$.

Let us first consider the more general problem of finding the function $y(x)$ that makes the integral

$$S = \int_a^b f\left(y, y', x\right) dx, \tag{A.2}$$

an extremum, when $y(x)$ is subjected to the conditions $y(a) = y_a$ and $y(b) = y_b$.

In June 1696, Johann Bernoulli, the father of Daniel Bernoulli, had used the pages of the *Acta Eruditorum Lipsidae* to pose a challenge to the international mathematical community: to find the form of the curve joining two fixed points so that a mass will slide down along it, under the influence of gravity alone, in the *minimum amount of time*. This problem was called the *brachistochrone problem*, from the Greek words brákhistos (shortest) and khrónos (time), and was solved by Newton based on the second law of mechanics.

In 1733, Euler elaborated on the brachistochrone problem using a geometrical approach but a few years later, in 1755, Joseph-Louis Lagrange, at the age of 19 years old, studying a problem named the tautochrone problem, which is a variation of the brachistochrone problem, published a paper explaining an analytical method for analyzing such a problem. Euler was so impressed that he drops his current method and switches to Lagrangian method which was purely analytical. In 1756, Euler pub-

P. C. Philippi, *Thermodynamics*, https://doi.org/10.1007/978-3-031-49357-7

lished a paper on the Calculus of Variations based on Lagrange's research, trying to extend it to a much wider range of problems. Then, in 1788, Lagrange published his *Mécanique Analytique*, a classical mechanics paper that, in contrast with Newton's mechanics, consider mechanical problems as minimization problems, proposing what is presently known as *Lagrangian Mechanics*. Fifty years later, in 1834, Willian Rowan Hamilton extended Lagrangian's approach to problems in electromagnetism and quantum theory, launching the foundations of the *principle of the least action*, considered by some physicists to be the most fundamental principle of physics.

Going back to Eq. (A.2), consider a family of curves

$$Y(x) = y(x) + \epsilon g(x),$$

in such a way that $g(a) = g(b) = 0$. Therefore, if we replace $y(x)$ by $Y(x)$ in Eq. (A.2), the action S becomes dependent on ϵ,

$$S(\epsilon) = \int_a^b f\left(Y, Y', x\right) dx,$$

and since we want to find the function $y(x)$ that makes S an extremum,

$$\frac{dS}{d\epsilon} = 0,$$

when $\epsilon = 0$.

On the other hand

$$\frac{dS}{d\epsilon} = \int_a^b \left(\frac{\partial f}{\partial Y}\frac{\partial Y}{\partial \epsilon} + \frac{\partial f}{\partial Y'}\frac{\partial Y'}{\partial \epsilon}\right) dx,$$

or

$$\frac{dS}{d\epsilon} = \int_a^b \left(\frac{\partial f}{\partial Y} g + \frac{\partial f}{\partial Y'} g'\right) dx.$$

The second term of the above equation may be integrated by parts

$$\int_a^b \frac{\partial f}{\partial Y'} g' dx = \frac{\partial f}{\partial Y'} g \Big]_a^b - \int_a^b \frac{d}{dx}\left(\frac{\partial f}{\partial Y'}\right) g dx,$$

and, since $g(a) = g(b) = 0$,

$$\int_a^b \frac{\partial f}{\partial Y'} g' dx = -\int_a^b \frac{d}{dx}\left(\frac{\partial f}{\partial Y'}\right) g dx.$$

Therefore

$$\frac{dS}{d\epsilon} = \int_a^b \left(\frac{\partial f}{\partial Y} - \frac{d}{dx}\left(\frac{\partial f}{\partial Y'}\right)\right) g dx.$$

When $\epsilon = 0$, $dS/d\epsilon = 0$ and $Y(x)$ becomes the function $y(x)$ we want to find

$$\int_a^b \left(\frac{\partial f}{\partial y} - \frac{d}{dx}\left(\frac{\partial f}{\partial y'}\right)\right) g dx = 0.$$

Since $g(x)$ is an arbitrary function, this integral is null when and only when

$$\frac{\partial f}{\partial y} - \frac{d}{dx}\left(\frac{\partial f}{\partial y'}\right) = 0 \qquad \text{(A.3)}$$

Equation (A.3) is the Euler–Lagrange equation, which is the basis of Lagrangian mechanics when the integrand f is replaced by the Lagrangian L, defined as the difference between the kinetic E_c and the potential energy E_p of a particle whose position and velocity are, respectively, given by the generalized coordinates x and $\dot{x} = dx/dt$ in the phase-space,

$$L = E_c\left(\dot{x}\right) - E_p(x),$$

and the independent variable is the time,

$$S = \int_a^b L\left(x, \dot{x}, t\right) dt.$$

In some cases we do not have an *explicite* dependence of the integrand $f\left(y, y'\right)$ on the coordinate x and the second term on the l.h.s. of Eq. (A.3) may be written as

$$\frac{d}{dx}\left(\frac{\partial f}{\partial y'}\right) = \frac{\partial^2 f}{\partial y \partial y'} y' + \frac{\partial^2 f}{\partial y''^2} y'', \tag{A.4}$$

because y and y' are dependent on x.

Replacing Eq. (A.4) into Eq. (A.3) we obtain

$$\frac{\partial f}{\partial y} - \frac{\partial^2 f}{\partial y \partial y'} y' - \frac{\partial^2 f}{\partial y'^2} y'' = 0. \tag{A.5}$$

Now, consider the function

$$h(x) = y' \frac{\partial f}{\partial y'} - f.$$

When this function is differentiated it gives

$$\frac{dh}{dx} = y' \underbrace{\left(\frac{\partial^2 f}{\partial y'^2} y'' + \frac{\partial^2 f}{\partial y \partial y'} y' - \frac{\partial f}{\partial y}\right)}_{=0} = 0.$$

Therefore, we get the following equation as a particular case of the Euler–Lagrange equation

$$y' \frac{\partial f}{\partial y'} - f = C,$$

named *Beltrami Identity* because it was firstly derived by Eugênio Beltrami in 1868 [120].

References

1. D. G. A. L. Aarts, H. N. W. Lekkerkerker, H. Guo, G. H. Wegdam, and D. Bonn, *Hydrodynamics of droplet coalescence*, Physical Review Letters, 95 (2005), p. 164503.
2. A. W. Adamson and A. P. Gast, *Physical Chemistry of Surfaces*, Wiley, 1997.

3. R. Adhikari and S. Succi, *Duality in matrix lattice Boltzmann models*, Physical Review E - Statistical, Nonlinear, and Soft Matter Physics, 78 (2008), pp. 1–9.
4. P. Adler, *Porous Media: Geometry and Transports*, Butterworth-Heinemann, 1992.
5. T. Andrews, *The Bakerian lecture: On the continuity of the gaseous and liquid states of matter*, Philosophical Transactions of the Royal Society of London, 159 (1869), pp. 575–590.
6. C. R. Appoloni, A. Macedo, C. P. Fernandes, and P. C. Philippi, *Characterization of porous microstructure by X-ray microtomography*, X-Ray Spectrometry, 31 (2002), pp. 124–127.
7. R. Aris, *Vectors, tensors, and the basic equations of fluid mechanics*, Dover Publications, 1989.
8. F. Bashfort, *An attempt to test the theories of capillary action by comparing the theoretical and measured forms of drops of fluid*, Cambridge University Press, 1883.
9. R. Bazarin, P. Philippi, A. Randles, and L. H. Junior, *Moments-based method for boundary conditions in the lattice Boltzmann framework: A comparative analysis for the lid driven cavity flow*, Computers & Fluids (2021), p. 105142.
10. P. L. Bhatnagar, E. P. Gross, and M. Krook, *A model for collision processes in gases. I. Small amplitude processes in charged and neutral one-component systems*, Physical Review, 94 (1954), pp. 511–525.
11. P. Bigdelou, C. Liu, P. Tarey, and P. Ramaprabhu, *An efficient ghost fluid method to remove overheating from material interfaces in compressible multi-medium flows*, Computers & Fluids, 233 (2022), p. 105250.
12. J. Black, *Lectures on the Elements of Chemistry delivered at the University of Edinburgh*, Mundell and Son, 1803.
13. N. Bogoliubov, *"Kinetic equations" (in Russian)*, Journal of Experimental and Theoretical Physics, 16 (1946), pp. 691–702.
14. L. Boltzmann, *Weitere studien uber das warmegleichgewicht unter gasmolekun*, Sitzungberichte der Akademie der Wissenschaften zu Wien, mathematisch-naturwissenschaftliche Klasse, 66 (1872), pp. 275–370.
15. L. Boltzmann, *Vorlesungen uber Gas Theory, Part I*, J.A. Barth, 1896.
16. L. Boltzmann, *Vorlesungen uber Gas Theorie. Part II.*, J.A. Barth, 1898.
17. M. Born and H. Green. *A general kinetic theory of liquids I. The molecular distribution functions*, Proc. Roy. Soc. A, 188 (1946), pp. 10–18.
18. C. Bosanquet, *Lv. on the flow of liquids into capillary tubes*, The London, Edinburgh, and Dublin Philosophical Magazine and Journal of Science, 45 (1923), pp. 525–531.
19. R. Boyle, *New experiments Physico-Mechanical touching the spring of the air and its effects*, H. Hall, 1660.
20. A. Braslau, M. Deutsch, P. S. Pershan, A. H. Weiss, J. Als-Nielsen, and J. Bohr, *Surface roughness of water measured by X-ray reflectivity*, Physical Review Letters, 54 (1985), pp. 114–117.
21. S. Brunauer, P. H. Emmett, and E. Teller, *Adsorption of gases in multimolecular layers*, Journal of the American Chemical Society, 60 (1938), pp. 309–319.
22. G. Buresti, *A note on Stokes' hypothesis*, Acta Mechanica, 226 (2015), pp. 3555–3559.
23. J. N. Butler and B. H. Bloom, *A curve-fitting method for calculating interfacial tension from the shape of a sessile drop*, Surface Science, 4 (1966), pp. 1–17.
24. J. W. Cahn, *Critical point wetting*, The Journal of Chemical Physics, 66 (1977), p. 3667.
25. J. W. Cahn and J. E. Hilliard, *Free energy of a nonuniform system. I. Interfacial free energy*, The Journal of Chemical Physics, 28 (1958), pp. 258–267.
26. C. Caratheodory, *Untersuchungen ber die grundlagen der thermodynamik*, Mathematische Annalen, 67 (1909), pp. 355–386.
27. S. Carnot, *Réflexions sur la puissance motrice du feu*, Bachelier, 1824.
28. H. Cavendish, *Three papers, containing experiments on factitious air, by the hon. Henry Cavendish, f. r. s.*, Philosophical Transactions of the Royal Society of London, 56 (1766), pp. 141–184.
29. C. Cercignani, *Mathematical Methods in Kinetic Theory*, London: Macmillan, 1969.

30. S. Chapman and T. Cowling, *The Mathematical Theory of Non-uniform Gases*, Cambridge University Press, 1970.
31. H. Chen, R. Zhang, I. Staroselsky, and M. Jhon, *Recovery of full rotational invariance in lattice Boltzmann formulations for high Knudsen number flows*, Physica A: Statistical Mechanics and its Applications, 362 (2006), pp. 125–131.
32. A. K. Chesters, *An analytical solution for the profile and volume of a small drop or bubble symmetrical about a vertical axis*, Journal of Fluid Mechanics, 81 (1977), pp. 609–624.
33. B. E. Clapeyron, *Memoire sur la puissance motrice de la chaleur*, Journal de l'Ecole Royale Polytechnique, 23 (1834), pp. 153–190.
34. R. Clausius, *Ueber die bewegende kraft der warme und die gesetze, welche sich daraus fur die warmelehre selbst ableiten lassen*, Annalen der Physik und Chemie, 155 (1850), pp. 500–524.
35. R. Clausius, *Ueber eine veranderte form des zweiten hauptsatzes der mechanischen warmetheorie*, Annalen der Physik und Chemie, 169 (1854), pp. 481–506.
36. R. Clausius, *Abhandlungen uber die mechanische Warmetheorie (On the mechanical theory of heat)*, John van Voorst, 1867.
37. J. P. Dalton, *Researches on the Joule–Kelvin effect, especially at low temperatures. I. Calculations for hydrogen*, Proceedings of the Royal Netherlands Academy of Arts and Sciences, 11 (1909), pp. 863–873.
38. J. H. de Boer and C. Zwikker, *Adsorption als folge von polarization*, Z. physik. Chem. B, 3 (1929), pp. 407–418.
39. S. de Groot and P. Mazur, *Non-Equilibrium Thermodynamics*, Dover Publications, 1984.
40. P. J. Dellar, *Incompressible limits of lattice Boltzmann equations using multiple relaxation times*, Journal of Computational Physics, 190 (2003), pp. 351–370.
41. D. D'Humieres, *Generalized lattice-Boltzmann equations*, 1 1994.
42. D. D'Humieres, I. Ginzburg, M. Krafczyk, P. Lallemand, and L.-S. Luo, *Multiple-relaxation-time lattice Boltzmann models in three dimensions*, Philosophical transactions. Series A, Mathematical, physical, and engineering sciences, 360 (2002), pp. 437–451.
43. L. O. E. dos Santos, F. G. Wolf, and P. C. Philippi, *Dynamics of interface displacement in capillary flow*, Journal of Statistical Physics, 121 (2005), pp. 197–207.
44. M. Dreyer, A. Delgado, and H.-J. Path, *Capillary rise of liquid between parallel plates under microgravity*, Journal of Colloid and Interface Science, 163 (1994), pp. 158–168.
45. P. Duhem, *Le potentiel thermodynamique et ses applications*, A. Hermann, 1886.
46. P. Duhem, *Josiah Willard Gibbs, á propos de la publication de ses mémoires scientifiques*, A. Hermann, 1908.
47. P. Duhem, *Traié d'énergétique ou de thermodynamique générale (2 vols.)*, Gauthier-Villars, 1911.
48. J. Dunn, *Interstitial working and a nonclassical continuum thermodynamics*, 1986.
49. C. Eckart, *The thermodynamics of irreversible processes. I. The simple fluid*, Physical Review, 58 (1940), pp. 267–269.
50. C. Eckart, *The thermodynamics of irreversible processes. II. Fluid mixtures*, Physical Review, 58 (1940), pp. 269–275.
51. S. Fechter, F. Jaegle, and V. Schleper, *Exact and approximate Riemann solvers at phase boundaries*, Computers & Fluids, 75 (2013), pp. 112–126.
52. S. Fechter, C.-D. Munz, C. Rohde, and C. Zeiler, *A sharp interface method for compressible liquid-vapor flow with phase transition and surface tension*, Journal of Computational Physics, 336 (2017), pp. 347–374.
53. S. Fechter, C.-D. Munz, C. Rohde, and C. Zeiler, *Approximate Riemann solver for compressible liquid vapor flow with phase transition and surface tension*, Computers & Fluids, 169 (2018), pp. 169–185.
54. R. P. Fedkiw, T. Aslam, and S. Xu, *The ghost fluid method for deflagration and detonation discontinuities*, Journal of Computational Physics, 154 (1999), pp. 393–427.
55. C. Fernandes, F. Magnani, P. Philippi, J. Daïan, C. Fernandas, F. Magnani, P. Philippi, and J. Daian, *Multiscale geometrical reconstruction of porous structures*, Physical Review E, 54 (1996), pp. 1734–1741.

56. J. S. Fernandes, C. R. Appoloni, and C. P. Fernandes, *Microstructural characterization of siltstone and sandstone pore space by X-ray microtomography*, Soils and Rocks, 42 (2019), pp. 281–288.
57. A. Fredenslund, R. L. Jones, and J. M. Prausnitz, *Group-contribution estimation of activity coefficients in nonideal liquid mixtures*, AIChE Journal, 21 (1975), pp. 1086–1099.
58. L. J. Gay-Lussac, *L'expansion des gaz par la chaleur*, Annales de Chimie, 43 (1802), p. 137. NULL.
59. J. Gibbs, *On the equilibrium of heterogeneous substances. Part i*, Transactions of the Connecticut Academy of Arts and Sciences, 3 (1875), pp. 108–248.
60. J. Gibbs, *On the equilibrium of heterogeneous substances. Part ii*, Transactions of the Connecticut Academy of Arts and Sciences, 3 (1877), p. 343–524.
61. J. W. Gibbs, *A method of geometrical representation of the thermodynamic properties of substances by means of surfaces*, Transactions of the Connecticut Academy of Arts and Sciences, 2 (1873), pp. 382–404.
62. E. P. Gross and E. A. Jackson, *Kinetic models and the linearized Boltzmann equation*, Physics of Fluids, 2 (1959), pp. 432–441.
63. F. B. Hanson and T. F. Morse, *Kinetic models for a gas with internal structure*, The Physics of Fluids, 10 (1967), pp. 345–353.
64. F. Hauksbee, *An account of an experiment touching the ascent of water between two glass planes, in an hyperbolick figure by Mr. Francis Hauksbee, f. r. s.*, Philosophical Transactions of the Royal Society of London, 27 (1710), pp. 539–540.
65. R. D. Hazlett, *Simulation of capillary-dominated displacements in microtomographic images of reservoir rocks*, Transport in Porous Media, 20 (1995), pp. 21–35.
66. X. He and G. D. Doolen, *Thermodynamic foundations of kinetic theory and lattice Boltzmann models for multi-phase flows*, Journal of Statistical Physics, 107 (2002), pp. 309–328.
67. X. He and L.-S. Luo, *Theory of the lattice Boltzmann method: From the Boltzmann equation to the lattice Boltzmann equation*, Physical Review E, 56 (1997), pp. 6811–6817.
68. H. Helmholtz, *On the conservation of force*, Transactions of the Physical Society of Berlin (1847), pp. 114–162.
69. W. Henry, *Experiments on the quantity of gases absorbed by water, at different temperatures, and under different pressures*, Philosophical Transactions of the Royal Society of London, 93 (1803), pp. 29–274.
70. D. J. Holdych, D. Rovas, J. G. Georgiadis, and R. O. Buckius, *An improved hydrodynamics formulation for multiphase flow lattice-Boltzmann models*, International Journal of Modern Physics C, 09 (1998), pp. 1393–1404.
71. I. K. Howard, *H is for enthalpy, thanks to Heike Kamerlingh Onnes and Alfred W. Porter*, Journal of Chemical Education, 79 (2002), p. 697.
72. H. Huang, M. Krafczyk, and X. Lu, *Forcing term in single-phase and Shan-Chen-type multiphase lattice Boltzmann models*, Physical Review E, 84 (2011), p. 046710.
73. F. S. V. Hurtado and P. C. Philippi, *Thermodynamic analysis of the heating-dissipation process inside a cavity (technical report)*, 2005.
74. T. Inamuro, N. Konishi, and F. Ogino, *A Galilean invariant model of the lattice Boltzmann method for multiphase fluid flows using free-energy approach*, Comput. Phys. Comm., 129 (2000), pp. 32–45.
75. G. Jaumann, *Die Grundlagen der Bewegungslehre*, Barth Verlag, 1905.
76. G. Jaumann, *Geschlossenes System physikalischer und chemischer Diffrerentialgesetze*, K. k. Hof- und Staatsdruckerei, 1911.
77. M. Y. Joshi, *A class of stochastic models for porous media*, 1974.
78. J. Joule, *On the changes of temperature produced by the rarefaction and condensation of air*, Philosophical Magazine Series 3, 26 (1845), pp. 369–383.
79. J. P. Joule, *On the heat evolved by metallic conductors of electricity, and in the cells of a battery during electrolysis*, Philosophical Magazine Series 3, 19 (1841), pp. 260–277.
80. J. P. Joule, *On the mechanical equivalent of heat*, Philosophical Transactions of the Royal Society of London, 140 (1850), pp. 61–82.

81. L. A. H. Junior, A. Scagliarini, M. Sbragaglia, K. K. Mattila, P. C. Philippi, D. F. Puleri, J. Gounley, and A. Randles, *High-Reynolds-number turbulent cavity flow using the lattice Boltzmann method*, Physical Review E, 98 (2018), p. 043302.
82. J. Jurin, *An account of some new experiments, relating to the action of glass tubes upon water and quicksilver by James Jurin, M. D. Reg. Soc. & Coll. Med. Lond. Soc.*, Philosophical Transactions of the Royal Society of London, 30 (1717), pp. 1083–1096.
83. I. V. Karlin, F. Bösch, and S. S. Chikatamarla, textitGibbs' principle for the lattice-kinetic theory of fluid dynamics, Physical Review E, 90 (2014), p. 031302.
84. I. V. Karlin, A. Ferrante, and H. C. Öttinger, *Perfect entropy functions of the lattice Boltzmann method*, Europhysics Letters (EPL), 47 (1999), pp. 182–188.
85. K. Kinosita and H. Yokota, *Temperature dependence of the optical surface thickness of water*, Journal of the Physical Society of Japan, 20 (1965), pp. 1086–1086.
86. J. G. Kirkwood, *The statistical mechanical theory of transport processes I. General theory*, The Journal of Chemical Physics, 14 (1946), p. 180.
87. J. G. Kirkwood, *The statistical mechanical theory of transport processes II. Transport in gases*, The Journal of Chemical Physics, 15 (1947), p. 72.
88. J. Kloubek, *Development of methods for surface free energy determination using contact angles of liquids on solids*, Advances in Colloid and Interface Science, 38 (1992), pp. 99–142.
89. G. M. Kontogeorgis, I. V. Yakoumis, H. Meijer, E. Hendriks, and T. Moorwood, *Multicomponent phase equilibrium calculations for water-methanol-alkane mixtures*, Fluid Phase Equilibria, 158–160 (1999), pp. 201–209.
90. K. G. Kornev and A. V. Neimark, *Spontaneous penetration of liquids into capillaries and porous membranes revisited*, Journal of Colloid and Interface Science, 235 (2001), pp. 101–113.
91. D. Korteweg, *Sur la forme que prennent les équations du mouvement des fluides si l'on tient compte des forces capillaires causées par des variations de densité considérables mais continues et sur la théorie de la capillarité dans l'hypothese d'une variation continue de la densité*, Archives Néerlandaises des Sciences Exactes et Naturelles, 2 (1901), pp. 1–24.
92. G. M. Kremer, *An Introduction to the Boltzmann Equation and Transport Processes in Gases*, Springer Berlin Heidelberg, 2010.
93. P. Kundu, I. Cohen, and D. Dowling, *Fluid Mechanics*, Academic Press, 6th ed., 2015.
94. A. Kupershtokh, D. Medvedev, and D. Karpov, *On equations of state in a lattice Boltzmann method*, Computers & Mathematics with Applications, 58 (2009), pp. 965–974.
95. A. J. C. Ladd, *Numerical simulations of particulate suspensions via a discretized Boltzmann equation. Part 1. Theoretical foundation*, Journal of Fluid Mechanics, 271 (1994), pp. 285–309.
96. P. Lallemand and L.-S. Luo, *Theory of the lattice Boltzmann method: Acoustic and thermal properties in two and three dimensions*, Physical Review E, 68 (2003), p. 036706.
97. I. Langmuir, *The constitution and fundamental properties of solids and liquids*, Journal of the American Chemical Society, 39 (1917), pp. 1848–1906.
98. I. Langmuir, *The adsorption of gases on plane surfaces of glass, mica and platinum*, Journal of the American Chemical Society, 40 (1918), pp. 1361–1403.
99. P. Laplace, *Traite de Mecanique Celeste*, Courcier, 1805.
100. J. Latt and B. Chopard, *Lattice Boltzmann method with regularized pre-collision distribution functions*, Mathematics and Computers in Simulation, 72 (2006), pp. 165–168.
101. A. Lavoisier, *Memoire sur la chaleur lu a l'academie royale des sciences, le 28 juin 1783*, Memoires de l' Academie Royale des Sciences (1783).
102. A. H. Lefebvre and V. G. McDonell, *Atomization and Sprays*, CRC Press, 3 2017.
103. A. M. Legendre, *Mémoire sur l'intégration de quelques équations aux différences partielles*, Mémoires de l' Académie des Sciences (1787), pp. 309–351.
104. S. Levine, J. Lowndes, E. J. Watson, and G. Neale, *A theory of capillary rise of a liquid in a vertical cylindrical tube and in a parallel-plate channel*, Journal of Colloid and Interface Science, 73 (1980), pp. 136–151.

105. G. N. Lewis, *The law of physico-chemical change*, Proceedings of the American Academy of Arts and Sciences, 37 (1901), p. 49.
106. Z. Liang, C. Fernandes, F. Magnani, and P. Philippi, *A reconstruction technique for three-dimensional porous media using image analysis and Fourier transforms*, Journal of Petroleum Science and Engineering, 21 (1998), pp. 273–283.
107. J. Liouville, *Sur la theorie de la variation des constantes arbitraires*, Journal de mathématiques pures et appliquées, 3 (1838), pp. 342–349.
108. L. Luo, *Unified theory of lattice Boltzmann models for nonideal gases*, Physical Review Letters, 81 (1998), pp. 1618–1621.
109. O. Malaspinas, *Increasing stability and accuracy of the lattice Boltzmann scheme: recursivity and regularization*, arXiv:1505.06900v1 [physics.flu-dyn] 26 May 2015 (2015), pp. 1–31.
110. V. Maltsev, M. Skote, and P. Tsoutsanis, *High-order methods for diffuse-interface models in compressible multi-medium flows: A review*, Physics of Fluids, 34 (2022), p. 021301.
111. E. Mariotte, *Essais de Physique, ou mémoires pour servir á la science des choses naturelles*, E. Michallet, 1679.
112. M. Matsumoto and Y. Kataoka, *Study on liquid-vapor interface of water. I. Simulational results of thermodynamic properties and orientational structure*, The Journal of Chemical Physics, 88 (1988), p. 3233.
113. K. K. Mattila, L. A. H. Junior, and P. C. Philippi, *Investigation of an entropic stabilizer for the lattice-Boltzmann method*, Physical Review E, 91 (2015), p. 063010.
114. K. K. Mattila, P. C. Philippi, and L. A. H. Junior, *High-order regularization in lattice-Boltzmann equations*, Physics of Fluids, 29 (2017), p. 046103.
115. J. C. Maxwell, *On the dynamical theory of gases*, Philosophical transactions of the Royal Society, 157 (1867), pp. 49–88.
116. J. C. Maxwell, *Molecules*, Nature (1873), pp. 437–441.
117. J. C. Maxwell, *On the dynamical evidence of the molecular constitution of bodies*, Nature, 11 (1875), pp. 357–359.
118. A. C. Moreira, C. P. Fernandes, M. V. de Oliveira, A. A. Ribeiro, I. F. Mantovani, L. R. Medeiros, K. K. Mattila, and R. Lyra, *Evaluating the performance of truncated Gaussian method for three-dimensional reconstruction of the pore microstructure of titanium scaffolds*, Journal of Porous Media, 24 (2021), pp. 31–44.
119. I. Muller, *A History of Thermodynamics*, Springer-Verlag Berlin Heidelberg, 2007.
120. P. J. Nahim, *When Least is Best*, Princeton University Press, 2003.
121. L. Natanson, *Uber die gezetze nicht umkehrbarer vorgange*, Z. phys. Chem. (1896), p. 193.
122. L. Onsager, *Reciprocal relations in irreversible processes. i*, Physical Review, 37 (1931), pp. 405–426.
123. L. Onsager, *Reciprocal relations in irreversible processes. ii*, Physical Review, 38 (1931), pp. 2265–2279.
124. H. Orbey and J. H. Vera, *Correlation for the third virial coefficient using T_c, P_c and ω as parameters*, AIChE Journal, 29 (1983), pp. 107–113.
125. S. Osher and J. A. Sethian, *Fronts propagating with curvature-dependent speed: Algorithms based on Hamilton-Jacobi formulations*, Journal of Computational Physics, 79 (1988), pp. 12–49.
126. J. F. Padday, *The profiles of axially symmetric menisci*, Philosophical Transactions of the Royal Society A: Mathematical, Physical and Engineering Sciences, 269 (1971), pp. 265–293.
127. R. Panton, *Incompressible Flow*, John Wiley and Sons, 2013.
128. D.-Y. Peng and D. B. Robinson, *A new two-constant equation of state*, Industrial & Engineering Chemistry Fundamentals, 15 (1976), pp. 59–64.
129. P. Philippi and R. Brun, *Kinetic modeling of polyatomic gas mixtures*, Physica A: Statistical Mechanics and its Applications, 105 (1981), pp. 147–168.
130. P. Philippi, L. H. Junior, R. Surmas, D. Siebert, and L. dos Santos, *From the Boltzmann to the lattice-Boltzmann equation: beyond BGK collision models*, International Journal of Modern Physics C, 18 (2007), pp. 556–565.

131. P. C. Philippi, L. O. E. dos Santos, L. A. H. Junior, C. E. P. Ortiz, D. N. Siebert, and R. Surmas, *Thermodynamic consistency in deriving lattice Boltzmann models for describing segregation in non-ideal mixtures*, Philosophical transactions of the Royal Society A, 369 (2011), pp. 2292–300.
132. P. C. Philippi, L. A. H. Júnior, L. D. Santos, and R. Surmas, *From the continuous to the lattice Boltzmann equation: The discretization problem and thermal models*, Physical Review E, 73 (2006), p. 056702.
133. P. C. Philippi, K. K. Mattila, D. N. Siebert, L. O. E. dos Santos, L. A. H. Júnior, and R. Surmas, *Lattice-Boltzmann equations for describing segregation in non-ideal mixtures*, Journal of Fluid Mechanics, 713 (2012), pp. 564–587.
134. K. S. Pitzer, *The volumetric and thermodynamic properties of fluids. I. Theoretical basis and virial coefficients 1*, Journal of the American Chemical Society, 77 (1955), pp. 3427–3433.
135. K. S. Pitzer, D. Z. Lippmann, R. F. Curl, C. M. Huggins, and D. E. Petersen, *The volumetric and thermodynamic properties of fluids. II. Compressibility factor, vapor pressure and entropy of vaporization 1*, Journal of the American Chemical Society, 77 (1955), pp. 3433–3440.
136. M. Planck, *Treatise on Thermodynamics*, Dover Publications, 1917.
137. B. E. Poling, J. Prausnitz, and J. P. O'Connell, *The properties of gases and liquids*, McGraw-Hill, fifth edition ed., 2001.
138. J. Prausnitz, R. Lichtenthaler, and E. Azevedo, *Molecular thermodynamics of fluid-phase equilibria*, Prentice Hall, 1986.
139. J. Priestley, *Considerations on the doctrine of phlogiston and the decomposition of water*, Thomas Dobson, 1796.
140. I. Prigogine, *Introduction to Thermodynamics of Irreversible Processes*, Interscience, 1968.
141. P. Proust and J. H. Vera, *PRSV: The Stryjek-Vera modification of the Peng-Robinson equation of state. Parameters for other pure compounds of industrial interest*, The Canadian Journal of Chemical Engineering, 67 (1989), pp. 170–173.
142. F.-M. Raoult, *Recherches expérimentales sur les tensions de vapeur des dissolutions*, Journal de Physique Théorique et Appliquée, 8 (1889), pp. 5–20.
143. L. Rayleigh, *On the stability, or instability, of certain fluid motions*, Proceedings of the London Mathematical Society, s1–11 (1879), pp. 57–72.
144. L. J. W. S. Rayleigh, *Investigation of the character of the equilibrium of an incompressible heavy fluid of variable density*, Proceedings of the London Mathematical Society, s1–14 (1882), pp. 170–177.
145. O. Redlich and J. N. S. Kwong, *On the thermodynamics of solutions. V. An equation of state. Fugacities of gaseous solutions*, Chemical Reviews, 44 (1949), pp. 233–244.
146. C. Rohde and C. Zeiler, *A relaxation Riemann solver for compressible two-phase flow with phase transition and surface tension*, Applied Numerical Mathematics, 95 (2015), pp. 267–279.
147. J. S. Rowlinson, *Cohesion: a Scientific History of Intermolecular Forces*, Cambridge University Press, 2004.
148. J. S. Rowlinson and B. Widom, *Molecular Theory of Capillarity (Dover Books on Chemistry)*, Dover Publications, 2003.
149. K. Sankarayanan, X. Shan, I. G. Kevrekidis, and S. Sundaresan, *Analysis of drag and virtual mass forces in bubbly suspensions using an implicit formulation of the lattice Boltzmann method*, Journal of Fluid Mechanics, 452 (2002), pp. 61–96.
150. M. Sbragaglia, R. Benzi, L. Biferale, S. Succi, K. Sugiyama, and F. Toschi, *Generalized lattice Boltzmann method with multirange pseudopotential*, Physical Review E, 75 (2007), p. 026702.
151. V. Schleper, *A HLL-type Riemann solver for two-phase flow with surface forces and phase transitions*, Applied Numerical Mathematics, 108 (2016), pp. 256–270.
152. X. Shan and H. Chen, *Lattice Boltzmann model for simulating flows with multiple phases and components*, Physical Review E, 47 (1993), pp. 1815–1819.
153. X. Shan and H. Chen, *Simulation of nonideal gases and liquid-gas phase transitions by the lattice Boltzmann equation*, Physical Review E, 49 (1994), pp. 2941–2948.

154. X. Shan, X.-F. Yuan, and H. Chen, *Kinetic theory representation of hydrodynamics: a way beyond the Navier-Stokes equation*, Journal of Fluid Mechanics, 550 (2006), pp. 413–441.
155. C. E. Shannon, *A mathematical theory of communication*, Bell System Technical Journal, 27 (1948), pp. 379–423.
156. D. Siebert, P. Philippi, and K. Mattila, *Consistent lattice Boltzmann equations for phase transitions*, Physical Review E, 90 (2014), p. 053310.
157. D. N. Siebert, *Discrete kinetic models for non-ideal fluids and phase transition (doctoral thesis, in Portuguese)*, 2013.
158. D. N. Siebert, L. A. Hegele, P. C. Philippi, L. A. H. Júnior, and P. C. Philippi, *Lattice Boltzmann equation linear stability analysis: Thermal and athermal models*, Physical Review E, 77 (2008), p. 026707.
159. G. Soave, *Equilibrium constants from a modified Redlich-Kwong equation of state*, Chemical Engineering Science, 27 (1972), pp. 1197–1203.
160. B. Song and J. Springer, *Determination of interfacial tension from the profile of a pendant drop using computer-aided image processing: 2. Experimental*, Journal of Colloid and Interface Science, 184 (1996), pp. 77–91.
161. M. Stange, M. E. Dreyer, and H. J. Rath, *Capillary driven flow in circular cylindrical tubes*, Physics of Fluids, 15 (2003), pp. 2587–2601.
162. G. G. Stokes, *Mathematical and Physical Papers, Vol IV*, Cambridge University Press, 1880.
163. R. Stryjek and J. H. Vera, *PRSV - an improved Peng-Robinson equation of state with new mixing rules for strongly nonideal mixtures*, The Canadian Journal of Chemical Engineering, 64 (1986), pp. 334–340.
164. R. Stryjek and J. H. Vera, *PRSV: An improved Peng-Robinson equation of state for pure compounds and mixtures*, The Canadian Journal of Chemical Engineering, 64 (1986), pp. 323–333.
165. R. Stryjek and J. H. Vera, *PRSV2: A cubic equation of state for accurate vapor-liquid equilibria calculations*, The Canadian Journal of Chemical Engineering, 64 (1986), pp. 820–826.
166. M. Sussman, P. Smereka, and S. Osher, *A level set approach for computing solutions to incompressible two-phase flow*, Journal of Computational Physics, 114 (1994), pp. 146–159.
167. M. M. R. Swift, E. Orlandini, W. R. W. Osborn, and J. M. Yeomans, *Lattice Boltzmann simulations of liquid-gas and binary fluid systems*, Physical Review E, 54 (1996), pp. 5041–5052.
168. J. Szekely, A. Neumann, and Y. Chuang, *The rate of capillary penetration and the applicability of the Washburn equation*, Journal of Colloid and Interface Science, 35 (1971), pp. 273–278.
169. R. S. Taylor, L. X. Dang, and B. C. Garrett, *Molecular dynamics simulations of the liquid/vapor interface of SPC/E water*, The Journal of Physical Chemistry, 100 (1996), pp. 11720–11725.
170. S. G. I. Taylor, *The instability of liquid surfaces when accelerated in a direction perpendicular to their planes. I*, Proceedings of the Royal Society of London. Series A. Mathematical and Physical Sciences, 201 (1950), pp. 192–196.
171. W. Thomson, *On an absolute thermometric scale founded on Carnot's theory of the motive power of heat, and calculated from Regnault's observations*, Proceedings of the Cambridge Philosophical Society, Philosophical Magazine (1848).
172. W. Thomson, *On the dynamical theory of heat, with numerical results deduced from Mr Joule's equivalent of a thermal unit, and M. Regnault's observations on steam*, Transactions of the Royal Society of Edinburgh, 20 (1851), pp. 261–288.
173. W. Thomson, *Mathematical and Physical Papers*, Cambridge University Press, 1890.
174. W. L. K. Thomson, *On the dynamical theory of heat*, Mathematical and Physical Papers, 1 (1882), pp. 174–332.
175. R. H. Thurston, *History of the Growth of the Steam-Engine*, Appleton, 1886.
176. C. Tsalis, *Nonextensive statistical mechanics and thermodynamics: Historical background and present status*, 2001.
177. C. Tsonopoulos, *An empirical correlation of second virial coefficients*, AIChE Journal, 20 (1974), pp. 263–272.

178. C. Tsonopoulos, *Second virial coefficients of polar haloalkanes*, AIChE Journal, 21 (1975), pp. 827–829.
179. C. Tsonopoulos, *Second virial coefficients of water pollutants*, AIChE Journal, 24 (1978), pp. 1112–1115.
180. C. Tsonopoulos and J. L. Heidman, *From the virial to the cubic equation of state*, Fluid Phase Equilibria, 57 (1990), pp. 261–276.
181. J. D. van der Waals, *Over de continuiteit van den gas en vloeistoftoestand*, 1873.
182. H. C. van Ness, *H is for enthalpy (letter)*, Journal of Chemical Education, 80 (2003), p. 486.
183. E. Vogle, *On the origins of water wetting terminology*, 2001.
184. A. Wagner, *Thermodynamic consistency of liquid-gas lattice Boltzmann simulations*, Physical Review E, 74 (2006), p. 056703.
185. C. Wang-Chang and C. Uhlenbeck, *University of Michigan engineering research report CM-681*, 1951.
186. F. G. Wolf, *Experimental investigation of the effect of the addition of surfactants on the water-oil interface tension*, 2007.
187. F. G. Wolf, L. O. E. dos Santos, and P. C. Philippi, *Capillary rise between parallel plates under dynamic conditions*, Journal of colloid and interface science, 344 (2010), pp. 171–179.
188. H. Xu and P. Sagaut, *Optimal low-dispersion low-dissipation LBM schemes for computational aeroacoustics*, Journal of Computational Physics, 230 (2011), pp. 5353–5382.
189. T. Young, *An essay on the cohesion of fluids*, Philosophical transactions of the Royal Society, 95 (1805), pp. 65–87.
190. P. Yuan and L. Schaefer, *Equations of state in a lattice Boltzmann model*, Physics of Fluids, 18 (2006), p. 042101.
191. J. Yvon, *Theorie statistique des fluides et l'equation d'etat*, Actes Scientifique et Industrie, 203 (1935).
192. R. Zhang, X. Shan, and H. Chen, *Efficient kinetic method for fluid simulation beyond the Navier-Stokes equation*, Physical Review E, 74 (2006), p. 046703.
193. B. Zhmud, F. Tiberg, and K. Hallstensson, *Dynamics of capillary rise*, Journal of Colloid and Interface Science, 228 (2000), pp. 263–269.